T0328627

ADVANCED DISTRIBUTED CONSENSUS FOR MULTIAGENT SYSTEMS

ADVANCED DISTRIBUTED CONSENSUS FOR MULTIAGENT SYSTEMS

MAGDI S. MAHMOUD
King Fahd University of Petroleum and Minerals
Systems Engineering Department
Dhahran, Saudi Arabia

MOJEED O. OYEDEJI
King Fahd University of Petroleum and Minerals
Systems Engineering Department
Dhahran, Saudi Arabia

YUANQING XIA
Beijing Institute of Technology
School of Automation
Beijing, China

ACADEMIC PRESS

An imprint of Elsevier

Academic Press is an imprint of Elsevier
125 London Wall, London EC2Y 5AS, United Kingdom
525 B Street, Suite 1650, San Diego, CA 92101, United States
50 Hampshire Street, 5th Floor, Cambridge, MA 02139, United States
The Boulevard, Langford Lane, Kidlington, Oxford OX5 1GB, United Kingdom

Library of Congress Cataloging-in-Publication Data
A catalog record for this book is available from the Library of Congress

British Library Cataloguing-in-Publication Data
A catalogue record for this book is available from the British Library

ISBN: 978-0-12-821186-1

For information on all Academic Press publications
visit our website at https://www.elsevier.com/books-and-journals

Publisher: Mara Conner
Acquisitions Editor: Sonnini R. Yura
Editorial Project Manager: Rafael G. Trombaco
Production Project Manager: Sojan P. Pazhayattil
Designer: Miles Hitchen

Typeset by VTeX

Working together
to grow libraries in
developing countries

www.elsevier.com • www.bookaid.org

This book is dedicated to our families. With tolerance, patience, and wonderful frame of mind, they have encouraged and supported us for many years.

Magdi S. Mahmoud, Mojeed O. Oyedeji, Yuanqing Xia

Contents

About the authors   *xiii*
Preface   *xvii*
Acknowledgments   *xxi*

1. An overview   **1**
    1.1. Introduction   1
    1.2. Notations   2
    1.3. Elements of graph theory   3
       1.3.1. Basic results   4
       1.3.2. Laplacian spectrum of graphs   4
       1.3.3. Properties of adjacency matrix   5
    1.4. Mathematical models for agent dynamics   9
       1.4.1. Single integrator model   9
       1.4.2. Double integrator model   10
       1.4.3. Uncertain fully actuated model   11
       1.4.4. Nonholonomic unicycle model   11
    1.5. Notes   12
    References   12

2. Consensus over fixed networks   **15**
    2.1. Introduction   15
    2.2. Leaderless consensus   17
       2.2.1. Preliminaries   17
       2.2.2. Problem formulation: linear dynamics   19
       2.2.3. Problem formulation: nonlinear dynamics   23
       2.2.4. Numerical example 2.1   24
    2.3. Leader–follower consensus   25
       2.3.1. Description   26
       2.3.2. Problem formulation   26
       2.3.3. Related results   29
       2.3.4. Numerical example 2.2   29
    2.4. Group consensus   36
       2.4.1. Initial considerations   39
       2.4.2. Problem layout   41
       2.4.3. Related work   46
       2.4.4. Numerical example 2.3   46
    2.5. Bipartite consensus   53
       2.5.1. Definitions   54
       2.5.2. Problem description   55

2.5.3. Numerical example 2.4   58

2.6. Scaled consensus   58

2.6.1. Basics   59

2.6.2. Problem statement   60

2.6.3. Related results   60

2.6.4. Numerical example 2.5   61

2.7. Pinning consensus   61

2.7.1. Main issues   62

2.7.2. Problem layout: linear dynamics   62

2.7.3. Problem layout: nonlinear dynamics   63

2.7.4. Related work   65

2.7.5. Numerical example 2.6   66

2.8. Notes   68

References   69

3. Consensus in multiagent systems over time-varying networks   73

3.1. Introduction   73

3.1.1. Definitions   73

3.2. Switching networks   76

3.2.1. Leaderless consensus problems   76

3.2.2. Numerical example 3.1   81

3.2.3. Leader–follower consensus   82

3.2.4. Group consensus   85

3.2.5. Numerical example 3.2   86

3.2.6. Bipartite consensus   94

3.2.7. Scaled consensus   95

3.2.8. Related studies   96

3.3. Random (Markovian) networks   101

3.3.1. Leaderless consensus   101

3.3.2. Leader–follower consensus   102

3.3.3. Group consensus   103

3.3.4. Bipartite consensus   104

3.3.5. Scaled consensus   104

3.3.6. Related studies   105

3.4. Time-varying delay networks   106

3.4.1. Leaderless consensus   106

3.4.2. Leader–follower consensus   106

3.4.3. Bipartite consensus   108

3.5. Notes   109

References   110

4. Distributed consensus of multiagent systems   117

4.1. Exponential consensus of stochastic delayed systems   117

4.1.1. Introduction   117
4.1.2. Model formulation   118
4.2. Asynchronous switching control   121
4.2.1. Numerical example 4.1   132
4.2.2. Numerical example 4.2   134
4.3. H_∞ and H_2 consensus in directed networks   137
4.3.1. Introduction   139
4.3.2. Notations   140
4.3.3. Graph theory   140
4.3.4. H_∞ consensus   141
4.4. H_2 consensus   147
4.4.1. Numerical example 4.3   150
4.4.2. Numerical example 4.4   151
4.5. Notes   153
References   156

5. Consensus over vulnerable networks   **161**
5.1. Introduction   161
5.2. Discrete-time resilient consensus algorithms   163
5.2.1. Mean subsequence reduced (MSSR) algorithm   164
5.2.2. Weighted mean subsequence reduced (W-MSSR) algorithm   164
5.2.3. Median-based consensus algorithm (MCA)   164
5.2.4. Synchronous DP-MSSR algorithm   165
5.2.5. Asynchronous DP-MSSR algorithm   167
5.2.6. Heterogeneous HP-MSSR algorithm   168
5.2.7. Resilient group consensus algorithm   170
5.3. Continuous-time resilient consensus algorithms   172
5.3.1. Adversarial robust consensus protocol (ARC-P)   173
5.3.2. k-connectivity-based robust algorithms   175
5.3.3. Absolute weighted MSR (AW-MSR) algorithm   178
5.3.4. Secure linear consensus protocols   179
5.4. Asymptotic consensus value   180
5.4.1. Introduction   180
5.4.2. Consensus over switching directed graphs   181
5.4.3. Mean analysis   182
5.4.4. Mean of consensus value   183
5.5. Variance of the asymptotic consensus value   185
5.5.1. Infinity norm $\|\mathbf{R} - \mathbf{Q}\|_\infty$ 186
5.5.2. Perturbation-based bound for the variance   188
5.5.3. Numerical example 5.1   189
5.5.4. Numerical example 5.2   191
5.6. Notes   192
References   194

6. Consensus on state-dependent fuzzy graphs   **197**
    6.1. Introduction   197
    6.2. Preliminaries   199
        6.2.1. Graph theory   199
    6.3. State-dependent graphs   200
        6.3.1. Eigenvalue optimization   201
        6.3.2. Consensus on state-dependent graphs   204
        6.3.3. Graph controllability   207
    6.4. Fuzzy graph theory   208
        6.4.1. Important definitions   208
        6.4.2. Connectivity in fuzzy graphs   209
    6.5. State-dependent fuzzy graphs   211
        6.5.1. Problem statement   212
        6.5.2. Fuzzy membership functions   215
        6.5.3. Fuzzy relations   224
    6.6. Notes   224
    References   225

7. Distributed consensus on state-dependent evolutionary graphs   **229**
    7.1. Introduction   229
        7.1.1. Games, payoffs and strategies   230
        7.1.2. The battle of the Bismarck Sea   231
        7.1.3. Matching pennies   231
        7.1.4. Nonzero-sum games   232
        7.1.5. Battle of sexes   233
        7.1.6. Matching pennies   233
        7.1.7. Extensive form games   234
        7.1.8. Cooperative games   234
    7.2. Evolutionary game theory   235
        7.2.1. Evolutionary games: hawk and dove game   235
        7.2.2. Evolutionary stability   237
    7.3. Evolutionary graph theory   238
        7.3.1. Moran process   238
        7.3.2. Evolutionary graph   238
        7.3.3. Fixation probability   239
        7.3.4. Time to fixation   241
    7.4. Games, coalitions, evolution, and consensus   242
        7.4.1. Games, graphs, and payoffs   242
        7.4.2. Replicator dynamics for games on graphs   247
    7.5. Consensus in evolutionary graphs   249
        7.5.1. Problem formulation   249
        7.5.2. Consensus: a coalition game   251
        7.5.3. Numerical example 7.1   256

7.6.   Notes   263
References   264

8. Multivehicle cooperative control   **267**
8.1.   Introduction   267
8.2.   Multivehicle architecture   268
    8.2.1.   Agent models   268
    8.2.2.   Attitude dynamics and kinematics   269
    8.2.3.   Information exchange   272
    8.2.4.   Behavior module   272
8.3.   Aerial vehicle systems   272
    8.3.1.   Vertical take-off landing aircrafts   273
8.4.   Ground vehicle systems   277
    8.4.1.   Differential drive vehicle   279
    8.4.2.   Car-like vehicle   280
    8.4.3.   Tractor-trailer systems   281
    8.4.4.   Off-road light autonomous vehicle   282
8.5.   Underwater vehicle systems   285
8.6.   Drone-assisted vehicular networks   286
8.7.   Load transportation   287
    8.7.1.   Single quadrotor with slung load   288
    8.7.2.   Dual quadrotor with slung load   291
    8.7.3.   Multiquadrotor systems with slung load 293
8.8.   Notes   294
References   294

9. Path planning in autonomous ground vehicles   **297**
9.1.   Introduction   297
9.2.   Vehicle model   299
9.3.   Global path planning   306
    9.3.1.   Graph-based planners   306
    9.3.2.   Sampling-based methods   309
    9.3.3.   Interpolating curve planners   311
    9.3.4.   Bio-inspired methods   312
9.4.   Multirobot path planning   319
    9.4.1.   Independent objectives   319
    9.4.2.   Mixed path planning   320
    9.4.3.   Distributed task assignment   321
9.5.   Some typical applications   323
9.6.   Notes   326
References   327

10. Path planning in autonomous aerial vehicles   **331**

10.1. Introduction   331
10.2. UAV and its environment   332
    10.2.1. UAV models   332
    10.2.2. Environment modeling   333
    10.2.3. Threat modeling   334
10.3. UAV path planning   335
    10.3.1. Potential fields   335
    10.3.2. Grid-based approaches   337
    10.3.3. Intelligent approaches   337
10.4. Multi-UAV path-planning problems   342
    10.4.1. Cooperative search–attack mission   343
    10.4.2. Coverage path planning   346
    10.4.3. Planning for cooperative sensing   349
10.5. Typical applications   351
    10.5.1. Simulation example   353
10.6. Notes   359
References   360

Index   363

About the authors

Magdi S. Mahmoud obtained BSc (Honors) in communication engineering, MSc in electronic engineering, and PhD in systems engineering, all from Cairo University in 1968, 1972, and 1974, respectively. He has been a Professor of Engineering since 1984. He is now a Distinguished Professor at King Fahd University of Petroleum and Minerals (KFUPM), Saudi Arabia. He was among the faculty at different universities worldwide including Egypt (CU, AUC), Kuwait (KU), UAE (UAEU), UK (UMIST), USA (Pitt, Case Western), Singapore (Nanyang), and Australia (Adelaide). He lectured in Venezuela (Caracas), Germany (Hanover), UK (Kent), USA (UoSA), Canada (Montreal), and China (BIT, Yanshan). He is the principal author of fifty-one (51) books, inclusive book-chapters, and the author/coauthor of more than 600 peer-reviewed papers. He is a fellow of the IEE, a senior member of the IEEE, the CEI (UK), and a registered consultant engineer of information engineering and systems (Egypt). He received the Science State Incentive Prize for outstanding research in engineering (1978, 1986), the State Medal For Science And Art, First Class (1978), and the State Distinction Award (1986), Egypt. He was awarded the Abdulhamed Showman Prize for Young Arab Scientists in the field of Engineering Sciences (1986), Jordan. In 1992, he received the Distinguished Engineering Research Award, College of Engineering and Petroleum, Kuwait University (1992), Kuwait. He is cowinner of the Most Cited Paper Award 2009, "Signal Processing", vol. 86, no. 1, 2006, pp. 140–152. His papers were selected among the 40 best papers in Electrical & Electronic Engineering by the Web of Science ISI in July 2012. He was interviewed for "People in Control", IEEE Control Systems Magazine, August 2010. He served as Guest Editor for the special issue "Neural Networks and Intelligence Systems in Neurocomputing" and Guest Editor for the 2015 International Symposium on Web-of-Things and Big Data (WoTBD 2015) 18–20 October 2015, Manama, Bahrain. He is a Regional Editor (Middle East and Africa) of International Journal of Systems, Control and Communications (JSCC), Interscience Publishers since 2007, member of the Editorial Board of the Journal of Numerical Algebra, Control and Optimization (NACO), Australia, since 2010, an Associate Editor of the International Journal of Systems Dynamics Applications (IJSDA), since 2011, member of the Editorial Board of the Journal of Engineering

Management, USA, since 2012, and an Academic Member of Athens Institute for Education and Research, Greece, since 2015. Since 2016, he is an Editor of the Journal Mathematical Problems in Engineering, Hindawi Publishing Company, USA. He is currently actively engaged in teaching and research in the development of modern methodologies to distributed control and filtering, networked control systems, fault-tolerant systems, cyber-physical systems and information technology.

Mojeed O. Oyedeji was born in Lagos, Nigeria. He received his BS degree in electrical and electronic engineering from Osun State University, Nigeria, in 2012 and MSc in Systems Engineering from King Fahd University of Petroleum and Minerals (KFUPM), Saudi Arabia. He is now a Lecturer B and working for his PhD degree in the Department of Systems Engineering at KFUPM, Saudi Arabia. His interests cover robust control theory, multiagent systems and cyber-physical systems.

Yuanqing Xia was born in Anhui Province, China, in 1971, and graduated from the Department of Mathematics, Chuzhou University, China, in 1991. He received his MSc degree in Fundamental Mathematics from Anhui University, China, in 1998, and PhD degree in control theory and control engineering from Beijing University of Aeronautics and Astronautics, China, in 2001. From 1991 to 1995, he was with Tongcheng Middle-School, China, where he worked as a teacher. From January 2002 to November 2003, he was a postdoctoral research associate in the Institute of Systems Science, Academy of Mathematics and System Sciences, Chinese Academy of Sciences, China, where he worked on navigation, guidance, and control. From November 2003 to February 2004, he was with the National University of Singapore as a Research Fellow, where he worked on variable structure control. From February 2004 to February 2006, he was with the University of Glamorgan, UK, as a research fellow, where he worked on networked control systems. From February 2007 to June 2008, he was a guest professor with Innsbruck Medical University, Austria, where he worked on biomedical signal processing. Since July 2004, he has been with the School of Automation, Beijing Institute of Technology, Beijing, first as an associate professor, then, since 2008, as a professor. And in 2012, he was appointed as Xu Teli distinguished professor at the Beijing Institute of Technology, then in 2016, as chair professor. In 2012, he obtained the National Science Foundation prize for Distinguished Young Scholars of China, and in 2016, he was honored as the

Yangtze River Scholar Distinguished Professor and was supported by National High Level Talents Special Support Plan ("Million People Plan") by the Organization Department of the CPC Central Committee. He is now the dean of School of Automation, Beijing Institute of Technology. He has published 8 monographs with Springer, John Wiley, and CRC, and more than 100 papers in international scientific journals. He is a deputy editor of Journal of Beijing Institute of Technology, an associate editor of Acta Automatica Sinica, Control Theory and Applications, International Journal of Innovative Computing, Information and Control, International Journal of Automation and Computing. He obtained the Second Award of the Beijing Municipal Science and Technology (No. 1) in 2010 and 2015, the Second National Award for Science and Technology (No. 2) in 2011, and the Second Natural Science Award of the Ministry of Education (No. 1) in 2012. His research interests include networked control systems, robust control and signal processing, active disturbance rejection control, and flight control.

Preface

Multiagent systems (MASs) is perhaps one of the most exciting and the fastest growing domains in the intelligent resource management and agent-oriented technology which deals with modeling of autonomous decision making entities. Recent developments have produced very encouraging results in its novel approach to handle multiplayer interactive systems. In particular, the multiagent system approach is adapted to model, control, manage, or test the operations and management of several system applications including multivehicles, microgrids, multirobots, where agents represent individual entities in the network. Each participant is modeled as an autonomous participant with independent strategies and responses to outcomes. It is able to operate autonomously and interact proactively with the environment. In view of the available results, it turns out that research avenues in multiagent systems offer great opportunities for further developments from theoretical, simulation, and implementation aspects.

This book aims at laying down the basic definitions, essential ingredients and providing some advanced developments in the fascinating area of multiagent systems (MASs) oriented toward dynamical systems.

Several multiagent models exist in the literature. The proposed book adopts the unique approach of structuring along four dimensions: *Agent, Environment, Interaction, and Organization*, while taking two main points of view: *global (system-centered) and local (agent-centered)*.

In the proposed book, the focus is on the wide-sense consensus problem in multiagent systems. We rely heavily in deriving our results on algebraic graph theory and topology. The main objective of writing this book is to contribute to the further development of advanced distributed consensus methods for different classes of multiagent methods. It will help in expanding the field of coordinated multiagent dynamic systems, including swarms, multivehicle and swarm robotics that has become increasingly popular in recent years. The book will establish new results based on rigorous math tools and implemented in efficient algorithmic procedures.

From this perspective, the pedagogical objectives of the book are as follows:

1. Introducing a coherent and unified framework for studying multiagent systems (MAS) with particular emphasis on different models;

2. Acquainting students with the system-theoretic background required to read and contribute to the research literature on MASs;

3. Helping expand the field of coordinated multiagent dynamical systems, including swarms, multivehicle and swarm robotics that have become increasingly popular in recent years; and

4. Providing a modest coverage of new results based on rigorous math tools and implemented in efficient algorithmic procedures.

The following is a short summary of the chapters of the book:

- **Chapter 1** (An overview)

  In this chapter, an extensive overview of multiagent distributed control will be carried out, detailing the current state-of-art research in this field. Detailed classifications will be provided highlighting existing solutions, problems and developments in this area.

- **Chapter 2** (Modeling of multiagent systems)

  This chapter presents a modeling framework and defines the intended problem formulation. In this chapter, we present the dynamics of some example practical systems where subsequently proposed algorithms will be applied. Formulations for different consensus problems will be discussed.

- **Chapter 3** (Consensus in multiagent systems over time-varying networks)

  Consensus control in multiagent systems is becoming increasingly popular among researchers due to its applicability in analyzing and designing coordination behaviors among agents in a multiagent framework. This chapter provides an extensive overview on consensus control in multiagent systems from the network perspective. Specifically, this chapter provides an overview of agent models (discrete and continuous) which have been studied by earlier researchers. A summary of different forms of consensus in multiagent systems will be provided, followed by an overview of recent results in consensus-related problems involving network phenomenon such as time-delay, actuator failures, switching, and random networks.

- **Chapter 4** (Adaptive cluster consensus for multiagent systems)

  In this chapter, we discuss adaptive cluster consensus for a class of multiagent systems (MASs) with unknown control coefficients. Specifically, we consider linear and nonlinear MASs with single and double integrator dynamics. It has been established in some previous studies that there is a relationship between coupling gains and eigenvalues of the Laplacian matrix, and based on the selection of the coupling gains, cluster consensus may or may not be achieved. Using Lyapunov theory, we propose some distributed adaptive algorithms based on relative errors between agents for adjusting intra- and intercluster coupling gains.

- **Chapter 5** (Consensus conditions under denial of services attacks)
  In this chapter, couple–group consensus of multiagent systems under denial-of-service (DoS) attacks will be studied. Specifically, couple–group consensus problem involving DoS attack within subgroups will be examined. We derive some necessary conditions for the multiagent system to remain in consensus using information about the eigenvalues of the Laplacian matrix representing the network, the Laplacian of the attack modes, and the coupling strengths. Based on simulation studies and derived conditions, we will deduce that the coupling strengths between and within subgroups can be adjusted after the influence of an attack for the multiagent system to remain in a consensus state or renegotiate a new consensus state.

- **Chapter 6** (A graph neural network approach to multiagent coordination)
  In this chapter, distributed coordination protocols will be designed for MASs using graph neural network theory. The interactions among the agents will be formed using neural networks function approximation based on the agents' and team objectives.

- **Chapter 7** (A fuzzy graph theory approach to multiagent coordination)
  In this chapter, distributed coordination protocols will be proposed for different consensus-type problems over fuzzy graphs. Much of the analysis discussed here will be largely based on the fuzzy graph theory. In this sense, the coordination among agents will be based on fuzzy graphs, that is, the links (edges) will be formed in a time-varying manner based on fuzzy laws.

- **Chapter 8** (An evolutionary graph approach to multiagent coordination)
  In this chapter, design of distributed coordination protocols will be presented for MASs via evolutionary graph theory approaches. Essentially, the graph topology is a switching topology where the switching rules are defined based on an evolutionary game.

- **Chapter 9** (Path planning in autonomous ground vehicles)
  Path planning is an important task in unmanned vehicle systems. This chapter will examine different path planning algorithms for unmanned ground vehicles such as grid-based, sampling based, and computational-intelligence based algorithms.

- **Chapter 10** (Path planning in autonomous aerial vehicles)
  Path planning is an important task in unmanned aerial vehicle systems. Path planning tasks are particularly difficult and cumbersome in UAV systems because the algorithms are developed in three-dimensional configuration spaces which may be affected with different forms of uncertainties. In this chapter, we discuss different path planning problems for UAVs.

Magdi S. Mahmoud, Mojeed O. Oyedeji, and Yuanqing Xia
KFUPM–Saudi Arabia, BIT–China
June 2019

Acknowledgments

Special thanks are due to Elsevier team; particularly Acquisitions Editor Sonnini R. Yura for guidance, assistance, and dedication throughout the publishing process. We are grateful to all the anonymous referees for carefully reviewing and selecting the appropriate topics for the final version during this process. Portions of this volume were developed and upgraded while offering the graduate courses **SCE-612-171, SCE-612-172, SCE-701-172, SCE-701-181, SCE-515-182** at KFUPM, Saudi Arabia. Dr. Mahmoud and Mr. Oyedeji acknowledge the support by the Deanship of Scientific Research (DSR) at KFUPM through distinguished research project no. **IN 161065**, and Dr. Xia acknowledges the National Key Research and Development Program of China under Grant **2018YFB1003700**.

Magdi S. Mahmoud, Mojeed O. Oyedeji, and Yuanqing Xia

March 2019

CHAPTER 1

An overview

1.1 Introduction

During the last two decades, cooperative control of multiagent systems (MASs) has received considerable attention due to its wide applications in many fields such as formation control, sensor networks, attitude of spacecraft alignment, and so on. Nowadays, multiagent systems (MASs) technology is growing to the point where the first multiagent systems are now being transferred from the laboratory to the utility, allowing industry to gain experience in the use of MASs and also to evaluate their effectiveness.

Consensus problems have a long history in computer science and form the foundation of the field of distributed computing. In networks of agents (or dynamical systems) "consensus" means reaching an agreement regarding a certain quantity of interest that depends on the state of all agents. A "consensus algorithm" (or protocol) is an interaction rule that specifies the information exchange between an agent and all of its neighbors in the network.

The problem of reaching consensus, that is, driving the state of a set of interconnected dynamical systems towards the same value, has received much attention due to its many applications in both the modeling of natural phenomena such as flocking and in the solution of several control problems involving synchronization or agreement between dynamical systems.

Cooperative collective behaviors in networks of autonomous agents, such as synchronization, consensus, swarming, and particularly flocking, have received considerable attention in recent years due to their broad applications to biological systems, sensor networks, unmanned air vehicle formations, robotic cooperation teams, mobile communication systems, and so on. In a flock, to coordinate with other dynamical agents, every individual needs to share information with each other, and they all need to agree on a common objective of interest. In this pursuit of scientific research, two strategies are commonly adopted: centralized control and distributed control. The centralized approach assumes that a central station is available and is powerful enough to communicate with and control the whole group of mobile agents. On the contrary, the distributed approach

Advanced Distributed Consensus for Multiagent Systems
https://doi.org/10.1016/B978-0-12-821186-1.00009-X

does not require such a central unit for control and management, at the cost of becoming more complicated in both network structure and organization of multiple agents. Although both approaches are practically depending on the situations and conditions of the applications at hand, the distributed approach is generally more attractive due to the existence of many inevitable physical constraints in practice such as only locally available information, limited resources and energy, distance decay in communications, and the large scale of agent systems. This section reviews some recent progress in distributed consensus and coordination control of mobile multiagent systems over complex communication networks.

The study of distributed coordination control of mobile multiagent systems was perhaps first motivated by the works in distributed computing, management science, and statistical physics, among others. Briefly stated, research studies on distributed coordination control of mobile multiagent systems include:

1. *Consensus.* This refers to the group behavior that all mobile agents asymptotically reach an agreement or alignment under a local distributed control protocol, with or without requiring some predefined common speed and orientation in their asymptotic motions.
2. *Formation control.* This refers to the group behavior that all mobile agents asymptotically form a predesigned geometrical configuration through local interactions, with or without a common reference such as a target state or convergence agreement.
3. *Distributed estimation and control.* This refers to designing distributed controllers for networked mobile systems, using local estimators to obtain the needed global information.

1.2 Notations

Throughout this book, the following notations are adopted in the different chapters. Let $\mathbb{R}^+$ and $\mathbb{R}^n$ respectively denote the set of nonnegative real numbers and the n-dimensional Euclidean space. Let $\mathbb{N}_+$ be the set of positive integers and let In denote an n-dimensional identity matrix. For $x \in \mathbb{R}^n$, x^T denotes its transpose. The vector norm is defined as $\|x\| = \sqrt{x^T x}$. For a matrix $A \in \mathbb{R}^{n \times n}$, $\|A\| = \sqrt{\lambda_{\max}(A^T A)}$, where $\lambda_{\max}(\cdot)$ (resp., $\lambda_{\min}(\cdot)$) represents the largest (resp., the smallest) eigenvalue, and the notation $A > 0$ means that A is a real symmetric and positive definite matrix. Moreover, let $(\Omega, \mathbb{F}, \{\mathbb{F}_t\}, \mathbb{P})$ be a complete probability space with filtration $\{\mathbb{F}_t\}_{t \geq t_0}$ satisfying the usual conditions (i.e., the filtration contains all $\mathbb{P}$-null sets and is right continuous). Denote by $C([-\tau, 0]; \mathbb{R}^n])$ the family of continuous functions φ from $[-\tau, 0]$ to $\mathbb{R}^n$ with the norm $\|\varphi\| = \sup_{-\tau \le \theta \le 0} |\varphi(\theta)|$. Let $L^2_{\mathbb{F}_{t_0}}([-\tau, 0]; \mathbb{R}^n)$ be the family of all $\mathbb{F}_{t_0}$-measurable $C([-\tau, 0]; \mathbb{R}^n)$-valued random variables $\xi = \{\xi(s) : -\tau \le s \le 0\}$ such that $\sup_{-\tau \le s \le 0} \mathbb{E}|\xi(s)|^2 < \infty$, where $\mathbb{E}\{\cdot\}$ stands for the mathematical expectation operator with respect to the given probability measure $\mathbb{P}$. The Dini derivative of $\psi(t)$ is defined as $D^+ \psi(t) = \lim \sup_{s \to 0^+} (\psi(t+s) - \psi(t))/s$.

1.3 Elements of graph theory

In this section, some preliminary knowledge of graph theory [1] is introduced to facilitate the subsequent analysis. For a system of n connected agents, its network topology can be modeled as a directed graph.

(A) Let $\mathbb{G} = (\mathbb{V}, \mathbb{E}, \mathbb{A})$ be a weighted directed graph of order n, where $\mathbb{V} = \{1, \ldots, n\}$ is the set of nodes, $\mathbb{E} \subseteq \mathbb{V} \times \mathbb{V}$ is the set of edges, and $\mathbb{A} = [a_{ij}] \in \mathbb{R}^{n \times n}$ is a nonnegative adjacency matrix. An edge of $\mathbb{G}$ is denoted by a pair of distinct nodes $(i, j) \in \mathbb{E}$, where nodes i and j are called the child and parent node, respectively. A path in a directed graph is a sequence $i_0, i_1, \ldots, i_f$ of different nodes such that (i_{j-1}, i_j) is an edge for $j = 1, 2, \ldots, f$, $f \in \mathbb{Z}^+$. Denote by $\mathbb{N}_i = \{j \mid (i, j) \in \mathbb{E})\}$ the set of neighbors of node i. The adjacency matrix $\mathbb{A} = [a_{ij}] \in \mathbb{R}^{n \times n}$ is defined such that a_{ij} is the nonnegative weight of edge (i, j).

(B) We assume $a_{ij} = 0$ if $(i, j) \notin \mathbb{E}$ and $a_{ii} = 0$ for all $i \in 1, \ldots, n$. The Laplacian matrix $\mathbb{L} = [l_{ij}] \in \mathbb{R}^{n \times n}$ is defined as $l_{ii} = \sum_{j=1, j \neq i}^n a_{ij}$ and $l_{ij} = -a_{ij}$ $(i \neq j)$. A *directed tree* is a directed graph, in which there is exactly one parent for every node except for a node called the root. A *directed spanning tree* is a directed tree, which consists of all of the nodes in $\mathbb{G}$. A directed graph contains a directed spanning tree if there exists a directed spanning tree as a subgraph of the graph. Let $\mathbb{G} = (\mathbb{V}, \mathbb{E}, \mathbb{A})$ be a directed graph of order n, where $\mathbb{V} = \{s_1, \ldots, s_n\}$ is the set of nodes, $\mathbb{E} \subseteq \mathbb{V} \times \mathbb{V}$ is the set of edges, and $\mathbb{A} = [a_{ij}] \in \mathbb{R}^{n \times n}$ is a weighted adjacency matrix. The node indexes belong to a finite index set $I = \{1, 2, \ldots, n\}$. An edge of $\mathbb{G}$ is denoted by $e_{ij} = (s_i, s_j)$, where the first element, s_i, of e_{ij} is said to be the tail of the edge and the other, s_j, the head. The adjacency elements associated with the edges are positive, that is, $e_{ij} \in \mathbb{E} \Longleftrightarrow a_{ij} > 0$. If a directed graph has the property that $a_{ij} = a_{ji}$ for any $i, j \in I$, then this directed graph is called undirected. The Laplacian of the directed graph is defined as $\mathbb{L} = \Delta - \mathbb{A} \in \mathbb{R}^{n \times n}$, where $\Delta = [\Delta_{ij}]$ is a diagonal matrix with $\Delta_{ii} = \sum_{j=1}^n a_{ij}$. An important fact of $\mathbb{L}$ is that all the row sums of $\mathbb{L}$ are zero and thus 1 is an eigenvector of $\mathbb{L}$

associated with the zero eigenvalue. The set of neighbors of node s_i is denoted by $\mathbb{N}_i = \{s_j \in \mathbb{V} : (s_i, s_j) \in \mathbb{E}\}$. A directed path is a sequence of ordered edges of the form $(s_{i1}, s_{i2}), (s_{i2}, s_{i3}), \ldots$, where $s_{ij} \in \mathbb{V}$ in a directed graph. A directed graph is said to be strongly connected, if there is a directed path from every node to every other node. Moreover, a directed graph is said to have spanning trees, if there exists a node such that there is a directed path from every other node to this node.

Let $\text{Re}(z)$, $\text{Im}(z)$, and $\|z\|$ be the real part, the imaginary part, and the modulus of a complex number z, respectively. Let $I_n (0_n)$ be the identity (zero) matrix of dimension n and 1_n be the $n \times 1$ column vector of all ones. Also $\otimes$ represents the Kronecker product.

1.3.1 Basic results

Lemma 1.1 ([7]). *If a graph $\mathbb{G}$ has a spanning tree, then its Laplacian $\mathbb{L}$ has the following properties:*

1. *Zero is a simple eigenvalue of $\mathbb{L}$, and $\mathbb{I}_n$ is the corresponding eigenvector, that is, $\mathbb{L}\mathbb{I}_n = 0$.*
2. *The other $n - 1$ eigenvalues all have positive real parts. In particular, if the graph G is undirected, then all these eigenvalues are positive and real.*

Lemma 1.2 ([5]). *Consider a directed graph $\mathbb{G}$. Let $\mathbb{D} \in \mathbb{R}^{n \times |\mathbb{E}|}$ be the 0–1-matrix with rows and columns indexed by the nodes and edges of $\mathbb{G}$, and $\mathbb{E} \in \mathbb{R}^{|\mathbb{E}| \times n}$ be the 0–1-matrix with rows and columns indexed by the edges and nodes of $\mathbb{G}$, such that*

$$D_{uf} = \begin{cases} 1 & \text{if the node } u \text{ is the tail of the edge } f, \\ 0 & \text{otherwise,} \end{cases} \tag{1.1}$$

$$E_{fu} = \begin{cases} 1 & \text{if the node } u \text{ is the head of the edge } f, \\ 0 & \text{otherwise,} \end{cases} \tag{1.2}$$

where $|\mathbb{E}|$ is the number of the edges. Let $\mathbb{Q} = \text{diag}\{q_1, q_2, \ldots, q_{|\mathbb{E}|}\}$, where q_p ($p = 1, \ldots, |\mathbb{E}|$) is the weight of the pth edge of $\mathbb{G}$ (i.e., the value of the adjacency matrix on the pth edge). Then the Laplacian of $\mathbb{G}$ can be transformed into $\mathbb{L} = \mathbb{D}\mathbb{Q}(\mathbb{D}^T - \mathbb{E})$.

1.3.2 Laplacian spectrum of graphs

This section is a concise review of the relationship between the eigenvalues of a Laplacian matrix and the topology of the associated graph. We refer the reader to [2] for a comprehensive treatment of the topic. We list a collection of properties associated with undirected graph Laplacians and adjacency matrices, which will be used in subsequent sections of the paper.

A graph $\mathbb{G}$ is defined as

$$\mathbb{G} = (\mathbb{V}, \mathbb{A}) \tag{1.3}$$

where $\mathbb{V}$ is the set of nodes (or vertices) $\mathbb{V} = \{1, \ldots, N\}$ and $\mathbb{A} \subseteq \mathbb{V} \times \mathbb{V}$ the set of edges (i, j) with $i \in \mathbb{V}$, $j \in \mathbb{V}$. The degree d_j of a graph vertex j is the number of edges which start from j. Let $d_{\max}(\mathbb{G})$ denote the maximum vertex degree of the graph $\mathbb{G}$.

1.3.3 Properties of adjacency matrix

We denote by $\mathbb{A}(\mathbb{G})$ the 0–1 adjacency matrix of the graph $\mathbb{G}$. Let $\mathbb{A}_{ij} \in \mathbb{R}$ be its (i, j)th element, then $\mathbb{A}_{i,i} = 0$, $\forall i = 1, \ldots, N$, $\mathbb{A}_{i,j} = 0$ if $(i, j) \notin \mathbb{A}$ and $\mathbb{A}_{i,j} = 1$ if $(i, j) \in \mathbb{A}$, $\forall i, j = 1, \ldots, N$, $i \neq j$. We will focus on *undirected* graphs, for which the adjacency matrix is symmetric.

Let $\mathbb{S}(\mathbb{A}(\mathbb{G})) = \{\lambda_1(\mathbb{A}(\mathbb{G})), \ldots, \lambda_N(\mathbb{A}(\mathbb{G}))\}$ be the spectrum of the adjacency matrix associated with an undirected graph $\mathbb{G}$ arranged in nondecreasing order.

- **Property 1.** $\lambda_N(\mathbb{A}(\mathbb{G})) \leq d_{\max}(\mathbb{G})$.

  This property, together with Proposition 2.1, implies

- **Property 2.** $\gamma_i \geq 0$, $\forall \gamma_i \in \mathbb{S}(d_{\max} I_N - \mathbb{A})$.

  We define the Laplacian matrix of a graph $\mathbb{G}$ in the following way:

$$\mathbb{L}(\mathbb{G}) = \mathbb{D}(\mathbb{G}) - \mathbb{A}(\mathbb{G}) \tag{1.4}$$

where $\mathbb{D}(\mathbb{G})$ is the diagonal matrix of vertex degrees d_i (also called the valence matrix). The eigenvalues of Laplacian matrices have been widely studied by graph theorists. Their properties are strongly related to the structural properties of their associated graphs. Every Laplacian matrix is a singular matrix. By Gershgorin theorem [4], the real part of each nonzero eigenvalue of $\mathbb{L}(\mathbb{G})$ is strictly positive.

For undirected graphs, $\mathbb{L}(\mathbb{G})$ is a symmetric, positive semidefinite matrix, which has only real eigenvalues. Let $\mathbb{S}(\mathbb{L}(\mathbb{G})) = \{\lambda_1(\mathbb{L}(\mathbb{G})), \ldots, \lambda_N(\mathbb{L}(\mathbb{G}))\}$ be the spectrum of the Laplacian matrix $\mathbb{L}$ associated with an undirected graph $\mathbb{G}$ arranged in nondecreasing order. Then,

- **Property 3.**
  1. $\lambda_1(\mathbb{L}(\mathbb{G})) = 0$ with corresponding eigenvector of all ones, and $\lambda_2(\mathbb{L}(\mathbb{G}))$ iff $\mathbb{G}$ is connected. In fact, the multiplicity of 0 as an eigenvalue of $\mathbb{L}(\mathbb{G})$ is equal to the number of connected components of $\mathbb{G}$.

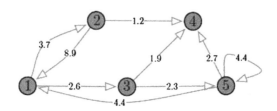

Figure 1.1 Sample graph.

2. The modulus of $\lambda_i(\mathbb{L}(\mathbb{G}))$, $i = 1, \ldots, N$ is less than N.

The second smallest Laplacian eigenvalue $\lambda_2(\mathbb{L}(\mathbb{G}))$ of graphs is probably the most important information contained in the spectrum of a graph. This eigenvalue, called the algebraic connectivity of the graph, is related to several important graph invariants, and it has been extensively investigated.

Let $\mathbb{L}(\mathbb{G})$ be the Laplacian of a graph $\mathbb{G}$ with N vertices and with maximal vertex degree $d_{\max}(\mathbb{G})$. Then properties of $\lambda_2(\mathbb{L}(\mathbb{G}))$ include

- **Property 4.**
  1. $\lambda_2(\mathbb{L}(\mathbb{G})) \leq (N/(N-1)) \min\{d(v), v \in \mathbb{V}\}$;
  2. $\lambda_2(\mathbb{L}(\mathbb{G})) \leq \nu(\mathbb{G}) \leq \eta(\mathbb{G})$;
  3. $\lambda_2(\mathbb{L}(\mathbb{G})) \geq 2\eta(\mathbb{G})(1 - \cos(\pi/N))$;
  4. $\lambda_2(\mathbb{L}(\mathbb{G})) \geq 2(\cos\frac{\pi}{N} - \cos\frac{2\pi}{N})\eta(\mathbb{G}) - 2\cos\frac{\pi}{N}(1 - \cos\frac{\pi}{N})d_{\max}(\mathbb{G})$,

where $\nu(\mathbb{G})$ is the vertex connectivity of the graph $\mathbb{G}$ (the size of a smallest set of vertices whose removal renders $\mathbb{G}$ disconnected) and $\eta(\mathbb{G})$ is the edge connectivity of the graph $\mathbb{G}$ (the size of a smallest set of edges whose removal renders $\mathbb{G}$ disconnected) [6].

Further relationships between the graph topology and Laplacian eigenvalue locations are discussed in [3] for undirected graphs. Spectral characterization of Laplacian matrices for directed graphs can be found in [4], see Fig. 1.1, where the corresponding graph Laplacian is given by

$$\mathbb{L}(\mathbb{G}) = \begin{bmatrix} 6.3 & -3.7 & -2.6 & 0 & 0 \\ -8.9 & 10.1 & 0 & -1.2 & 0 \\ 0 & 0 & 4.2 & -1.9 & -2.3 \\ 0 & 0 & 0 & 0 & 0 \\ -4.4 & 0 & 0 & -2.7 & 6.7 \end{bmatrix}. \tag{1.5}$$

A lemma about Laplacian $\mathbb{L}$ associated with a balanced digraph $\mathbb{G}$ is given hereafter:

Lemma 1.3. *If $\mathbb{G}$ is balanced, then there exists a unitary matrix*

$$
V = \begin{bmatrix} \frac{1}{\sqrt{n}} & * & \cdots & * \\ \frac{1}{\sqrt{n}} & * & \cdots & * \\ \vdots & \vdots & \ddots & \vdots \\ \frac{1}{\sqrt{n}} & * & \cdots & * \end{bmatrix} \in \mathbb{C}^{n \times n} \tag{1.6}
$$

such that

$$
V^* \mathbb{L} V = \begin{bmatrix} 0 & \\ & H \end{bmatrix} = \Lambda \in \mathbb{C}^{n \times n}, \quad H \in \mathbb{C}^{(n-1) \times (n-1)}. \tag{1.7}
$$

Moreover, if $\mathbb{G}$ has a globally reachable node, $H + H^*$ is positive definite.

Proof. Let $V = [\zeta_1, \zeta_2, \ldots, \zeta_n]$ be a unitary matrix where $\zeta_i \in \mathbb{C}^n$ ($i = 1, \ldots, n$) are the column vectors of $\mathbb{V}$ and

$$
\zeta_1 = (1/\sqrt{n})1_n = (1/\sqrt{n}, 1/\sqrt{n}, \ldots, 1/\sqrt{n})^T.
$$

Notice that if $\mathbb{G}$ is balanced, then $\zeta_1^* \mathbb{L} = 0$. Thus we have

$$
V^* \mathbb{L} V = V^* \mathbb{L} [\zeta_1, \zeta_2, \ldots, \zeta_n]
$$

$$
= \begin{bmatrix} \zeta_1^* \\ \zeta_2^* \\ \vdots \\ \zeta_n^* \end{bmatrix} [0_n, \mathbb{L}\zeta_2, \ldots, \mathbb{L}\zeta_n]
$$

$$
= \begin{bmatrix} 0 & 0_{n-1}^T \\ 0_{n-1} & H \end{bmatrix}.
$$

Furthermore, if $\mathbb{G}$ has a globally reachable node, then $\mathbb{L} + \mathbb{L}^T$ is positive semi-definite, see Theorem 7 in [8]. Hence, $V^*(\mathbb{L} + \mathbb{L}^T)V$ is also positive semidefinite. Furthermore, we know that 0 is a simple eigenvalue of $\mathbb{L}$ and therefore $H + H^*$ is positive definite. $\square$

As a closing remark, the Laplacian matrix satisfies the property $\mathbb{L} = CC^T$. It is well-known fact that this property holds regardless of the choice of the orientation of $\mathbb{G}$. Let x_i denote a scalar real value assigned to v_i. Then $x = [x_1, \ldots, x_n]^T$ denotes the state of the graph $\mathbb{G}$. We define the *Laplacian potential* of the graph as follows:

$$\Psi_{\mathbb{G}}(x) = \frac{1}{2}x^T \mathbb{L}x. \tag{1.8}$$

From this definition, the following property of the Laplacian potential of the graph follows:

Lemma 1.4 ([9]). *The Laplacian potential of a graph is positive definite and satisfies the following identity:*

$$x^T \mathbb{L}x = \sum_{j \in \mathbb{N}_i}(x_j - x_i)^2. \tag{1.9}$$

Moreover, given a connected graph, $\Psi_{\mathbb{G}}(x) = 0$ if and only if $x_i = x_j$, $\forall i, j$.

It follows from Lemma 1.4 that the Laplacian potential of the graph $\Psi_{\mathbb{G}}(x)$ is a measure of the *total disagreement* among all nodes. If at least two neighboring nodes of $\mathbb{G}$ disagree, then $\Psi_{\mathbb{G}} > 0$. Hence, minimizing $\Psi_{\mathbb{G}}$ is equivalent to reaching a consensus, which signifies a fundamental key in the design of consensus protocols.

Remark 1.1. It well know [1] that for a connected undirected graph, the following property holds:

$$\min_{x \neq 0,\, 1_n^T x = 0} \frac{x^T \mathbb{L}x}{||x||^2} = \lambda_2(\mathbb{L}). \tag{1.10}$$

The proof follows from a special case of **Courant–Fischer theorem** in [10]. A connection between $\lambda_2(\hat{\mathbb{L}})$, with $\hat{\mathbb{L}} = \frac{1}{2}(\mathbb{L} + \mathbb{L}^T)$, called the Fiedler eigenvalue of $\hat{\mathbb{L}}$ [11] and the performance (that is, the worst case speed of convergence) of a protocol on digraphs is established in [2].

Remark 1.2. Consider a simple digraph $\mathbb{G} = (\mathbb{V}, \mathbb{E}, \mathbb{A})$ with $\mathbb{V} = \{v_1, v_2, \ldots, v_n\}$ a nonempty finite set of nodes or vertices, a set of edges or arcs $\mathbb{E} \subseteq \mathbb{V} \times \mathbb{V}$, and an adjacency matrix $\mathbb{A} = [a_{ij}]$ with weights $a_{ij} > 0$ if $(v_j, v_i) \in \mathbb{E}$ and $a_{ij} = 0$ otherwise. Let $(v_i, v_i) \notin \mathbb{E}$, $\forall i$, so that there are no self-loops and no multiple edges in the same direction between the same pair of nodes. Thus, $a_{ii} = 0$. Define the in-degree of node v_i as the ith row sum of $\mathbb{A}$, $d_{\text{in}}(v_i) = \sum_{j=1}^{n} a_{ij}$, and the out-degree of node v_i as the ith column sum of $\mathbb{A}$, $d_{\text{out}}(v_i) = \sum_{j=1}^{n} a_{ji}$. The node of a digraph is balanced if and only if its in-degree and out-degree are equal, i.e., $d_{\text{in}}(v_i) = d_{\text{out}}(v_i)$. A graph $\mathbb{G}$

is called balanced if and only if all of its nodes are balanced. Define the diagonal in-degree matrix $\mathbb{D} = \text{diag}\{d_{\text{in}}(v_i)\}$ and the graph Laplacian matrix $\mathbb{L} = \mathbb{D} - \mathbb{A}$. The set of neighbors of a node v_i is $\mathbb{N}_i = \{v_j : (v_j, v_i) \in \mathbb{E}\}$, the set of nodes with edges incoming to v_i. A directed path is a sequence of nodes $v_1, v_2, \ldots, v_r$ such that $(v_i, v_{i+1}) \in \mathbb{E}$, $i \in \{1, 2, \ldots, r-1\}$. A semipath is a sequence of nodes $v_1, v_2, \ldots, v_r$ such that $(v_i, v_{i+1}) \in \mathbb{E}$, or $(v_{i+1}, v_i) \in \mathbb{E}$, $i \in \{1, 2, \ldots, r-1\}$. Node v_i is said to be connected to node v_j if there is a directed path from v_i to v_j. Node v_i is called a root node if it has a directed path to all other nodes. Graph $\mathbb{G}$ is said to be strongly connected if there is a directed path from every node to every other node, and weakly connected if any two different nodes are connected by a semipath. A subgraph of $\mathbb{G}$ is a digraph whose vertices and edges respectively belong to $\mathbb{V}$ and $\mathbb{E}$. A spanning subgraph of $\mathbb{G}$ is a subgraph of $\mathbb{G}$ with vertices $\mathbb{V}$. A directed tree is a connected digraph where every node has in-degree equal to 1, except for one with in-degree of 0. A spanning tree of a digraph is a directed tree formed by graph edges that connects all the nodes of the graph. A graph is said to have a spanning tree if a subset of the edges forms a directed tree. This is equivalent to saying that all nodes in the graph are reachable from a single (root) node.

1.4 Mathematical models for agent dynamics

In this section, we focus on a particular element, the agents, and on modeling of their dynamics. We briefly summarize some of the mathematical models for agent/vehicle dynamics considered in the systems and control literature on multiagent dynamic systems (or swarms). We consider a swarm consisting of N individuals/agents moving in an n-dimensional Euclidean space and, unless otherwise stated, denote with $x_i \in \mathbb{R}^n$ the state vector and with $m_i \in \mathbb{R}^m$, $m \le n$ the control input of agent i. Depending on the context, the state vector x_i may denote (a collection of) the position, orientation, synchronization frequency, information to be agreed upon, etc. The dimensions of the state and control spaces (the values of n and m) change depending on the context as well.

1.4.1 Single integrator model

The simplest mathematical model considered in the literature for studying MASs or swarm behavior is the so-called higher-level, or kinematic, or single integrator model in which the agent motions are given by

$$\dot{x}_i(t) = u_i(t), \quad i = 1, \ldots, N, \tag{1.11}$$

where x_i is the state of agent i, u_i is its control input, and the dot represents the derivative (the change) with respect to time. As mentioned above, depending on the context the state x_i can represent the position p_i, the orientation angle or synchronization frequency θ_i, or other variables (or collection of those).

We refer to this model as a higher-level or kinematic model since it ignores the lower-level vehicle dynamics of the individual agents (e.g., robots). However, it is a relevant and useful model since it can be used for studying higher-level algorithms independent of the agent/vehicle dynamics and obtaining "proof of concept"-type results for swarm behavior. Moreover, in certain control tasks involving path planning, the trajectories generated using the higher-level agent models can be used as reference trajectories for the actual agents to track. Furthermore, (1.11) is a realistic simplified kinematic model for a class of omnidirectional mobile robots with so-called universal (or Swedish) wheels.

1.4.2 Double integrator model

Another dynamic model which is commonly used in the multiagent coordination and control literature is the point mass, or double integrator, model given by

$$\dot{p}_i(t) = v_i(t),$$
$$\dot{v}_i(t) = \frac{1}{m_i} u_i(t), \quad 1 \leq i \leq N, \tag{1.12}$$

where p_i is the position, v_i is the velocity, m_i is the mass of the agent, and u_i is the force (control) input, and the state of the systems can be defined as

$$x_i = [p_i^T, v_i^T]^T.$$

The higher-level model in (1.11) can be viewed also as a special case of the point mass model (1.12) under the assumption that the motion environment is very viscous, that $m_i \approx 0$ (as is the case for some bacteria), and the control input is taken as

$$u_i(t) = -k_v v_i + \bar{u}_i$$

with the velocity damping coefficient $k_v = 1$, and the control term $\bar{u}_i$ corresponding to u_i of (1.11). However, in general, for many biological and engineering systems, this assumption is not satisfied and the point mass model in (1.12) becomes more relevant.

1.4.3 Uncertain fully actuated model

A more realistic model for agent/vehicle dynamics (compared to the higher-level and the point mass models) is the fully actuated model:

$$M_i(p_i)\ddot{p}_i + f_i(p_i, \dot{p}_i) = u_i(t), \quad 1 \leq i \leq N, \tag{1.13}$$

where p_i represents the position or configuration (recall that $x_i = [p_i^T, v_i^T]^T$), $M_i(p_i) \in \mathbb{R}^{n \times n}$ is the mass or inertia matrix, $f_i(p_i, \dot{p}_i) \in \mathbb{R}^n$ represents the centripetal, Coriolis, gravitational effects, and additive disturbances. It is a realistic model for fully actuated omnidirectional mobile robots or for some fully actuated manipulators. What makes the model even more realistic is that it is assumed that (1.13) contains uncertainties and disturbances. In particular, it is assumed that

$$f_i(p_i, \dot{p}_i) = f_i^k(p_i, \dot{p}_i) + f_i^u(p_i, \dot{p}_i), \quad 1 \leq i \leq N, \tag{1.14}$$

where $f_i^k(p_i, \dot{p}_i)$ represents the known part and $f_i^u(p_i, \dot{p}_i)$ represents the unknown part. The latter is assumed to be bounded with a known bound, that is,

$$\|f_i^u(p_i, \dot{p}_i)\| \leq \bar{f}_i(p_i, \dot{p}_i), \quad 1 \leq i \leq N, \tag{1.15}$$

where $\bar{f}_i(p_i, \dot{p}_i)$ are all known. Moreover, besides the additive disturbances and uncertainties, it is assumed that for all i, the mass/inertia matrix is unknown, but nonsingular, and lower and upper bounded by known bounds. This means that the matrices $M_i(p_i)$ satisfy

$$\underline{M}_i(p_i) \|\gamma\|^2 \leq \gamma^T M_i(p_i)\gamma \leq \overline{M}_i(p_i)\|\gamma\|^2, \quad 1 \leq i \leq N, \tag{1.16}$$

where $\gamma \in \mathbb{R}^n$ is arbitrary and $\underline{M}_i(p_i)$, $\overline{M}_i(p_i)$ are known and satisfy $0 < \underline{M}_i(p_i) < \overline{M}_i(p_i) < \infty$. These uncertainties provide an opportunity for developing algorithms that are robust with respect to the above type of realistic uncertainties and disturbances.

1.4.4 Nonholonomic unicycle model

$$\dot{p}_{ix}(t) = v_i(t)\cos(\theta_i),$$

$$\dot{p}_{iy}(t) = v_i(t)\sin(\theta_i),$$
$$\dot{\theta}_i(t) = \omega_i(t),$$
$$\dot{v}_i(t) = \frac{1}{m_i}F_i,$$
$$\dot{\omega}_i(t) = \frac{1}{J_i}\tau_i(t), \quad 1 \le i \le N, \tag{1.17}$$

where p_{ix} and p_{iy} are the Cartesian (x and y, respectively) coordinates (on the 2-dimensional motion space), θ_i is the steering angle (or orientation), v_i is the translational (linear) speed, and ω_{ix} is the rotational (angular) speed of each agent i. The quantities m_i and J_i are positive constants and represent the mass and the moment of inertia of each agent, respectively. The control inputs to the system are the force input F_i and the torque input τ_i. Many mobile robots used for experimentation in the laboratories (e.g., robots with one castor and two differentially driven wheels) obey the model in (1.17).

It must be emphasized that the main mathematical tools used in representing swarms, beside differential or difference equations describing agent dynamics, are directed and undirected graphs and their geometric representations in the particular motion space [1]. See Section 1.2 for a concise introduction on these tools.

1.5 Notes

This chapter provided an extensive overview on consensus control in multiagent systems from the network perspective. Such systems are becoming increasingly popular among researchers due to their applicability in analyzing and designing coordination behaviors among agents in a multiagent framework. Specifically, the material presented agent models (discrete and continuous) as studied by earlier researchers. This was followed by a summary of different forms of consensus in multiagent systems and recent results in consensus-related problems involving network phenomena such as time-delay, actuator failures, switching, and random networks. Suggestions for future work towards designing better consensus protocols that address real-life problems in autonomous multiagent systems were outlined.

References

[1] C. Godsil, G. Royle, Algebraic Graph Theory, Graduate Texts in Mathematics, vol. 207, Springer-Verlag, Berlin, Germany, 2001.

[2] R. Olfati-Saber, J. Alex Fax, R.M. Murray, Consensus and cooperation in networked multi-agent systems, Proc. IEEE 95 (2007) 215–233.

[3] M.S. Mahmoud, Distributed Control and Filtering for Industrial Systems, IET Press, UK, December 2012.

[4] D. Angeli, P.-A. Bliman, Convergence speed of unsteady distributed consensus: decay estimate along the settling spanning-trees, SIAM J. Control Optim. 48 (1) (2009) 1–2.

[5] P. Lin, Y.M. Jia, L. Li, Distributed robust $\mathcal{H}_\infty$ consensus control in directed networks of agents with time-delay, Syst. Control Lett. 57 (8) (2008) 643–653.

[6] J. Zhou, Q. Wang, Convergence speed in distributed consensus over dynamically switching random networks, Automatica 45 (6) (2009) 1455–1461.

[7] W. Ren, R.W. Beard, Consensus seeking in multi-agent systems under dynamically changing interaction topologies, IEEE Trans. Autom. Control 50 (5) (May 2005) 655–661.

[8] R. Olfati-Saber, R.M. Murray, Consensus problems in networks of agents with switching topology and time delays, IEEE Trans. Autom. Control 49 (9) (2004) 1520–1533.

[9] R. Olfati-Saber, R.M. Murray, Consensus problems in networks of agents with switching topology and time-delays, IEEE Trans. Autom. Control 49 (9) (2004) 1520–1533.

[10] R.A. Horn, C.R. Johnson, Matrix Analysis, Cambridge Univ. Press, Cambridge, UK, 1987.

[11] M. Fiedler, Algebraic connectivity of graphs, Czechoslov. Math. J. 23 (98) (1973) 298–305.

CHAPTER 2

Consensus over fixed networks

2.1 Introduction

Consider a team of n robots described by a graph $\mathbb{G}(\mathbb{V}, \mathbb{E})$ where $\mathbb{V} = \{v_1, \ldots, v_n\}$ represents the set of vehicles and $\mathbb{E}$ captures the interaction in terms of communication, perception or trust level among the agents over directed or undirected networks. Matrices such as adjacency $\mathbb{A} = [a_{ij}] \in \mathbb{R}^{n \times n}$ and Laplacian matrix $\mathbb{L} = [\ell_{ij}] \in \mathbb{R}^{n \times n}$ introduced earlier provide useful and intuitive information about the interaction between the agents such that results from spectral analysis of these matrices can be applied in deriving useful theorems and lemmas for governing consensus in multiagent systems. Mathematical models representing agent dynamics can be homogeneous or heterogeneous. Homogeneous MASs have similar mathematical models defining the dynamics of each agent, implying that they are of the same model. In heterogeneous MASs, the agent dynamics vary, for example, in a team of multivehicle robot varying in shape, make, and parameters, but having similar state definitions, that is, displacement and velocity.

Consensus in a multirobot system generally means that the states of each robot (or agent) reach an agreement. However, agreement has different forms depending on the nature of the problem and consensus objectives. Different consensus forms including average, asymptotic, mean–square, fixed (or finite)-time, and max consensus have been studied in literature. Consider a multiagent system defined by the following dynamics:

$$\dot{x}_i = f(x_i, u_i) \tag{2.1}$$

where $x_i \in \mathbb{R}^n$ and $u_i \in \mathbb{R}^m$ are state and control inputs respectively, $f(x_i, u_i)$ represents any linear or nonlinear state function for each agent. Formal definitions for consensus are given for system (2.1) as follows

Definition 2.1 (Asymptotic consensus). An MAS consisting of n agents described by (2.1) reaches an asymptotic consensus locally if for any $x_i(0)$ in the neighborhood $|x_i(0) - x_j(0)| < \alpha$, where $\mu > 0$, condition (2.2) is satisfied:

$$\lim_{t \to \infty} ||x_i(t) - x_j(t)|| = 0 \ \forall \, i = 1, 2, \ldots, n. \tag{2.2}$$

Advanced Distributed Consensus for Multiagent Systems
https://doi.org/10.1016/B978-0-12-821186-1.00010-6

A global asymptotic consensus is reached if for any initial condition $x_i(0)$, $i = 1, 2, \ldots, N$, condition (2.2) is fulfilled.

Definition 2.2 (Average consensus). An average consensus problem is solved locally for any MAS consisting of n agents described by (2.1) if there exists a constant μ such that for any initial state $x_i(0)$, in the neighborhood $||x_i(0) - x_j(0)|| < \mu$, there exists a finite time $T > 0$ satisfying

$$\lim_{t \to \infty} ||x_i(t)|| = \frac{1}{n} \sum_{i=1}^{n} x_i(0) \ \forall i = 1, 2, \ldots, n. \tag{2.3}$$

A global average consensus is reached if for any initial condition $x_i(0)$, $i = 1, 2, \ldots, N$, condition (2.3) is fulfilled.

Definition 2.3 (Finite-time consensus). A finite-time consensus problem is solved locally for any MAS consisting of n agents described by (2.1) if there exists a constant μ such that for any initial state $x_i(0)$, in the neighborhood $||x_i(0) - x_j(0)|| < \mu$, there exists a finite time $T > 0$ satisfying

$$\lim_{t \to T} ||x_i(t) - x_j(t)|| = 0. \tag{2.4}$$

Definition 2.4 (Mean-square average consensus). A mean-square average consensus problem is solved for MAS described by a stochastic differential equation, or a normal differential equation over a random graph, if for any initial condition in the set $\mathbb{X}(0) \in \mathbb{R}^n$ the following conditions are satisfied:

$$\mathbb{E}(x^*) = \frac{1}{N} \sum_{j=1}^{N} x_j(0), \tag{2.5}$$

$$\mathrm{Var}(x^*) < \infty, \tag{2.6}$$

$$\lim_{t \to \infty} E(x_i(t) - x^*)^2 = 0 \ \forall i = 1, 2, \ldots, N. \tag{2.7}$$

Definition 2.5 (Max-consensus). A max consensus is reached for MAS consisting of n agents described by (2.1) if there exists a control protocol $\mathbb{U} \in \mathbb{R}^n$ such that given a sequence of initial conditions $\mathbb{X}(0) \in \mathbb{R}^n$, convergence as defined by (2.8) is achieved

$$x_i(t) = \max\{x_1(0), x_2(0), \ldots, x_n(0)\} \ \forall i \in N. \tag{2.8}$$

A **strong** max-consensus is achieved if (2.8) is attainable for all possible initial conditions in $x(0) \in \mathbb{R}^n$, otherwise, if (2.8) is feasible for only a subset of $x(0) \in \mathbb{R}^n$, then a **weak** max-consensus is achieved.

Having provided these important definitions, now we proceed in subsequent sections to examine different types of consensus problems and algorithms over fixed network topologies.

2.2 Leaderless consensus

Leaderless consensus is perhaps the most basic form of a consensus problem in multiagent systems where the control protocols are strictly designed based on relative state information between agents. All types of consensus problems in the absence of a leader agent are classified as *leaderless*.

2.2.1 Preliminaries

Consider an MAS with the following single-integrator agent dynamics, that is,

$$\dot{x}_i(t) = u_i(t). \tag{2.9}$$

For system (2.9), a popular leaderless distributed consensus algorithm is based on relative state information is defined as:

$$u_i(t) = -\sum_{j=1}^{n} a_{ij}(x_i(t) - x_j(t)), \tag{2.10}$$

where a_{ij} represents the adjacency relationship between agent i and j, where $a_{ij} = 0$ if $i = j$ and $a_{ij} \geq 0$ if $i \neq j$. The overall system dynamics under control protocol (2.10) is given as

$$\dot{\mathbf{x}}(t) = \mathbb{L}\,\mathbf{x}(t) \tag{2.11}$$

where $\mathbb{L} = \mathbb{D} - \mathbb{A}$ is the graph Laplacian. Some spectral properties of the graph Laplacian have been used by to derive important theorems for reaching consensus.

Definition 2.6 (Agreement subspace [1]). An agreement set $\boldsymbol{\Gamma} \subset \mathbb{R}^n$ is defined on the subspace span$\{\mathbb{I}\}$, that is,

$$\boldsymbol{\Gamma} = \{x \in \mathbb{R}^n \mid x_i = x_j, \ \forall \ i, j\}. \tag{2.12}$$

That is, the agreement space lies in an invariant set where the coordination states of the agents are equal.

For the MAS defined by (2.1), with an undirected graph $\mathbb{G}$, the states of the agent reach asymptotic consensus provided that the graph is connected. The solution of the state-space system (2.11) for an arbitrary initial condition $x(0)$ is defined as

$$x(t) = e^{-\mathbb{L}t}x(0). \tag{2.13}$$

The Laplacian $\mathbb{L}$ can be rewritten via diagonalizing similarity transformation in terms of its eigenvalues,

$$\mathbb{L} = M\Lambda M^T, \tag{2.14}$$

where M satisfies the property $M^T = M^{-1}$ and $\Lambda = \mathrm{diag}[\lambda_1(\mathbb{L}), \ldots, \lambda_n(\mathbb{L})]$. Therefore it is possible to write

$$\begin{aligned} x(t) &= e^{-M\Lambda M^T t}x(0) \\ &= Me^{-\Lambda t}M^T x(0) \\ &= e^{-\lambda_1 t}(m_1^T x(0))m_1 + e^{-\lambda_2 t}(m_2^T x(0))m_2 + \cdots + e^{-\lambda_n t}(m_n^T x(0))m_n. \end{aligned} \tag{2.15}$$

Hereinafter, we review some theorems and lemmas relevant to leaderless consensus problems.

Theorem 2.1 ([1]). *For an undirected network $\mathbb{G}$, the eigenvalues of the corresponding Laplacian matrix are ordered as*

$$0 = \lambda_1(\mathbb{L}) < \lambda_2(\mathbb{L}) \le \cdots \le \lambda_n(\mathbb{L}), \tag{2.16}$$

provided that the undirected graph is connected.

Theorem 2.2 ([1]). *The rate of convergence for system (2.11) is dictated by $\lambda_2(\mathbb{L})$ in a connected graph.*

Proof. As stated earlier, in a connected (undirected) graph, the eigenvalues are ordered as in (2.16), hence λ_2 is the smallest positive eigenvalue of $\mathbb{L}$ and dictates the speed of convergence of (2.15), and thus, $x(t)$ converges,

$$x(t) \to \frac{\mathbb{I}^T x(0)\mathbb{I}}{n}, \tag{2.17}$$

that is, the states of the agents converge to the average values of the initial conditions, where $\mathbb{I}$ stands for a column vector of ones, that is, $\mathbb{I} = [1, 1, \ldots, 1]^T$. $\qquad\square$

Lemma 2.1 ([9]). *Consider a multiagent system consisting of n agents with states denoted by x on a graph with Laplacian $\mathbb{L}$ satisfying $l_{ij} \geq 0 \; \forall \; i \neq j$ and $\sum_{j=1}^{N} \ell_{ij} = 0$, $i = 1, 2, \ldots, n$. The following conditions are equivalent:*

- *$\mathbb{L}$ has a simple zero eigenvalue and all other eigenvalues have positive real parts.*
- *$\mathbb{L}x = 0$ implies that $x_1 = x_2 = \cdots = x_n$.*
- *Consensus is reached asymptotically for the system described by $\dot{x} = -\mathbb{L}x$.*
- *The directed graph of $\mathbb{L}$ has a directed spanning tree.*
- *The rank of $\mathbb{L}$ is $n - 1$.*

2.2.2 Problem formulation: linear dynamics

Consider the following MAS (2.18) on a graph $\mathbb{G}$:

$$\dot{x}_i(t) = a_i x_i(t) + b_i u_i(t), \tag{2.18}$$

where a_i and b_i are state and control coefficients, respectively, which may be unknown and time-varying. When $a_i = 0$, we have a network of single-integrator agents. A problem of interest is to derive distributed consensus protocols such that in the MAS modeled by (2.18) leaderless consensus is achieved with respect to any of the earlier defined consensus forms. Consider the following distributed control protocol u_i for an MAS modeled by a graph $\mathbb{G}$ described by the single-integrator dynamics in (2.9),

$$u_i(t) = k \sum_{\substack{j \in \mathbb{N}_i}}^{n} a_{ij}(x_j(t) - x_i(t)). \tag{2.19}$$

A typical research problem can be posed as follows:

Under what conditions on k and the eigenvalues of the corresponding graph Laplacian, does the system (2.9) solve the leaderless consensus problem as defined by (2.2)?

Consider the following MAS on $\mathbb{G}$:

$$\dot{x}_i(t) = u_i(t - \tau_i), \tag{2.20}$$

where τ_i represents input delay. The following problems are of interest:

- What is the maximum tolerable delay by the system such that the MAS reaches leaderless consensus?
- Is it possible to design state estimators for each agent to compensate the effect of the delay?

For the consensus protocol in (2.10), a time-delayed version is given by

$$u_i(t) = -\sum_{j=1}^{n} a_{ij}(x_i(t - \tau_{ij}) - x_j(t - \tau_{ij})), \qquad (2.21)$$

where τ_{ij} denotes a time-delay parameter. In time-delayed consensus problems, the primary concern is finding the bounds on τ such that consensus is achieved.

Theorem 2.3 ([1]). *A network of single-integrator multiagent system described by a fixed, undirected and connected communication graph with constant communication delay $\tau_{ij} = \tau$ over the edges globally asymptotically solves the average consensus problem, provided one of the following equivalent conditions is satisfied:*

- *$\tau \in (0, \tau^*)$ with $\tau^* = \dfrac{\pi}{2\lambda_n}$, $\lambda_n = \lambda_{\max}(\mathbb{L})$,*

- *The Nyquist plot of $\Gamma(s) = \dfrac{e^{\tau s}}{s}$ has a zero encirclement around $\dfrac{-1}{\lambda_k} \ \forall \, k > 1$.*

Proof. See [1]. $\qquad\qquad\qquad\qquad\qquad\qquad\qquad\qquad\qquad\qquad$ □

Consider a network of agents described by the following double-integrator dynamics:

$$\dot{x}_i(t) = v_i(t),$$
$$\dot{v}_i(t) = ax_i(t) + bv_i(t) + u_i(t). \qquad (2.22)$$

For this system, a distributed consensus protocol is designed as [4]:

$$u_i(t) = k_1 \sum_{j=1}^{n} a_{ij}[x_j(t) - x_i(t)] + k_2 \sum_{j=1}^{n} a_{ij}[v_j(t) - v_i(t)]. \qquad (2.23)$$

Theorem 2.4 ([4]). *The MAS system described by (2.22) achieves consensus asymptotically under protocol (2.23), provided the following conditions are satisfied simultaneously:*

$$k_2^2 \mathrm{Im}^2(\lambda_i) - 4(a - k_1 \mathrm{Re}(\lambda_i)) > 0, \qquad (2.24)$$

$$\frac{k_1^2 \mathrm{Im}^2(\lambda)}{(b - k_2 \mathrm{Re}(\lambda_i))^2} + \frac{k_1 k_2 \mathrm{Im}^2(\lambda)}{b - k_2 \mathrm{Re}(\lambda)} < k_1 \mathrm{Re}(\lambda_i) - a, \qquad (2.25)$$

$$ab - (ak_2 + bk_1)\mathrm{Re}(\lambda_i) + k_1 k_2 |\lambda_i|^2 > 0, \qquad (2.26)$$

$$b - k_2 \mathrm{Re}(\lambda_i) \neq 0, \qquad (2.27)$$

where $\mathrm{Re}(S)$, $\mathrm{Im}(S)$ stand for the real part and imaginary part of a complex quantity S, respectively.

Corollary 2.1 ([4]). *If $a = 0$, $b = 0$, (2.22) reduces to a double-integrator system which achieves consensus asymptotically under protocol (2.23), provided the following conditions are satisfied simultaneously:*

$$k_1 > 0, \quad k_2 > 0, \tag{2.28}$$

$$\frac{k_2^2}{k_1} > \max_{i=1,2,3,\dots,n} \frac{\text{Im}^2(\lambda_i)}{\text{Re}(\lambda_i)|\lambda_i|^2}. \tag{2.29}$$

Consider the distributed control protocol (2.31) for an MAS modeled by a graph $\mathbb{G}$ described by the double-integrator dynamics (2.30),

$$\dot{x}_i = v_i,$$
$$\dot{v}_i = u_i, \tag{2.30}$$

$$u_i = k_1 \sum_{j \in \mathbb{N}_i}^{n} a_{ij}(x_j - x_i) + k_2 \sum_{j \in \mathbb{N}_i}^{n} a_{ij}(v_j - v_i). \tag{2.31}$$

Under what conditions on k and the eigenvalues of the corresponding graph Laplacian, does the system (2.9) solve the leaderless consensus problem as defined by (2.2)? Consider a network of n agents over a graph $\mathbb{G}$ in Fig. 2.1 with the following double-integrator model:

$$\dot{x}_i(t) = v_i(t),$$
$$\dot{v}_i(t) = u_i(t). \tag{2.32}$$

For system (2.32), a distributed control protocol is designed as [3]:

$$u_i(t) = \sum_{j=1}^{n} a_{ij}[(x_j(t) - x_i(t)) + \gamma(v_i(t) - v_j(t))]. \tag{2.33}$$

Applying distributed control protocol (2.33) to system (2.32), we get

$$\dot{x}_i(t) = v_i(t),$$
$$\dot{v}_i(t) = \sum_{j=1}^{n} a_{ij}[(x_j(t) - x_i(t)) + \gamma(v_i(t) - v_j(t))]. \tag{2.34}$$

System (2.34) reaches a global asymptotic consensus, provided that for all $x_i(0)$, $v_i(0)$, $i = 1, 2, \dots, n$,

$$\lim_{t \to \infty} ||x_i - x_j|| = 0, \tag{2.35}$$

$$\lim_{t \to \infty} ||v_i - v_j|| = 0. \tag{2.36}$$

System (2.34) can be rewritten as

$$\begin{bmatrix} \dot{x} \\ \dot{v} \end{bmatrix} = \begin{bmatrix} 0_{n \times n} & I_n \\ -\mathbb{L} & -\gamma \mathbb{L} \end{bmatrix} \begin{bmatrix} x \\ v \end{bmatrix}. \tag{2.37}$$

To derive useful results on the stability of system (2.37), let us examine the nature of its eigenvalues. To compute the eigenvalues of system, we need to solve the eigenvalue problem:

$$\det(\lambda I_{2n} - \Pi) = 0, \tag{2.38}$$

where Π is defined as

$$\Pi = \begin{bmatrix} 0_{n \times n} & I_n \\ -\mathbb{L} & -\gamma \mathbb{L} \end{bmatrix}. \tag{2.39}$$

Then

$$\det(\lambda I_{2n} - \Pi) = \det\left(\begin{bmatrix} \lambda I_n & -I_n \\ \mathbb{L} & \lambda I_n + \gamma \mathbb{L} \end{bmatrix} \right), \tag{2.40}$$

$$\det(\lambda I_{2n} - \Pi) = \det\left(\begin{bmatrix} \lambda^2 I_n + (1 + \gamma \lambda) \mathbb{L}_n \end{bmatrix} \right). \tag{2.41}$$

But

$$\det(\lambda I_n + \mathbb{L}_n) = \prod_{i=1}^{n} (\lambda - \mu_i), \tag{2.42}$$

where μ_i is the ith eigenvalue of $-\mathbb{L}_n$.

Therefore,

$$\det\left(\begin{bmatrix} \lambda^2 I_n + (1 + \gamma \lambda) \mathbb{L}_n \end{bmatrix} \right) = \prod_{i=1}^{n} (\lambda^2 - (1 + \gamma \lambda) \mu_i). \tag{2.43}$$

The eigenvalues of Π is thus computed as

$$\lambda_{i\pm} = \frac{\gamma \mu_i \pm \sqrt{\gamma^2 \mu_i^2 + 4\mu_i}}{2}, \tag{2.44}$$

where $\lambda_{i\pm}$ are eigenvalues of Π associated with μ_i.

Lemma 2.2 ([3]). *The distributed consensus algorithm (2.33) reaches consensus asymptotically if and only if Π has exactly two zero eigenvalues and all other eigenvalues have negative real parts.*

Lemma 2.3 ([3]). *If $-\mathbb{L}_n$ has a simple zero eigenvalue and all other eigenvalues are real and hence negative, then algorithm (2.33) achieves consensus asymptotically for any $\gamma > 0$.*

Consider a network of agents with the following dynamical representation:

$$\dot{x}_i(t) = v_i(t),$$
$$\dot{v}_i(t) = ax_i(t) + bv_i(t) + u_i(t).$$

For this system, a distributed time-delayed consensus protocol is designed as

$$u_i(t) = k_1 \sum_{j=1}^{n} a_{ij}[x_j(t-\tau) - x_i(t-\tau)] + k_2 \sum_{j=1}^{n} a_{ij}[v_j(t-\tau) - v_i(t-\tau)].$$
$$(2.45)$$

Theorem 2.5 ([4]). *The MAS system described by (2.22) achieves consensus asymptotically under protocol (2.45), provided*
* *the system is guaranteed to reach consensus in the delay-free case,*
* *the delay parameter τ is bounded as $0 < \tau < \tau^*$,*

where $\tau^* = \min_r \tau_r$ and τ_r is defined as

$$\tau_r = (k\pi + \tan^{-1} \psi_r)\omega_r^{-1},$$
$$(2.46)$$

ω_r is the root of the polynomial

$$(\omega_r^2 + a)^2 + b^2\omega_r^2 + (k_1^2 + k_2^2\omega_r^2)|\lambda_r|^2 = 0,$$
$$(2.47)$$

ϕ_r, ψ_r, and φ_r are defined as follows:

$$\psi_r = \frac{\phi_r(\omega_r^2 + a) + \phi_r\omega_r}{\phi_r b\omega_r - \varphi_r(\omega_r^2 + a)},$$
$$(2.48)$$
$$\phi_r = k_2\omega_r\mathrm{Re}(\lambda_r) + k_1\mathrm{Im}(\lambda_r),$$
$$(2.49)$$
$$\varphi_r = k_2\omega_r\mathrm{Im}(\lambda_r) - k_1\mathrm{Re}(\lambda_r),$$
$$(2.50)$$

$r = 2, 3, \ldots, n$.

2.2.3 Problem formulation: nonlinear dynamics

An average leaderless consensus problem was studied in [5] for MAS (2.51) exhibiting unknown nonlinear dynamics,

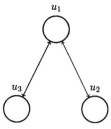

Figure 2.1 Graph showing interconnections between agents.

$$\dot{x}_i = f_i(t, x_i) + g_i(t, x_i)u_i + \omega_i(t), \tag{2.51}$$

where $x_i(t)$ is the state of the agent; $u_i(t)$ is the control input, f_i and g_i represent state and state-dependent controlling functions, while ω_i denotes unknown external disturbance. In [5], the following bounds were given for f_i, g_i, ω_i:

$$|f_i(t, x_i)| \leq \rho_i(t, x_i), \tag{2.52}$$

$$g_i(t, x_i) \geq g_0 > 0, \tag{2.53}$$

$$|\omega_i(t)| \leq \bar{\omega}, \tag{2.54}$$

where $\rho_i(t, x_i), g_0, \bar{\omega}$ are known nonnegative quantities.

Theorem 2.6 ([5]). *The nonlinear MAS (2.51) whose functions are bounded as in (2.52), (2.53), and (2.54) solves the average leaderless consensus problem asymptotically under the following distributed consensus algorithm:*

$$u_i = -\frac{1}{g_0}\mathrm{sgn}(x_i - \zeta_i)\left[\rho_i(t, x_i) + \bar{\omega} + \left\|\sum_{j\in\mathbb{N}} a_{ij}(x_i(t) - x_j(t))\right\| + \epsilon_i\right]. \tag{2.55}$$

2.2.4 Numerical example 2.1

Here, we examine consensus in a network of agents with single integrator dynamics. Consider a multirobot system described by (2.9) with communication topology $\mathbb{G}$ in Fig. 2.1. For the graph $\mathbb{G}$, we define the following adjacency and degree matrices:

$$\mathbb{A} = \begin{bmatrix} 0 & 1 & 1 \\ 0 & 0 & 1 \\ 0 & 1 & 0 \end{bmatrix}, \quad \mathbb{D} = \begin{bmatrix} 2 & 0 & 0 \\ 0 & 1 & 0 \\ 0 & 0 & 1 \end{bmatrix}. \tag{2.56}$$

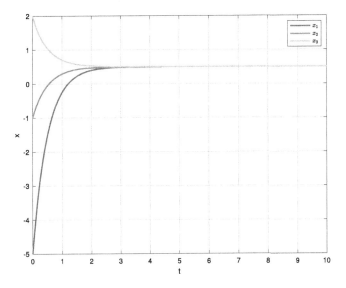

Figure 2.2 Leaderless consensus.

The corresponding Laplacian matrix is given as:

$$\mathbb{L} = \begin{bmatrix} 2 & -1 & -1 \\ 0 & 1 & -1 \\ 0 & -1 & 1 \end{bmatrix}, \tag{2.57}$$

and $\lambda(\mathbb{L}) = 0, 1, 2$. A simulation of the MAS (2.11), with initial conditions $x(0) = [-5 \ -1 \ 2]$ is demonstrated in Fig. 2.2.

2.3 Leader–follower consensus

Leader–follower consensus problems are quite common in MAS literature due to their real-life applications. In the communication/perception graph for a leader–follower network, there exists a root or leader node whose dynamics serves as a reference or tracking point for the followers. Leader–follower problems often appear in alignment, rendezvous, and containment control problems. In the preceding discussions we provide a summary of some recent advances in a leader–follower consensus problem citing relevant references which discuss problems related to time-delay, switching networks, random networks, and actuator failures. In leader–follower consensus problems, there is a leader node in the MAS network. Such consensus problems find practical applications of MAS where it is desired that other (follower) agents track the trajectory of a leader agent, especially in scenarios where only the leader agent has full information about the consensus objective. Here, distributed consensus protocols are such that the follower agents' states track the leader states with almost exact precision.

2.3.1 Description

Again, consider an MAS described by a single-integrator dynamics in (2.9), assuming the presence of a leader node also described by

$$\dot{x}_0(t) = f(x_0(t), u_0(t)), \tag{2.58}$$

where $x_0 \in \mathbb{R}^n$ represents the states of the leader agent and $u_0 \in \mathbb{R}^m$ represents the reference input to the leader. A simple distributed leader–follower protocol is defined for systems (2.58) and (2.9) as

$$u_i(t) = \sum_{j=1}^{n} a_{ij}(x_j(t) - x_i(t)) + b_i(x_i(t) - x_0(t)). \tag{2.59}$$

Definition 2.7 (Leader–follower consensus). The network of MAS described by (2.9) and (2.58) under distributed protocol (2.59) solves the leader–follower consensus problem asymptotically if

$$\lim_{t \to \infty} ||x_i(t) - x_0(t)|| = 0. \tag{2.60}$$

2.3.2 Problem formulation

We focus our attention on linear systems. The leader–follower consensus was studied for a class of second-order MAS in [11] under directed topology. Consider the following MAS:

$$\begin{aligned} \dot{x}_i(t) &= v_i(t) + u_i^1(t), \\ \dot{v}_i(t) &= u_i^2(t), \end{aligned} \tag{2.61}$$

where x_i and v_i represent position and velocity, respectively, of the follower agent; u_i^1, u_i^2 are control inputs of the follower agent. Also

$$\begin{aligned} \dot{x}_0(t) &= v_0(t), \\ \dot{v}_0(t) &= u_0(t), \end{aligned} \tag{2.62}$$

where $x_0(t)$ and $v_0(t)$ represent position and velocity, respectively, of the leader agent; $u_0(t)$ is the control input of the leader agent. In [11], the following distributed control protocol was designed for $u_i^1(t)$, $u_i^2(t)$:

$$u_i^1(t) = -\alpha \left[\sum_{j \in \mathbb{N}_i} a_{ij}(x_i(t) - x_j(t)) + b_i(x_i(t) - x_0(t)) \right], \tag{2.63}$$

$$u_i^2(t) = a_0(t) - \beta \left[\sum_{j \in \mathbb{N}_i} a_{ij}(v_i(t) - v_j(t)) + b_i(v_i(t) - v_0(t)) \right]. \tag{2.64}$$

2.3.2.1 Nonlinear dynamics

Adaptive leaderless consensus was proposed for first- and second-order nonlinear dynamics with unknown control coefficients in [6] under undirected and connected communication topologies. In the first-order case, the authors considered a nonlinear MAS with follower (2.65) and leader (2.66) dynamics:

$$\dot{x}_i = a_i u_i + b_i \phi_i(x_i), \tag{2.65}$$

$$\dot{x}_0 = u_0, \tag{2.66}$$

where a_i and b_i represent unknown coefficients and $\phi_i(x_i)$ is a known continuous function. The following distributed adaptive consensus algorithms were proposed to guarantee convergence to the leader–follower consensus:

$$u_i = -c_i \hat{\theta}_i \zeta_i - \hat{\theta}_i \hat{\theta}_i' \phi_i(x_i) + \hat{\lambda}_i, \tag{2.67}$$

$$\dot{\hat{\theta}}_i = \mathrm{sgn}(a_i)(c_i \zeta_i^2 + \zeta_i \hat{\theta}_i' \phi_i(x_i)), \tag{2.68}$$

$$\dot{\hat{\theta}}_i' = \zeta_i \phi_i(x_i), \tag{2.69}$$

$$\dot{\hat{\lambda}}_i = -\mathrm{sgn}(a_i)\zeta_i. \tag{2.70}$$

Also, distributed consensus protocols (2.73)–(2.75) were proposed for nonlinear second-order MAS (2.71):

$$\dot{x}_i = v_i, \tag{2.71}$$

$$\dot{v}_i = a_i u_i + b_i \phi_i(x_i, v_i), \tag{2.72}$$

$$u_i = -\mu_i \hat{\theta}_i z_i - \hat{\theta}_i(\zeta_i + \dot{\zeta}_i) - \hat{\theta}_i \hat{\theta}_i' \phi_i(x_i, v_i), \tag{2.73}$$

$$\dot{\hat{\theta}}_i = \mathrm{sgn}(a_i)(\mu_i z_i^2 + z_i(\zeta_i + \dot{\zeta}_i) + z_i \hat{\theta}_i' \phi_i(x_i, v_i)), \tag{2.74}$$

$$\dot{\hat{\theta}}_i' = z_i \phi_i(x_i). \tag{2.75}$$

The second-order consensus problem was investigated for a nonlinear MAS with a virtual leader in [8], that is, consider

$$\dot{x}_i = v_i,$$
$$\dot{v}_i = f(x_i) + u_i, \tag{2.76}$$

where x_i, v_i represent position and velocity of agent i; $f(x_i)$ is a nonlinear function, and u_i is the control input. The virtual leader dynamics is defined as:

$$\dot{x}_0 = v_0,$$
$$\dot{v}_0 = f(x_0). \tag{2.77}$$

The following Lipschitz assumption was made on the nonlinear function f:

$$(x - y)^T[f(x) - f(y)] \le (x - y)^T\Gamma(x - y), \quad \forall x, y \in \mathbb{R}^n, \tag{2.78}$$

where Γ is a positive definite diagonal matrix. A distributed adaptive linear protocol was proposed for system (2.76) and (2.77) as [8]

$$u_i = -\sum_{j \in N_i} \nabla_{q_i} \psi(\|x_i - x_j\|) - \sum_{j \in N_i} m_{ij}(v_i - v_j) - h_i c_i(x_i - x_0)$$
$$- h_i c_{2i}(v_i - v_0), \tag{2.79}$$

where m_{ij} and c_{2i} are adaptive gains defined as:

$$m_{ij} = k_{ij}(v_i - v_j)^T(v_i - v_j), \tag{2.80}$$
$$c_{2i} = k_i(v_i - v_j)^T(v_i - v_j). \tag{2.81}$$

Stability and convergence results were then derived based on connectivity assumptions and Lyapunov analysis. In some cases the nonlinear function f does not satisfy the Lipschitz property; problems of this nature were addressed in [12], where a nonlinear adaptive distributed control protocol (2.82) was designed for (2.76) and (2.77) as

$$u_i = -5\hat{\theta}_i \gamma_i(\zeta_i)^2 \zeta_i - \kappa \mathrm{sgn}(\zeta_i), \quad i = 1, \ldots, N, \tag{2.82}$$
$$\dot{\hat{\theta}}_i = \chi_i(\zeta_i)\zeta_i^2. \tag{2.83}$$

The leader–follower consensus problem for MAS (2.84) was studied in [7] with the assumption that only some of the system states are measurable:

$$\dot{x}_{k,i} = x_{k,i+1} + f_i(t, \bar{x}_{k,i}), \quad i = 1, 2, \ldots, n - 1,$$

$$\dot{x}_{k,n} = u_k + f_n(t, \bar{x}_{k,n}), \tag{2.84}$$

$$y_k = x_{k,1}, \quad k = 0, 1, \ldots N,$$

where $\bar{x}_{k,i} = (x_{k,i}, x_{k,2}, \ldots, x_{k,i})$, u_k, and y_k represent state, control input, and output of agent k, respectively. Index 0 is used to reference the states of the leader agent. The function f is assumed to satisfy the Lipschitz property:

$$|f_i(t, \bar{x}_{k,i} - f_i(t, \bar{x}_{l,i}))| \leq \eta(t) \sum_{j=1}^{i} |x_{k,j} - x_{l,j}|, \tag{2.85}$$

where $\eta(t) = c_1 e^{c_2 t}$, with c_1 and c_2 being known nonnegative numbers. Output feedback controllers were designed using relative state errors connectivity assumptions, Lipschitz property, and state estimators.

2.3.3 Related results

The mean-square leader–follower consensus problem was investigated in [9] over fixed undirected communication topologies under measurement noises and time-varying delays. The leader–follower tracking problem was investigated in [10] for an MAS with general linear dynamics and unknown disturbances using observer-based consensus algorithms. The leader–follower consensus problem for a nonlinear second-order MAS under denial-of-service attacks was investigated in [13] using event-triggering control strategy. The leader–follower consensus for a class of discrete-time multiagent systems with large-delay sequences under directed topology was studied in [48]. Some novel Lyapunov functionals have been proposed to analyze consensus problems under two different types of time-varying delays. Sufficient conditions for ensuring a leader–follower consensus were established using LMIs.

2.3.4 Numerical example 2.2

In what follows, we consider modeling and analysis of the battery energy storage systems (BESS) and the associated energy storage systems (ESS). The dynamic model of the state of charge (SoC) of an energy storage system (ESS), according to [17], is

$$s_i(t) = s_i(0) - \frac{\rho_i}{q_i} \int_0^t I_i(t) dt, \quad i \in \mathbb{V}, \tag{2.86}$$

where $s_i(t)$, I_i, q_i are state-of-charge, output current, and capacity of the ith ESS; ρ_i represent Coulomb efficiency defined as

$$\rho_i = \begin{cases} 1 & \text{when discharging,} \\ \eta_i & \text{when charging.} \end{cases} \tag{2.87}$$

Taking the derivative of (2.86), we get

$$\dot{s}_i(t) = -\frac{\rho_i}{q_i} I_i(t). \tag{2.88}$$

But $I_i = \frac{\sigma_i p_i}{v_i}$, therefore,

$$\dot{s}_i(t) = -\frac{\sigma_i \rho_i p_i}{q_i v_i}, \tag{2.89}$$

where σ_i is related to the converter efficiencies and defined as

$$\sigma_i = \begin{cases} \dfrac{1}{\tau_i^d} & \text{when discharging,} \\ \tau_i^c & \text{when charging,} \end{cases} \tag{2.90}$$

τ_i^d and τ_i^c are charging and discharging efficiencies.

Let μ_i be defined as follows:

$$\mu_i = \frac{\sigma_i \rho_i}{q_i v_i} = \begin{cases} \dfrac{1}{\tau_i^d Q_i V_i} = \mu_i^d & \text{when discharging,} \\ \dfrac{\rho_i \tau_i^c}{Q_i V_i} = \mu_i^c & \text{when charging.} \end{cases} \tag{2.91}$$

We can rewrite $\mu_i(t)$ as

$$\mu_i(t) = 0.5\,\text{sgn}(P_m(t))(\text{sgn}(P_m(t)) + 1)\mu_i^d \\ + 0.5\,\text{sgn}(P_m(t))(\text{sgn}(P_m(t) - 1))\mu_i^c, \tag{2.92}$$

where $P_m(t)$ is the power imbalance in the network between the generation and loads. Therefore, (2.89) becomes

$$\dot{s}_i = -\mu_i(t)p_i. \tag{2.93}$$

According to [18], the reference power generator of the ith ESS can be modeled with a first-order integrator as

$$\dot{p}_i = u_i. \tag{2.94}$$

Therefore, the overall dynamic model of the ith ESS unit can be represented as:

$$\dot{s}_i(t) = -\mu_i(t)p_i,$$
$$\dot{p}_i(t) = u_i. \qquad (2.95)$$

Further, each battery package in the BESS network is subject to some SoC (2.96) and power (2.97) constraints:

$$s_i^{\min} \le s_i \le s_i^{\max}, \qquad (2.96)$$
$$p_i^{\min} \le p_i \le p_i^{\max}. \qquad (2.97)$$

Energy dispatch objective

In the literature, based on the dynamic model of the BESS and safety constraints, the control objective for a distributed BESS is maintaining SoC and power levels, that is,

$$\lim_{t \to \infty} |s_i(t) - s_j(t)| = 0 \quad \forall i, j = 1, 2, \ldots, n,$$
$$\lim_{t \to \infty} |p_i(t) - p_{av}(t)| \approx 0 \quad \forall i, j = 1, 2, \ldots, n,$$

where p_{av} represents average power defined as a function of power mismatch:

$$p_{av}(t) = \frac{P_m(t)}{n}. \qquad (2.98)$$

The battery model (2.95) is modified to include an energy co-coordinator state,

$$\dot{s}_i(t) = -\mu_i \phi_i(z_i)p_i,$$
$$\dot{z}_i(t) = -z_i + v_i,$$
$$\dot{p}_i(t) = u_i. \qquad (2.99)$$

The extended state z_i is considered an energy co-coordinator according to [18], u_i and v_i are control inputs to the energy co-coordinator and power reference dynamics.

Assumption 2.3.1. There exists a charge controller in the network responsible for dispatching average power reference for the battery packages. The charge controller will be referred to as the "leader" in the multiagent system. At least one battery package is connected via the communication graph to the leader.

Under Assumption 2.3.1, we define the leader dynamics as

$$\dot{p}_0(t) = k_l(p_{av} - p_0), \tag{2.100}$$

where k_l is a control gain and p_{av} is the average power mismatch defined in (2.98). For the system defined by (2.99), the distributed control protocol u_i and v_i are designed as:

$$u_i = \omega_1 \sum_{j=1}^{n} a_{ij}(s_i - s_j), \tag{2.101}$$

$$v_i = \omega_2 \sum_{j=1}^{n} a_{ij}(p_i - p_j) + m_i(p_i - p_0), \tag{2.102}$$

where $\omega_1 > 0$ for discharging mode, $\omega_1 < 0$ for charging mode, and $\omega_2 > 0$. The closed-loop dynamics under protocols (2.101) and (2.102) is:

$$\dot{s}_i(t) = -\mu_i \phi_i(z_i) p_i,$$
$$\dot{z}_i(t) = -z_i + \omega_1 \sum_{j=1}^{n} a_{ij}(s_i - s_j),$$
$$\dot{p}_i(t) = \omega_2 \sum_{j=1}^{n} a_{ij}(p_i - p_j) + m_i(p_i - p_0). \tag{2.103}$$

Next, we consider the effect of model uncertainties in battery parameters on the battery packages and proceed to design an adaptive distributed controller.

Assumption 2.3.2. The battery parameter μ_i is subject to modeling uncertainties and parameter variations.

The dynamics in (2.99) is rewritten as:

$$\dot{s}_i(t) = -\theta_i \phi_i(z_i) p_i,$$
$$\dot{z}_i(t) = -z_i + v_i,$$
$$\dot{p}_i(t) = u_i, \tag{2.104}$$

where the battery parameter μ_i has been replaced with an uncertain parameter θ_i. To derive the control law v_i, we use the following Lyapunov analysis.

Consider the Lyapunov function,

$$V = \frac{1}{2} s^T \mathbb{L} s + \frac{1}{2} z^T z + \frac{1}{2} \sum_{i=1}^{N} \theta_i^2. \qquad (2.105)$$

The time-derivative of (2.105) along the trajectories of (2.104) is

$$\dot{V} = -\sum_{i=1}^{N} \eta_i \theta_i \phi_i(z_i) + \sum_{i=1}^{N} z_i(-z_i + u_i) + \sum_{i=1}^{N} \theta_i^2,$$

where η_i is defined as

$$\eta_i = \sum_{i=1}^{N} a_{ij}(s_i - s_j).$$

Let $\phi_i = 1 + z_i$, then

$$\dot{V} = -\sum_{i=1}^{N} \eta_i \theta_i p_i + \sum_{i=1}^{N} \eta_i \theta_i p_i z_i - \sum_{i=1}^{N} z_i^2 \qquad (2.106)$$
$$+ \sum_{i=1}^{N} z_i u_i + \sum_{i=1}^{N} \dot{\theta}_i \theta_i.$$

Collecting like terms gives

$$\dot{V} = -\sum_{i=1}^{N} z_i^2 + \sum_{i=1}^{N} \theta_i(\dot{\theta}_i - \eta_i p_i) + \sum_{i=1}^{N} z_i(u_i + \theta_i \eta_i p_i).$$

Choose the adaptive control laws:

$$u_i = -\theta_i \eta_i p_i, \qquad (2.107)$$
$$\dot{\theta}_i = \eta_i p_i. \qquad (2.108)$$

The Lyapunov function then satisfies

$$\dot{V} = -\sum_{i=1}^{N} z_i^2 < 0. \qquad (2.109)$$

For simulation purposes, we consider a BESS consisting of 10 battery packages with Coulombic and converter efficiencies given in Table 2.1.

Table 2.1 Battery parameters.

Batt	ζ_{di}	ζ_{ci}	η_i
1	0.942	0.921	0.981
2	0.941	0.923	0.983
3	0.945	0.926	0.984
4	0.947	0.924	0.985
5	0.944	0.929	0.982
6	0.912	0.866	0.963
7	0.912	0.863	0.961
8	0.914	0.867	0.967
9	0.917	0.868	0.966
10	0.918	0.861	0.962

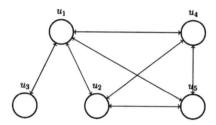

Figure 2.3 Graph showing interconnections between battery packages.

Fig. 2.3 shows the communication graph of the BESS system. The function $\phi(z)$ is defined as $\phi(z) = 1 + az$ with $a = 1$. The SoC limits of the each of the battery packages are $10\% \leq s_i \leq 90\%$. The system parameters of battery packages are $Q = 120$ Ah and $V = 220$.

The simulation results are presented in two categories: leader–follower consensus and adaptive leader–follower consensus. Figs. 2.4–2.7 present simulation plots for the leader–follower case in both charging and discharging operation. For this case, the control gains are chosen as $\omega_1 = 100$ and $\omega_2 = 100$. Fig. 2.4 shows the SoC profiles of each battery in charging mode. Fig. 2.5 shows the power profile of the batteries tracking the leader power profile in charging mode. Likewise, Figs. 2.6 and 2.7 present the SoC profile and power response in discharging modes.

Figs. 2.8–2.11 present simulation results for the adaptive leader–follower consensus instance. In Figs. 2.8 and 2.9, we show the SoC and power response in charging mode under control protocol (2.107). Figs. 2.10 and 2.11 present simulation responses in the discharging mode under adaptive protocol (2.107). Judging from simulation plots, we conclude that the adap-

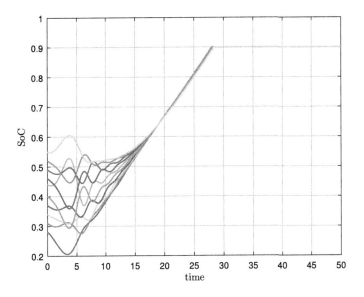

Figure 2.4 SoC profile during charging operation: leader–follower consensus.

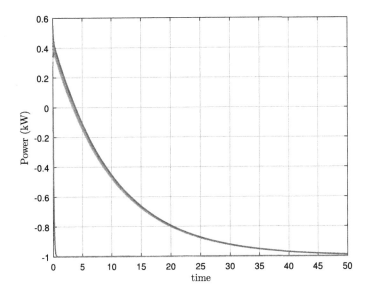

Figure 2.5 Power response during charging mode: leader–follower consensus.

tive leader–follower consensus protocol improves the overall performance of the BESS.

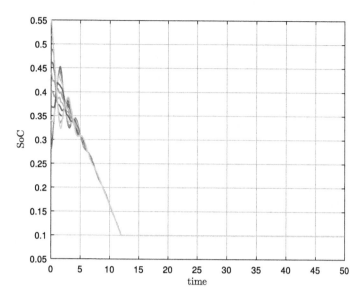

Figure 2.6 SoC profile during discharging operation: leader–follower consensus.

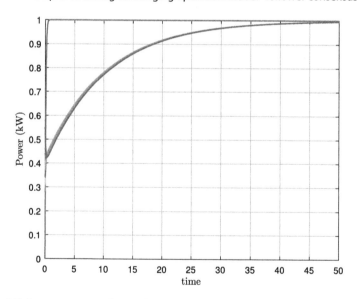

Figure 2.7 Power response during discharging mode: leader–follower consensus.

2.4 Group consensus

Group (or cluster) consensus problems emerge in cases where it is desired to have more than one agreement or consensus value. Practical applications

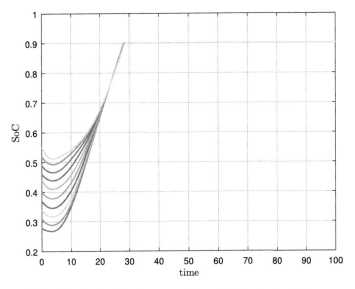

Figure 2.8 SoC profile during charging operation: adaptive leader–follower consensus.

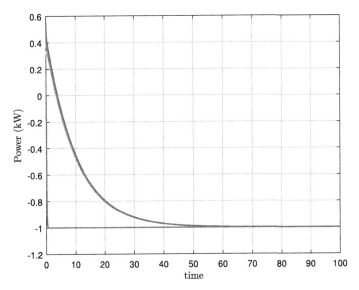

Figure 2.9 Power response during charging mode: adaptive leader–follower consensus.

of group consensus can be found in distributed robotics where agents need to converge to distinct agreement states to achieve a common coordination objective, for example, a group of fire-fighting robots may find it necessary

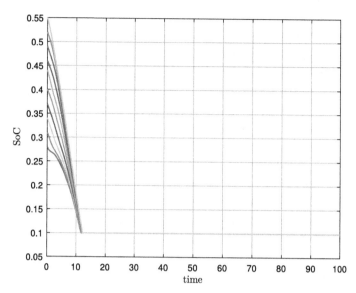

Figure 2.10 SoC profile during discharging operation: adaptive leader–follower consensus.

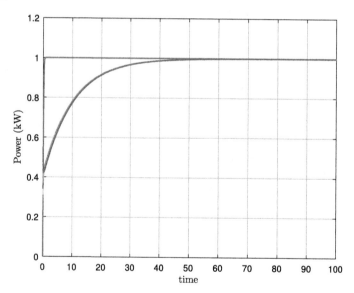

Figure 2.11 Power response during discharging mode: leader–follower consensus.

to be concentrated at multiple locations to minimize the time it will take to extinguish a wildfire. Group consensus theory relies mainly on established results in basic consensus problems.

2.4.1 Initial considerations

Here we provide some formal definitions of some terminologies in relation to group consensus:

Definition 2.8 (Couple-group consensus). A network of multiagent systems described by a graph $\mathbb{G}$ reaches couple-group consensus asymptotically, provided the states of the agent converge to two consistent values, that is,

$$\lim_{t \to \infty} ||x_i(t) - x_j(t)|| = 0 \quad \forall\, i, j \in \mathbb{N}_r, r = 1, 2. \tag{2.110}$$

Definition 2.9 (p-group consensus). A network of multiagent systems described by a graph $\mathbb{G}$ reaches p-group consensus asymptotically if the states of the agent converge to p consistent values, that is,

$$\lim_{t \to \infty} ||x_i(t) - x_j(t)|| = 0 \quad \forall\, i, j \in \mathbb{N}_r, r = 1, \dots, p. \tag{2.111}$$

Definition 2.10 (Average couple-group consensus). A network of multiagent systems described by a graph $\mathbb{G}$ solves an average couple-group consensus problem asymptotically if the states of the agent converge as follows:

$$\lim_{t \to \infty} ||x_i(t) - x_j(t)|| = \frac{1}{n_r} \sum_{j \in \mathbb{N}_1} x_i(0) \quad \forall\, i, j \in \mathbb{N}_r, r = 1, 2, \tag{2.112}$$

where n_r is the number of agents in each subgroup $\mathbb{N}_r$.

Definition 2.11 (Average group consensus). A network of multiagent systems described by a graph $\mathbb{G}$ solves an average group consensus problem asymptotically provided that the states of the agent converge as follows:

$$\lim_{t \to \infty} ||x_i(t) - x_j(t)|| = \frac{1}{n_r} \sum_{j \in \mathbb{N}_1} x_i(0) \quad \forall\, i, j \in \mathbb{N}_r, r = 1, \dots, p, \tag{2.113}$$

where n_r is the number of agents in each subgroup $\mathbb{N}_r$.

Consider the following distributed consensus protocol $u_i(t)$ for an MAS with single-integrator agents:

$$u_i(t) = \begin{cases} \displaystyle\sum_{j \in \mathbb{N}_1} a_{ij}(x_j(t) - x_i(t)) + \sum_{j \in \mathbb{N}_2} a_{ij}x_j(t), \\ \displaystyle\sum_{j \in \mathbb{N}_1} a_{ij}x_j(t) + \sum_{j \in \mathbb{N}_2} a_{ij}(x_j(t) - x_i(t)). \end{cases} \tag{2.114}$$

For the corresponding interaction graph $\mathbb{G}$ which describes the topology of the MAS under protocol (2.114), the following *balance-of-effect* assumptions are made [14]:

$$\sum_{j \in \mathbb{N}_2} a_{ij} = 0 \quad \forall \ i \in \mathbb{N}_1, \tag{2.115}$$

$$\sum_{j \in \mathbb{N}_1} a_{ij} = 0 \quad \forall \ i \in \mathbb{N}_2. \tag{2.116}$$

Applying the consensus protocol (2.114) to (2.9), the resulting system can also be written in closed-form of (2.117) as a function of the graph Laplacian:

$$\dot{x} = -\mathbb{L}x. \tag{2.117}$$

Lemma 2.4 ([14]). *The corresponding Laplacian $\mathbb{L}$ for the consensus protocol (2.117) has a zero eigenvalue with geometric multiplicity of two.*

Proposition 2.1 ([14]). *Protocol (2.114) solves the average couple-group consensus problem asymptotically if the following conditions are satisfied:*
- *Assumptions (2.115) and (2.116) are satisfied;*
- *$\mathbb{L}$ has exactly two simple zero eigenvalues and the rest of the eigenvalues have positive real parts;*
- *$\mathbb{G}_1$ and $\mathbb{G}_2$ are balanced and form a balanced couple.*

Consider the following distributed consensus protocol $u_i(t)$ for an MAS with single-integrator agents:

$$u_i(t) = \sum_{j \in \mathbb{N}_r} a_{ij}(x_j(t) - x_i(t)) + \sum_{j \notin \mathbb{N}_r} a_{ij}x_j(t) \quad \forall \ i \in \mathbb{N}_r. \tag{2.118}$$

Applying the consensus protocol (2.118) to (2.9), the resulting system can also be written in closed-form of (2.117) as a function of the graph Laplacian. The following lemmas are proposed for reaching group consensus in [14].

Lemma 2.5 ([14]). *The corresponding Laplacian $\mathbb{L}$ for the consensus protocol (2.118) has p simple zero eigenvalues.*

Proposition 2.2 ([14]). *Protocol (2.118) solves the average group consensus problem asymptotically provided that the following conditions are satisfied:*
- *Assumptions (2.115) and (2.116) are satisfied;*

- $\mathbb{L}$ *has exactly p simple zero eigenvalues and the rest of the eigenvalues have positive real parts;*
- $\mathbb{G}_i$ *is balanced for any $i \in 1, 2, \ldots, p$, and G_i and G_j are a balanced couple $\forall$ $i \neq j$ with $i, j \in 1, 2, \ldots, s$.*

2.4.2 Problem layout

Here, we start with linear dynamics and examine some typical group consensus problems for agents with single integrator dynamics. Consider an MAS on a graph $\mathbb{G}$ modeled by the following dynamics, where a_i and b_i are state and control coefficients, respectively, which may be unknown and time-varying. Some necessary and sufficient conditions were proposed in [15] for MAS with dynamics in (2.18). Consider the following distributed consensus protocol $u_i(t)$:

$$u_i(t) = \begin{cases} \alpha_1 \sum_{j \in \mathbb{N}_1} a_{ij}(x_j(t) - x_i(t)) + \beta_1 \sum_{j \in \mathbb{N}_2} a_{ij}x_j(t) & \forall i \in \mathbb{N}_1, \\ \alpha_2 \sum_{j \in \mathbb{N}_2} a_{ij}(x_j(t) - x_i(t)) + \beta_2 \sum_{j \in \mathbb{N}_1} a_{ij}x_j(t) & \forall i \in \mathbb{N}_2, \end{cases} \tag{2.119}$$

where α_1, α_2 are considered as intragroup coupling gains and β_1, β_2 are considered as intergroup coupling gains. System (2.18) under protocol (2.119) can be written in closed-loop form of

$$\dot{x} = \tilde{\mathbb{L}}x. \tag{2.120}$$

where $\tilde{\mathbb{L}}$ is defined as

$$\tilde{\mathbb{L}} = \begin{bmatrix} aI_n - b\alpha_1 \mathbb{L}_{11} & b\beta_1 \mathbb{L}_{12} \\ b\beta_2 \mathbb{L}_{21} & aI_n - b\alpha_2 \mathbb{L}_{22} \end{bmatrix}. \tag{2.121}$$

Theorem 2.7 ([15]). *The MAS described by $\mathbb{G}$ with dynamics (2.18) consisting of subgraphs $\mathbb{G}_1$ and $\mathbb{G}_2$ under protocol (2.119) achieves group consensus if and only if each component contains a spanning tree and the following inequalities are satisfied simultaneously:*

$$b^2[(\gamma_2 + a\gamma_1)^2 + 4\alpha_1\alpha_2\gamma_3] + 4a^2 > 4b^2\beta_1\beta_2\mu_l, \tag{2.122}$$

$$2a - b\gamma_1 \neq 0, \tag{2.123}$$

where $\gamma_1 = \alpha_1 \mathrm{Re}(\mu_i^1) + \alpha_2 \mathrm{Re}(\mu_j^2)$, $\gamma_2 = \alpha_1 \mathrm{Im}(\mu_i^1) + \alpha_2 \mathrm{Im}(\mu_j^2)$, $\gamma_3 = \mathrm{Re}(\mu_i^1)\mathrm{Re}(\mu_j^2) - \mathrm{Im}(\mu_j^1)\mathrm{Im}(\mu_j^2)$, μ_i^1, μ_j^2 are the eigenvalues of $\hat{\mathbb{L}}_{11}$ and $\hat{\mathbb{L}}_{22}$, respectively, and μ_l is the maximum eigenvalue of $\hat{\mathbb{L}}_{12}\hat{\mathbb{L}}_{21}$.

In the case where $a = 0$ and $b = 1$, we have a single integrator and we have the following

Corollary 2.2 ([15]). *The MAS described by $\mathbb{G}$ with dynamics (2.9) consisting of subgraphs $\mathbb{G}_1$ and $\mathbb{G}_2$ under protocol (2.119) achieves group consensus if and only if each component contains a spanning tree and the following inequalities are satisfied simultaneously:*

$$\gamma_2 + 4\alpha_1\alpha_2\gamma_3 > 4\beta_1\beta_2\mu_l, \tag{2.124}$$

$$\gamma_1 \neq 0. \tag{2.125}$$

The study was also extended to the case where the consensus protocol has some communication delays, that is, one can consider the distributed consensus protocol [15]:

$$u_i(t) = \begin{cases} \alpha_1 \sum_{j\in\mathbb{N}_1} a_{ij}(x_j(t-\tau) - x_i(t-\tau)) + \beta_1 \sum_{j\in\mathbb{N}_2} a_{ij}x_j(t-\tau) & \forall i \in \mathbb{N}_1, \\ \alpha_2 \sum_{j\in\mathbb{N}_2} a_{ij}(x_j(t-\tau) - x_i(t-\tau)) + \beta_2 \sum_{j\in\mathbb{N}_1} a_{ij}x_j(t-\tau) & \forall i \in \mathbb{N}_2. \end{cases}$$

$$\tag{2.126}$$

The authors investigated conditions on the coupling gains and the delay parameter such that consensus is achieved, leading to the following theorems and corollaries.

Theorem 2.8 ([15]). *If control protocol (2.117) makes system (2.18) reach couple-group consensus in the delay-free case, then protocol (2.126) does the same for system (2.18) with fixed coupling gains $\alpha_1 = \alpha_2 = \beta_1 = \beta_2 = \alpha$, if there exists a $\tau < \tau^*$, where τ^* is defined as*

$$\tau^* = \min_{i=1,\dots,n+m} \tau_i \tag{2.127}$$

and τ_i is defined as

$$\tau_i = \frac{1}{\omega_i} \tan^{-1}\left(\frac{a\mathrm{Im}(\lambda_i) - \omega_i\mathrm{Re}(\lambda_i)}{\omega_i\mathrm{Im}(\lambda_i) + a\mathrm{Re}(\lambda_i)}\right), \tag{2.128}$$

where $\omega_i = \sqrt{\alpha_2 b^2 |\lambda_i|^2 - a^2}$.

Corollary 2.3 ([15]). *Assuming control protocol (2.117) makes system (2.18) reach couple-group consensus in the delay-free case, then protocol (2.126) does the* *same for system (2.18) with fixed coupling gains $\alpha_1 = \alpha_2 = \beta_1 = \beta_2 = \alpha$, if there exists a $\tau < \tau^*$, where τ^* is defined as*

$$\tau^* = \min_{i=1,\ldots,n+m} \tau_i,$$

$$\tau_i = \begin{cases} \dfrac{1}{\omega_i} \tan^{-1}\left[-\dfrac{\kappa(\lambda_i)}{\nu(\lambda_i)} \right], & \nu(\lambda_i) \neq 0, \\[2mm] \dfrac{1}{\omega_i} \sin^{-1}(1), & \nu(\lambda_i) = 0, \end{cases} \tag{2.129}$$

$$\omega_i = \alpha|\lambda_i|, \tag{2.130}$$

where $\omega_i = \sqrt{\alpha_2 b^2 |\lambda_i|^2 - a^2}$.

Corollary 2.4 ([15]). *Suppose MAS (2.9) achieves group consensus under protocol (2.117) with fixed coupling gains $\alpha_1 = \alpha_2 = \alpha$, $\beta_1 = \beta_2 = \beta$. Then the MAS (2.9) under protocol (2.126) achieves group consensus if and only if $\tau < \tau^*$, where τ^* is*

$$\tau^* = \min_{i=1,\ldots,n+m} \tau_i,$$

$$\tau_i = \begin{cases} \dfrac{1}{\omega_i} \tan^{-1}\left[-\dfrac{\kappa(\lambda_i)}{\nu(\lambda_i)} \right], & \nu(\lambda_i) \neq 0, \\[2mm] \dfrac{1}{\omega_i} \sin^{-1}(1), & \nu(\lambda_i) = 0, \end{cases} \tag{2.131}$$

$$\omega_i = \sqrt{(\alpha + \beta)|\lambda_i|^2}. \tag{2.132}$$

Corollary 2.5 ([15]). *Suppose MAS (2.18) achieves group consensus under protocol (2.117) with fixed coupling gains $\alpha_1 \neq \alpha_2 > 0$, $\beta_1 \neq \beta_2 > 0$. Then the MAS (2.18) under protocol (2.126) achieves group consensus if and only if $\tau < \tau^*$, where τ^* is*

$$\tau^* = \min_{i=1,\ldots,n+m} \tau_i,$$

$$\tau_i = \frac{1}{\omega_i} \tan^{-1}\left[\frac{a\nu(\lambda_i) - \omega_i\kappa(\lambda_i)}{\omega_i\nu(\lambda_i) + a\kappa(\lambda_i)} \right],$$

$$\omega_i = \sqrt{(\alpha + \beta)b^2|\lambda_i|^2 - a^2}, \tag{2.133}$$

$$\alpha = \max[\alpha_1, \alpha_2], \ \beta = \max[\beta_1, \beta_2].$$

Corollary 2.6 ([15]). *Suppose MAS (2.9) achieves group consensus under protocol (2.117) with fixed coupling gains $\alpha_1 \neq \alpha_2 > 0$, $\beta_1 \neq \beta_2 > 0$. Then the MAS*

(2.18) *under protocol (2.126) achieves group consensus if and only if* $\tau < \tau^*$, *where* τ^* *is*

$$\tau^* = \min_{i=1,\ldots,n+m} \tau_i,$$

$$\tau_i = \begin{cases} \dfrac{1}{\omega_i} \tan^{-1}\left[-\dfrac{\kappa(\lambda_i)}{\nu(\lambda_i)}\right], & \nu(\lambda_i) \neq 0, \\ \dfrac{1}{\omega_i} \sin^{-1}(1), & \nu(\lambda_i) = 0, \end{cases} \quad (2.134)$$

$$\omega_i = \sqrt{(\alpha + \beta)|\lambda_i|^2}, \quad (2.135)$$

$$\alpha = \max[\alpha_1, \alpha_2], \quad \beta = \max[\beta_1, \beta_2].$$

Next, we direct attention to the case of nonlinear dynamics.

Fixed-time group consensus tracking control was studied in [16] for an MAS with nonlinear dynamics (2.136):

$$\dot{x}_i = f(t, x_i) + u_i, \quad (2.136)$$

$$\dot{x}_{\theta_k} = f(t, x_{\theta_k}) + u_{\theta_k}, \quad (2.137)$$

where x_i and x_{θ_k} represent follower and leader states, respectively, u_i and u_{θ_k} are the control inputs of agent i and leader belonging to group k. The following assumption was made on the nonlinear function f:

$$\|f(t, x_1) - f(t, x_2)\| \leq l_1 + l_2\|x_1 - x_2\|^2 + l_3\|x_1 - x_2\|^4. \quad (2.138)$$

In this study, cases involving both the presence and absence of intergroup balance conditions are considered. In the presence of intergroup balance condition, the following distributed control protocol was proposed for system (2.136):

$$u_i = \alpha\left[\sum_{j\in\mathbb{G}_k\cup\theta_k} a_{ij}(x_j - x_i)\right]^2 + \left[\sum_{k'\neq k}\sum_{j\in\mathbb{G}'_k} a_{ij}x_j\right]^2$$

$$+ \beta\left(\sum_{j\in\mathbb{G}_k\in\theta_k} a_{ij}(x_j - x_i)\right) + \left(\sum_{k'\neq k}\sum_{j\in\mathbb{G}'_k} a_{ij}x_j\right)$$

$$+ \gamma\operatorname{sgn}\left(\sum_{j\in\mathbb{G}_k\cup\theta_k} a_{ij}(x_j - x_i))\right) + \operatorname{sgn}\left(\sum_{k'\neq k}\sum_{j\in\mathbb{G}'_k} a_{ij}x_j\right). \quad (2.139)$$

Theorem 2.9. *Under Assumption 2.3.1, the consensus protocol (2.139) drives the distributed tracking errors of MAS (2.136) to zero with the following control gains α, β, γ:*

$$\alpha = \frac{(\bar{a}N^3 + \sqrt{l_3 N}(\lambda_{\max}^{3/2(W)})}{\lambda_{\min}^{3/2(W)}}, \tag{2.140}$$

$$\beta = \frac{\bar{a}N + \sqrt{l_2}}{\lambda_{\min}(W)}, \tag{2.141}$$

$$\gamma = 1 + W + \sqrt{l_1 \max_{1 \le k \le K} |\mathbb{G}_k|} + \frac{\rho}{\sqrt{2}\lambda_{\min}(W)}, \tag{2.142}$$

$$\bar{a} = \max_{i \in \mathbb{G}_k', j \in \mathbb{G}_k', k \ne k'} |a_{ij}|, \tag{2.143}$$

$$u_i = \alpha \left[\sum_{j \in \mathbb{G}_k \cup \theta_k} a_{ij}(x_j - x_i) \right]^2 + \left[\sum_{k' \in \phi_i^0} \sum_{j \in \mathbb{G}_k'} a_{ij}x_j \right]^2$$

$$+ \left[\sum_{k' \in \phi_i^1} \sum_{j \in \mathbb{G}_k'} a_{ij}(x_j - x_{\theta_{k'}}) \right]^2 + \beta \left(\sum_{j \in \mathbb{G}_k \cup \theta_k} a_{ij}(x_j - x_i) \right)$$

$$+ \left(\sum_{k' \in \phi_i^0} \sum_{j \in \mathbb{G}_k'} a_{ij}x_j \right) + \left(\sum_{k' \in \phi_i^1, j \in \mathbb{G}_k'} a_{ij}(x_j - x_{\theta_{k'}}) \right)$$

$$+ \gamma \operatorname{sgn}\left(\sum_{j \in \mathbb{G}_k \cup \theta_k} a_{ij}(x_j - x_i)) \right) + \operatorname{sgn}\left(\sum_{k' \ne k} \sum_{j \in \mathbb{G}_k'} a_{ij}x_j \right)$$

$$+ \operatorname{sgn}\left(\sum_{k' \in \phi_i^1, j \in \mathbb{G}_k'} a_{ij}(x_j - x_{\theta_{k'}}) \right). \tag{2.144}$$

Theorem 2.10. *Under Assumption 2.3.1, the consensus protocol (2.144) drives the distributed tracking errors of MAS (2.136) to zero with the following control gains α, β, γ:*

$$\alpha = \frac{(\bar{a}^2 N^3 + \sqrt{l_3 N})(\lambda_{\max}^{3/2(W)})}{\lambda_{\min}^{3/2(W)}} + \frac{\rho\sqrt{N}}{2\sqrt{2}\lambda_{\min}^{3/2}(W)}, \tag{2.145}$$

$$\beta = \frac{\bar{a}N + \sqrt{l_2}}{\lambda_{\min}(W)}, \tag{2.146}$$

$$\gamma = 1 + W + \sqrt{l_1 \max_{1 \le k \le K} |\mathbb{G}_k|} + \frac{\rho}{\sqrt{2}\lambda_{\min}(W)}, \tag{2.147}$$

$$\bar{a} = \max_{i \in \mathbb{G}_k', j \in \mathbb{G}_k', k \ne k'} |a_{ij}|. \tag{2.148}$$

2.4.3 Related work

The problem of reaching cluster consensus over directed interaction topology, regardless of the magnitude of coupling strengths, was investigated in [32] for generic linear multiagent systems with results depending solely on the topological conditions. Markov chains and nonnegative matrix analysis were used to derive and analyze two cluster consensus algorithms for an MAS exhibiting discrete-time dynamics over fixed and switching topologies in [33].

Novel algorithms for reaching group consensus were proposed in [34] for heterogeneous agents modeled by an Euler–Lagrange system and double integrator dynamics. Sufficient conditions were proposed based on algebraic graph theory and Barbalat's lemma. An adaptive group consensus problem was investigated in [35] for networked mechanical systems modeled by Lagrangian dynamics over directed acyclic networks with communication delays. An adaptive group consensus problem was investigated in [37] for an MAS with agents modeled by Euler–Lagrange dynamics with inherent parametric uncertainties. Necessary conditions for reaching group consensus were derived for topologies with directed acyclic and balanced couple structures.

A couple-group consensus problem was investigated for a class of MASs with heterogeneous agents exhibiting first- and second-order dynamics over networks with input and communication delays. A cluster consensus problem was investigated in [38] for general linear agents with heterogeneous dynamics considering both semi- and fully heterogeneous cases. The results suggest that in the semi-heterogeneous case, cluster consensus is achieved when agents are able to suppress the influence of intercluster couplings, while in the fully heterogeneous case, agents must be able to suppress both inter- and intracluster couplings. Frequency-domain analysis was used in [59] to derive sufficient conditions for reaching a group consensus for a network of a heterogeneous MAS with input time-delays.

2.4.4 Numerical example 2.3

Consider the MAS described by the graph in Fig. 2.12 with the Laplacian matrix (2.149):

$$\mathbb{L} = \begin{bmatrix} 4 & -3 & -1 & 1 & -1 \\ -3 & 3 & 0 & -1 & 1 \\ -1 & 0 & 1 & 0 & 0 \\ 1 & -1 & 0 & 3 & -3 \\ -1 & 1 & 0 & -3 & 3 \end{bmatrix}. \tag{2.149}$$

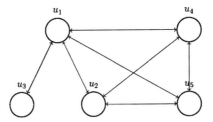

Figure 2.12 Graph showing interconnections between agents.

By simple computation, we obtain the eigenvalues of the Laplacian matrix as $\lambda = 0, 0, 1.3263, 4.3457, 8.3280$. The system comprises two subgroups $\mathbb{G}_1$ and $\mathbb{G}_2$ with 3 and 2 agents, respectively. The following Laplacian matrices $\mathbb{L}_{11}, \mathbb{L}_{22}, \mathbb{L}_{12} = \mathbb{L}_{21}^T$ represent the links between each of the subgroups and the interconnection amongst the two subgroups, respectively:

$$\mathbb{L}_{11} = \begin{bmatrix} 4 & -3 & -1 \\ -3 & 3 & 0 \\ -1 & 0 & 1 \end{bmatrix}, \quad \mathbb{L}_{22} = \begin{bmatrix} 3 & -3 \\ -3 & 3 \end{bmatrix}, \quad (2.150)$$

$$\mathbb{L}_{12} = \begin{bmatrix} 1 & -1 \\ -1 & 1 \\ 0 & 0 \end{bmatrix}, \quad (2.151)$$

$\mu_i = 0.0000, 1.3542, 6.6458$ and $\mu_j = 0, 6$ and $\mu_{\max}(\mathbb{L}_{12}\mathbb{L}_{21}) = 4$.

Next, we direct attention to the class of delay-free first-order MAS.

In what follows, we present some simulations based on the derived consensus conditions for the no-delay case. Fig. 2.13 shows the simulation behavior of MAS (2.18) with coupling gains $\alpha_1 = \alpha_2 = 1$ and $\beta_1 = \beta_2 = 1$. Since $\mathbb{L}$ has real eigenvalues, conditions (2.8) reduce to $\alpha_1\alpha_2\mu_i\mu_j > \beta_1\beta_2\mu_{\max}(\hat{\mathbb{L}}_{12}\hat{\mathbb{L}}_{21})$ and $\alpha_1\mu_i + \alpha_2\mu_j - \beta_1\beta_2 \neq 0$. Since these conditions are satisfied by $\alpha_1, \alpha_2, \beta_1$ and β_2, as shown in Fig. 2.13, a group consensus is achieved. Furthermore, we examine the case where $\alpha_1 = \alpha_2 = 1$ and $\beta_1 = \beta_2 = 2$. In this case also, the group consensus conditions established based on the eigenvalues of the subgroups are satisfied, therefore as shown in Fig. 2.14, a group consensus is achieved. However, in Fig. 2.15, the coupling gains used are $\alpha_1 = \alpha_2 = 1$ and $\beta_1 = \beta_2 = 4$. Here, the condition that $\alpha_1\alpha_2\mu_i\mu_j > \beta_1\beta_2\mu_{\max}(\hat{\mathbb{L}}_{12}\hat{\mathbb{L}}_{21})$ is violated and the system becomes unstable. However, for $\alpha_1 = \alpha_2 = 2$, $\beta_1 = \beta_2 = 4$, a choice which satisfies this condition, we see that in Fig, 2.16, group consensus is achieved.

Now, we consider the class of delayed first-order MASs.

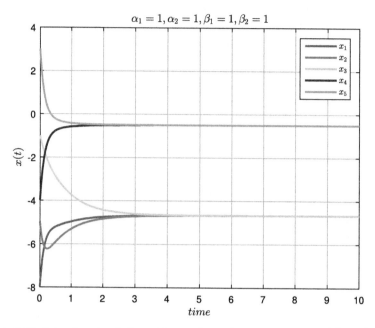

Figure 2.13 Delay-free simulation with coupling gains $\alpha_1 = \alpha_2 = 1$ and $\beta_1 = \beta_2 = 1$, showing that consensus is achieved.

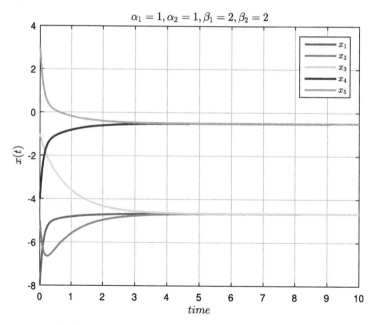

Figure 2.14 Delay-free simulation with coupling gains $\alpha_1 = \alpha_2 = 1$ and $\beta_1 = \beta_2 = 2$, showing that consensus is achieved.

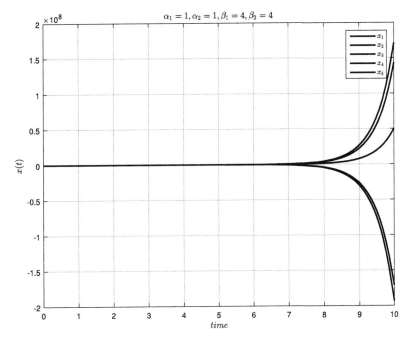

Figure 2.15 Delay-free simulation with coupling gains $\alpha_1 = \alpha_2 = 1$ and $\beta_1 = \beta_2 = 4$, showing that consensus is not achieved.

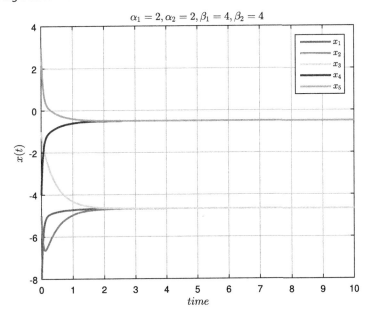

Figure 2.16 Delay-free simulation with coupling gains $\alpha_1 = \alpha_2 = 2$ and $\beta_1 = \beta_2 = 4$, showing that consensus is achieved.

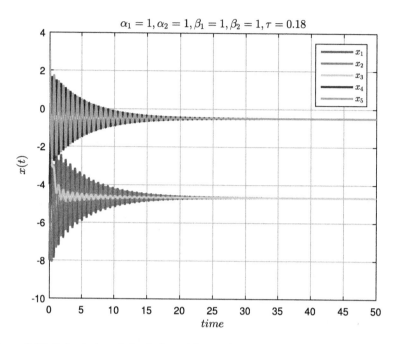

Figure 2.17 Delay-induced simulation with coupling gains $\alpha_1 = \alpha_2 = 1$, $\beta_1 = \beta_2 = 1$, and delay parameter $\tau = 0.18$, showing that consensus is achieved.

Consider the MAS (2.18) having Laplacian $\mathbb{L}$ in (2.149) with the eigenvalues $\lambda_i = 0, 0, 1.3263, 4.3457, 8.3280$. Since all the eigenvalues of $\mathbb{L}$ are real, ω_i becomes $\omega_i = \alpha\lambda_i$. Clearly, the maximum time delay τ^* required for group consensus is related to the maximum nonzero eigenvalue of $\mathbb{L}$ which in this case is $\lambda_{\max} = 8.3280$. For the case $\alpha_1 = \alpha_2 = 1$, $\beta_1 = \beta_2 = 1$, $\omega_{\max} = \lambda_{\max} = 8.3280$, and $\tau^* = 0.1886$. Fig. 2.17 shows the response of MAS (2.18) when $\tau = 0.18 < \tau^*$, while Fig. 2.18 shows the response of the system when $\tau = 0.19 > \tau^*$. Next we examine the case where $\alpha_1 = \alpha_2 = 2$, $\beta_1 = \beta_2 = 4$. From (2.131), $\omega_{\max} = (\sqrt{\alpha + \beta})\lambda_{\max} = 20.399$ and $\tau^* = 0.0754$. Fig. 2.19 shows the case where $\tau = 0.07 < \tau^*$ where the system achieves group consensus. Again, we simulate the instance where $\tau = 0.08 > \tau^*$ in Fig. 2.20, and it can be observed that the system becomes unstable. Finally, we present the case where the intra- and intercoupling gains α and β are nonuniform. Consider the case where $\alpha_1 = 1$, $\alpha = 3$, $\beta = 4$, and $\beta = 2$. By Corollary 2.4, we get $\tau^* = 0.0754$. Figs. 2.21 and 2.22 show the case where $\tau = 0.07 < \tau^*$ and $\tau = 0.08 > \tau^*$, respectively. Observe that group consensus is achieved in Fig. 2.21, and the system is unstable in Fig. 2.22.

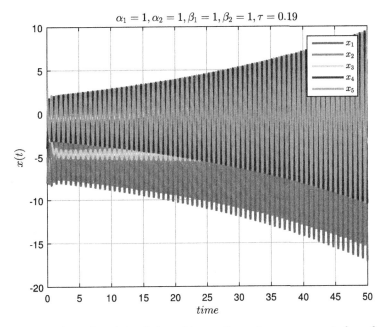

Figure 2.18 Delay-induced simulation with coupling gains $\alpha_1 = \alpha_2 = 1$, $\beta_1 = \beta_2 = 1$, and delay parameter $\tau = 0.19$, showing that consensus is not achieved.

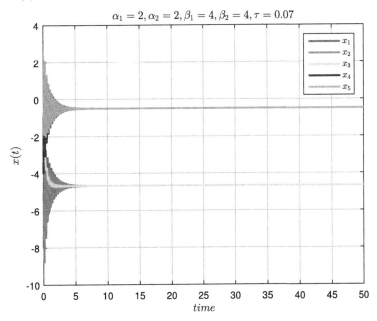

Figure 2.19 Delay-induced simulation with coupling gains $\alpha_1 = \alpha_2 = 2$, $\beta_1 = \beta_2 = 4$, and delay parameter $\tau = 0.07$, showing that consensus is achieved.

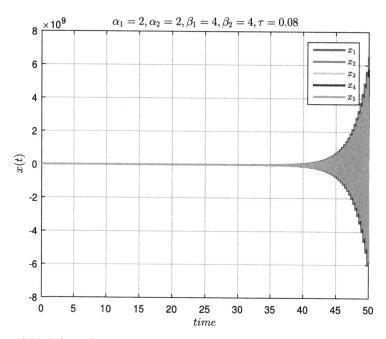

Figure 2.20 Delay-induced simulation with coupling gains $\alpha_1 = \alpha_2 = 2$, $\beta_1 = \beta_2 = 4$, and delay parameter $\tau = 0.08$, showing that consensus is not achieved.

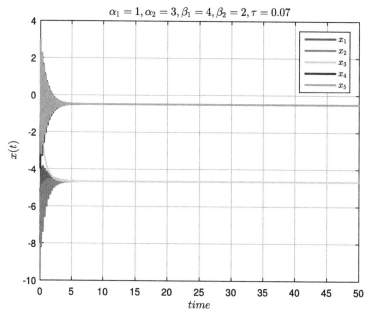

Figure 2.21 Delay-induced simulation with coupling gains $\alpha_1 = 1$, $\alpha_2 = 3$, $\beta_1 = 4$, $\beta_2 = 2$, and delay parameter $\tau = 0.07$, showing that consensus is achieved.

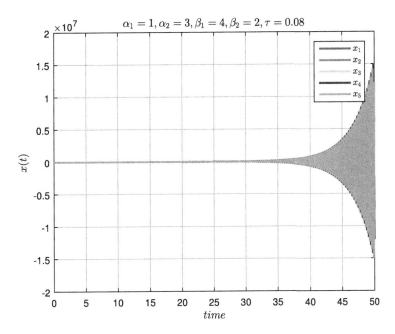

Figure 2.22 Delay-induced simulation with coupling gains $\alpha_1 = 1$, $\alpha_2 = 3$, $\beta_1 = 4$, $\beta_2 = 2$, and delay parameter $\tau = 0.08$, showing that consensus is not achieved.

2.5 Bipartite consensus

Bipartite consensus problems are often concerned with the study of multiagent interactions involving cooperative–competitive dynamics among agents. In this type of problem, the agents reach two distinct agreement states of equal magnitude but opposite signs. Generally, the interaction graph is described by a bipartite graph consisting of two teams where agents belonging to the same team are cooperating, while agents in the opposing teams are competing. Many results have been reported, especially in problems involving opinion forming, antagonistic interactions, social network interactions, and competition.

Pioneering studies in bipartite consensus were carried out by [2], towards solving the problem of reaching consensus over antagonistic interaction networks. This study models antagonistic interactions over *signed* graphs. A signed graph $\mathbb{G}_s$ is a triple consisting of a vertex set $\mathbb{V} = \{v_1, v_2, \ldots, v_n\}$, edge set $\mathbb{E} \subset \mathbb{V} \times \mathbb{V}$, and an adjacency matrix $\mathbb{A} \in \mathbb{R}^{n \times n}$ consisting of signed weights. Much of the research in [2] was concerned with deriving necessary and sufficient conditions. One of such conditions is that the signed graph is structurally balanced.

2.5.1 Definitions

Definition 2.12 (Structurally balanced graph [2]). A signed graph consisting of a bipartition of nodes $\mathbb{V}_1$ and $\mathbb{V}_2$ such that $\mathbb{V}_1 \cup \mathbb{V}_2 = \mathbb{V}$ and $\mathbb{V}_1 \cap \mathbb{V}_2 = \varnothing$ is structurally balanced provided that $a_{ij} \geq 0 \ \forall \ v_i, v_j \in \mathbb{V}_q, q \in \{1, 2\}$ and $a_{ij} \leq 0 \ \forall \ v_i, \in \mathbb{V}_q, v_j \in \mathbb{V}_r, q \neq r \ (q, r \in \{1, 2\})$, and it is *structurally unbalanced* otherwise. In most cases, structural balance is a necessary condition for reaching bipartite consensus.

Definition 2.13 (Homogeneous network). A competition network is said to be homogeneous if all the interactions are cooperative or competitive.

Definition 2.14 (Heterogeneous network). A competition network is said to be heterogeneous if the interactions are cooperative and competitive.

Definition 2.15 (Bipartite and interventional bipartite consensus). Consider a network of multiagent systems described by the following dynamics:

$$\dot{x}_i = f(x_i, u_i). \tag{2.152}$$

A simple distributed consensus protocol is designed for this system over a signed communication graph $\mathbb{G}_s$ as

$$u_i(t) = \sum_{j \in \mathbb{N}_i} |a_{ij}|(x_i(t) - \text{sgn}(a_{ij})x_j(t)). \tag{2.153}$$

The overall system under protocol (2.153) can also be written in closed form as

$$\dot{x} = -\mathbb{L}_s x. \tag{2.154}$$

However, the Laplacian $\mathbb{L}_s$ is defined as

$$l_{ik}^s = \begin{cases} \sum_{j \in \mathbb{N}_i} |a_{ij}|, & k = i, \\ -a_{ik}, & k = i. \end{cases} \tag{2.155}$$

A bipartite consensus problem is solved by a network of agents modeled by (2.9) under (2.153) provided that

$$\lim_{t \to \infty} |x_i| = c, \tag{2.156}$$

and an interventional bipartite consensus problem is solved provided that

$$\lim_{t \to \infty} |x_i - x_0| = 0, \tag{2.157}$$

$$\lim_{t \to \infty} |x_i + x_0| = 0, \tag{2.158}$$

where x_0 represents the states of a leader node modeled by

$$\dot{x}_0 = f(x_0, u_0). \tag{2.159}$$

Lemma 2.6 ([2]). *For a structurally balanced connected signed graph $\mathbb{G}_s$, the following conditions are equivalent:*
- *All cycles of $\mathbb{G}_s$ are positive;*
- *There exists a matrix $\mathbb{M}$ such that $\mathbb{MAM}$ has only positive entries, where $\mathbb{A}$ is the weighted adjacency matrix of $\mathbb{G}_s$;*
- *The smallest eigenvalue of $\mathbb{L}$ is 0.*

Corollary 2.7 ([2]). *For a structurally unbalanced connected signed graph $\mathbb{G}_s$, the following conditions are equivalent:*
- *At least one cycle of $\mathbb{G}_s$ is negative;*
- *There exists no matrix $\mathbb{M}$ such that $\mathbb{MAM}$ has only positive entries, where $\mathbb{A}$ is the weighted adjacency matrix of $\mathbb{G}_s$;*
- *All eigenvalues of $\mathbb{L}$ are positive.*

2.5.2 Problem description

In this subsection, some bipartite consensus problems from different references are revisited. Focusing on the linear dynamics case, consider the following MAS on a graph $\mathbb{G}$ modeled by the following first-order dynamics:

$$\dot{x}_i = a_i x_i + b_i u_i, \tag{2.160}$$

where a_i and b_i are scalar coefficients for each of the agents. A simple distributed bipartite consensus algorithm is defined as

$$u_i = k_i \sum_{j \in \mathbb{N}_i} |a_{ij}| (x_i(t) - \mathrm{sgn}(a_{ij}) x_j(t)). \tag{2.161}$$

Some bipartite consensus problems are mostly about investigating necessary and sufficient conditions on the resulting signed graph $\mathbb{G}_s$ and the gain k_i such that system (2.160) solves either bipartite or interventional bipartite consensus problem. In some cases, the network is affected by communication delays. Consider the following distributed bipartite consensus protocol with communication delays:

$$u_i = k_i \sum_{j \in \mathbb{N}_i} |a_{ij}|(x_i(t - \tau_{ij}) - \text{sgn}(a_{ij})x_j(t - \tau_{ij})), \tag{2.162}$$

where τ_{ij} denotes a time-delay parameter. Here, research investigations are conducted to determine bounds on τ_{ij} such that bipartite consensus is achieved. Consider a network of agents described by the following double-integrator dynamics:

$$\dot{x}_i(t) = v_i(t),$$
$$\dot{v}_i(t) = ax_i(t) + bv_i(t) + u_i(t). \tag{2.163}$$

A typical bipartite consensus protocol for MAS with agents exhibiting this second-order dynamics is given as:

$$u_i(t) = k_i^1 \sum_{j \in \mathbb{N}_i} |a_{ij}|(x_i(t) - \text{sgn}(a_{ij})x_j(t))$$
$$+ k_i^2 \sum_{j \in \mathbb{N}_i} |a_{ij}|(v_i(t) - \text{sgn}(a_{ij})v_j(t)). \tag{2.164}$$

Again, we are interested in properties that should be satisfied by the control gains in conjugation with the nature of the graph. A variant of this problem was investigated in [42] for an MAS with unknown disturbances and some leader nodes, that is, the authors considered (2.165)

$$\dot{x}_i(t) = v_i(t),$$
$$\dot{v}_i(t) = u_i(t) + \delta_i(t), \tag{2.165}$$

where $\delta_i(t)$ represents unknown disturbances. The corresponding leader node was modeled by (2.166)

$$\dot{x}_0(t) = v_0(t),$$
$$\dot{v}_0(t) = u_0(t). \tag{2.166}$$

Here, distributed adaptive control laws were derived based on Lyapunov theory and graph theoretic analysis via local neighborhood errors (2.167) and (2.168) for each of the states:

$$\zeta_i(t) = \sum_{j \in \mathbb{N}_i} |a_{ij}|(x_i(t) - \text{sgn}(a_{ij})x_j(t)) + a_{i0}(x_i - s_i x_0), \tag{2.167}$$

$$\eta_i(t) = \sum_{j \in \mathbb{N}_i} |a_{ij}|(x_i(t) - \mathrm{sgn}(a_{ij})x_j(t)) + a_{i0}(x_i - s_i x_0). \tag{2.168}$$

Bipartite consensus for higher-order linear multiagent systems was studied in [44].

Hereinafter, we treat the cases of nonlinear dynamics.

Bipartite consensus problems have also been investigated for agents with nonlinear dynamics. In [43], the authors consider interventional bipartite tracking consensus problem for an MAS exhibiting Lipschitz-type dynamics. Consider an MAS with leader and follower agents described by the following dynamics:

$$\dot{x}_i(t) = Ax_i(t) + Bu_i(t) + f(x_i, t), \tag{2.169}$$

$$\dot{x}_0(t) = Ax_0(t) + Bu_0(t) + f(x_0, t), \tag{2.170}$$

where $x_i(t), x_0(t) \in \mathbb{R}^n$ represent states of the follower and leader agent, respectively, where $u_i(t), u_0(t) \in \mathbb{R}^m$ represent control inputs of the follower and leader agent, respectively; $f : \mathbb{R} \times \mathbb{R} \times \mathbb{R}^+$ is a vector-valued function satisfying the following Lipschitz condition:

$$\|f(x, t) - f(y, t)\| < k(x - y) \quad \forall x, y \in \mathbb{R}^n, t \geq 0. \tag{2.171}$$

A distributed bipartite consensus protocol proposed in [43] is

$$u_i(t) = -K\left[\sum_{j \in \mathbb{N}_i} |a_{ij}|(x_i(t) - \mathrm{sgn}(a_{ij})x_j(t)) + a_{i0}(x_i - s_i x_0(t))\right] + s_i u_0(t).$$

$$\tag{2.172}$$

Theorem 2.11 ([43]). *An interventional bipartite consensus problem is solved for an MAS with dynamics (2.169) under protocol (2.172) if there exists a matrix $P > 0$ satisfying the following Riccati equation:*

$$A^T P + PA - \alpha PBB^t P + \gamma P = 0, \tag{2.173}$$

where $\alpha \leq 2\lambda(\mathbb{L}_r)$, $\gamma \geq 2k$, $K = B^T P$, and the following assumptions are fulfilled:

- *The interaction graph $\mathbb{G}_s$ satisfies the structural balance property;*
- *The function f satisfies the Lipschitz property;*
- *The pair A, B is stabilizable.*

Further, the authors of [43] also studied the case where the leader has zero control input and proposed the following distributed consensus algorithm:

$$u_i(t) = -K \left[\sum_{j \in \mathbb{N}_i} |a_{ij}|(x_i(t) - \mathrm{sgn}(a_{ij})x_j(t)) + a_{i0}(x_i - s_i x_0(t)) \right]$$
$$- \mathrm{sgn}\left(B^T P \sum_{j \in \mathbb{N}_i} |a_{ij}|(x_i(t) - \mathrm{sgn}(a_{ij})x_j(t)) + a_{i0}(x_i - s_i x_0(t)) \right). \quad (2.174)$$

Theorem 2.12 ([43]). *An interventional bipartite consensus problem is solved for an MAS with dynamics (2.169) under protocol (2.174) if there exists a matrix $P > 0$ satisfying the following Riccati equation:*

$$A^T P + PA - \alpha PBB^T P + \gamma P = 0, \quad (2.175)$$

where $\alpha \le 2\lambda(\mathbb{L}_r)$, $\gamma \ge 2k$, $K = B^T P$, and the following assumptions are fulfilled:
- *The interaction graph $\mathbb{G}_s$ satisfies the structural balance property;*
- *The function f satisfies the Lipschitz property;*
- *The pair A, B is stabilizable.*

2.5.3 Numerical example 2.4

Consider a network of three multiagent systems described by (2.9). The following bipartite consensus protocol u_i is applied to the system:

$$\dot{u}_i = \sum_{j \in \mathbb{N}_i} |a_{ij}|(\mathrm{sgn}(a_{ij})x_j(t) - \alpha_i x_i(t)). \quad (2.176)$$

Fig. 2.23 shows the response of the system.

2.6 Scaled consensus

Scaled consensus involves reaching a consensus defined for specified weights of the states of each agent. In other consensus problems, reaching agreement is strictly by neighborhood errors. Contrary to other consensus problems, the neighborhood errors are defined over a scale of constant values for each agent. The notion of a scaled consensus was first introduced in [45].

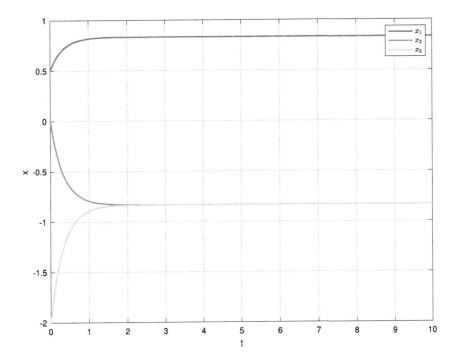

Figure 2.23 Bipartite consensus of multiagent systems.

2.6.1 Basics

Definition 2.16 (Scaled consensus [45]). Let $\beta_1, \beta_2, \ldots, \beta_n$ represent nonzero scalar values for each of the agent states $x_1, x_2, \ldots, x_n$. The MAS (2.1) consisting of n agents achieves a scaled consensus if

- The origin is not globally asymptotically stable;
- $\lim_{t \to \infty}(\beta_1 x_1(t) - \beta_j x_j(t)) = 0, j = 2, \ldots, n \ \forall \ x_i(0)$.

Definition 2.17 (Scaled consensus [46]). The MAS (2.9) with n agents exponentially reaches scaled consensus under nearest-neighbor protocol if there exist positive constants $\alpha, \beta > 0$ such that for any initial state $x_i(t_0)$, $i \in \mathbb{G}$, the following condition is satisfied:

$$||x_i(t) - \omega_i x_f|| \leq \alpha e^{-\beta(t-t_0)}, \quad \forall t_1 > t_0, \tag{2.177}$$

where $x_f \in \mathbb{R}^n$ is a vector of constants.

Lemma 2.7 ([45]). *For any strongly connected graph with Laplacian $\mathbb{L}_s$, scaled consensus is achieved with respect to $(\alpha_1, \ldots, \alpha_n)$.*

2.6.2 Problem statement

Consider the following distributed scaled consensus protocol for the MAS (2.9):

$$u_i(t) = \text{sgn}(\beta_i) \sum_{j \in N_i} a_{ij}(\beta_j x_j(t) - \beta_i x_i(t)) + \frac{1}{\alpha_i} \dot{r}_i(t). \qquad (2.178)$$

Stability analysis was carried out in [47] for a scaled consensus problem of this nature using Lyapunov–Krasovskii analysis, and the following theorem was proposed based on LMI formalisms.

Theorem 2.13 (Theorem 2.5.x, [47]). *Given any constant time delay* $\tau \in (0, \bar{\tau})$, *the MAS (2.9) solves the scaled consensus problem if there exist* $P, Z > 0$, *and* $Q = Q^T$ *such that the following LMIs are satisfied:*

$$\begin{bmatrix} P + Z & -Z \\ -Z & \bar{\tau}Q + Z \end{bmatrix} > 0, \qquad (2.179)$$

$$\begin{bmatrix} Q - \frac{1}{\bar{\tau}}Z & PA_1 + \frac{1}{\bar{\tau}}Z & 0 \\ A_1^T + \frac{1}{\bar{\tau}}Z & -Q - \frac{1}{\bar{\tau}}Z & \bar{\tau}A_1^T Z \\ 0 & \bar{\tau}ZA_1 & -\bar{\tau}Z \end{bmatrix} < 0. \qquad (2.180)$$

2.6.3 Related results

Scaled consensus was investigated in [48] for a class of heterogeneous agents under a leader–follower architecture. Using frequency domain analysis, some consensus conditions were derived with and without communication delay under undirected and directed topologies. Scaled consensus was investigated for a single-integrator MAS with output saturation in [49] under both undirected and strongly connected topologies. Necessary and sufficient conditions were derived via Lyapunov analysis. Also H_∞ sliding mode adaptive control was designed for reaching scaled consensus in [50] with some LMI conditions guaranteeing some sufficient conditions. Scaled group consensus was studied for first- and second-order MASs under continuous- and discrete-time setting in [51] using some necessary and sufficient conditions were derived via algebraic analysis.

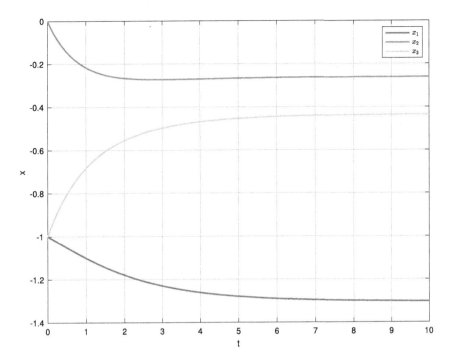

Figure 2.24 Scaled consensus of multiagent systems.

2.6.4 Numerical example 2.5

Consider a network of three multiagent systems described by (2.9). The following scaled consensus protocol u_i is applied to the system:

$$\dot{u}_i = \text{sgn}(\alpha_i) \sum_{j \in \mathbb{N}_i} a_{ij} (\alpha_j x_j(t) - \alpha_i x_i(t)). \tag{2.181}$$

Fig. 2.24 shows the response of the system.

2.7 Pinning consensus

Pinning consensus is mostly concerned with reaching an agreement with respect to a fixed point (that is, it is desired to *pin* agent states to a fixed point) with numerous applications in high-dimensional systems such as plasma instabilities, turbulence, multimode lasers, and reaction–diffusion systems [58]. The notion of pinning consensus was introduced in [3] in the context of a static reference tracking problem in MAS. Some practical applications of the pinning consensus strategy have been proposed in [52,53] for power grids. The authors of [53] studied pinning group distributed consensus control for multimicrogrids, while [52] proposed new algorithms based on state estimators for ensuring stability under network phenomena like noise and malicious attacks in a distributed power system.

2.7.1 Main issues

Here, some definitions related to pinning consensus are formally introduced.

Definition 2.18 (Pinning consensus). A network of MAS (2.11) is said to solve the pinning consensus problem asymptotically provided that

$$\lim_{t \to \infty} x_i(t) = \bar{x} \quad \forall i = 1, 2, \ldots, n. \tag{2.182}$$

Definition 2.19 (Exponential pinning consensus [54]). A network of MAS (2.11) reaches a pinning consensus exponentially with respect to a desired state x_d if there exist positive constants $\gamma > 0$ and $\nu > 0$ such that for any initial state $x(0) \in \mathbb{R}^n$ and $t \in \mathbb{R}_+$, the following condition is satisfied:

$$\left\| x_{(t)} - x_d \mathbf{1}_n \right\| \leq \gamma \left\| x(0) - x_d \mathbf{1}_n \right\| e^{-\nu t}, \tag{2.183}$$

where ν denotes the consensus speed of the MAS. Note that ν is directly related to the value of the smallest eigenvalue of the graph Laplacian describing the communication topology.

Definition 2.20 (Group pinning consensus). Consider an MAS (2.11) defined over $\mathbb{G}$ with k subgroups. It is assumed that each subgroup has an equilibrium state, denoted by x_{σ_k}. A network of MAS (2.11) is said to solve the pinning consensus problem asymptotically provided that

$$\lim_{t \to \infty} |x_i(t) - x_{\sigma_k}| = 0 \quad \forall i \in \mathbb{G}_k, k = 1, \ldots, K. \tag{2.184}$$

2.7.2 Problem layout: linear dynamics

Consider a single-integrator MAS (2.9) over a communication topology $\mathbb{G}$. It is desired to pin the states of all the agents to a desired state x_d, which represents the desired equilibrium of the system. The following pinning consensus algorithm (2.185) is proposed for (2.9):

$$u_i = -\sum_{j \in \mathbb{N}_i} a_{ij}(x_i(t) - x_j(t)) + k_i(x_d - x_i(t)). \tag{2.185}$$

A major advantage of the pinning consensus scheme is that not all the agents require explicit knowledge of the desired pinning state x_d. At least one agent is required to be *pinned*, that is, $k_i > 0$ for at least one agent. Through distributed exchange of state information, all agents eventually converge to the pinned node. Finite-time pinning consensus control was investigated in [55] for a second-order MAS given by

$$\dot{x}_i(t) = v_i(t),$$
$$\dot{v}_i(t) = u_i(t) + g_i(t), \tag{2.186}$$

with virtual leader

$$\dot{x}_0(t) = v_0(t),$$
$$\dot{v}_0(t) = f_0(t), \tag{2.187}$$

where $|f_0| < f_l$ and f_l is a known constant.

2.7.3 Problem layout: nonlinear dynamics

Consider a network of n MASs with the following nonlinear scale-free dynamical system [56]:

$$\dot{x}_i = f(x_i) + c\sum_{j=1}^{N} a_{ij}\Gamma x_j, \quad i = 1, 2, \ldots, N, \tag{2.188}$$

where $x_i \in \mathbb{R}^n$ represents the state of node i; $c > 0$ denotes the coupling strength, $\gamma \in \mathbb{R}^{n \times n}$ represents a matrix defining the state coupling relationship among the agents, and $A = (a_{ij}) \in \mathbb{R}^{N \times N}$ represents the scale-free coupling of the network. The stability analysis of system (2.188) was discussed in [56] under control protocol (2.189), and Lemma 2.8 was proposed for reaching stability. Here

$$u_i = \begin{cases} -cd\Gamma(x_i - \bar{x}), & i = 1, 2, \ldots, l, \\ 0, & i = l+1, l+2, \ldots, N, \end{cases} \tag{2.189}$$

where l is the maximum number of pinned nodes.

Lemma 2.8. *If there exists a constant $\rho > 0$ such that the matrix $Df(\bar{x}) + \rho\Gamma$ is Hurwitz, then the homogeneous stationary state $\bar{x}$ of the controlled system is locally exponentially stable provided that*

$$c\lambda_1 \leq \rho, \tag{2.190}$$

where λ_1 is the largest eigenvalue of the matrix $A - D$.

A distributed adaptive pinning consensus protocol (2.191) was proposed in [57] for a scale-free coupled dynamical system (2.188) as

$$
u_i = \begin{cases} -p_i e_i, \ p_i = k_i \|e_i\|^2, & i = 1, 2, \ldots, l, \\ 0, & i = l+1, l+2, \ldots, N, \end{cases} \tag{2.191}
$$

where p_i denotes adaptive control gains and k_i represents static control gains which dictates the speed of convergence of p_i. The convergence properties of controlled system (2.188) under control protocol (2.191) were guaranteed by Theorem 2.14.

Theorem 2.14. *If there exists a constant $\rho > 0$ such that the matrix $Df(\bar{x}) + \rho r$ is Hurwitz, then the homogeneous stationary state $\bar{x}$ of the controlled system is locally exponentially stable provided that*

$$
c\lambda_1 \leq \rho, \tag{2.192}
$$

where λ_1 is the largest eigenvalue of the matrix $A - D$.

Pinning consensus was investigated in [58] for an MAS with nonlinear coupling functions:

$$
\dot{x}_i(t) = \sum_{j \in \mathbb{N}_i} h_{ij}(x_j(t) - x_i(t)) + u_i(t), \tag{2.193}
$$

where h_{ij} is a nonlinear coupling function satisfying a Lipschitz condition

$$
\|h_{ij}(x_j - x_i)\| \leq k_{ij} \|x_j - x_i\|. \tag{2.194}
$$

The control input u_i was designed as

$$
u_i = d_i(x_i - \bar{x}), \tag{2.195}
$$

where d_i is a control gain.

Theorem 2.15. *If the Lipschitz condition (2.194) is satisfied and $K - D > 0$, then consensus is reached about $\bar{x}$ where K is defined by*

$$
K = \begin{bmatrix} \sum_{j \in \mathbb{N}_i} a_{1j} k_{1j} & \cdots & \sum_{j \in \mathbb{N}_i} a_{1n} k_{1,n} \\ \vdots & \ddots & \vdots \\ \sum_{j \in \mathbb{N}_i} a_{n1} k_{nn} & \cdots & \sum_{j \in \mathbb{N}_i} a_{nj} k_{nj} \end{bmatrix}. \tag{2.196}
$$

Theorem 2.16. *If h_{ij} is continuously differentiable and F is negative definite, then consensus is reached asymptotically around $\bar{x}$ for*

$$F = \frac{1}{2}\left(\frac{\partial h^T}{\partial e} + \frac{\partial h}{\partial e} - 2(D \otimes I_m)\right) \tag{2.197}$$

and $\dfrac{\partial h}{\partial e}$ defined by

$$\frac{\partial h}{\partial e} = \begin{bmatrix} \dfrac{\partial h_1}{\partial e_1} & \cdots & \dfrac{\partial h_1}{\partial e_n} \\ \vdots & \ddots & \vdots \\ \dfrac{\partial h_n}{\partial e_1} & \cdots & \dfrac{\partial h_n}{\partial e_n} \end{bmatrix}. \tag{2.198}$$

Theorem 2.17. *Any node can be controlled with arbitrarily small gain such that consensus is reached asymptotically around $\bar{x}$, if the graph topology for (2.193) is strongly connected, $m = 1$, and $\dfrac{\partial h_{ij}}{\partial x}(x) > 0 \; \forall \, (i,j) \in \mathbb{E}$.*

2.7.4 Related work

A dynamic pinning consensus problem was introduced in [62] for a single-integrator MAS based on frequency-domain approaches using a minimum-order observer and an integrator to boost consensus speed. Pinning group consensus was studied in [59] for a heterogeneous MAS with input saturation via a pinning scheme with both first- and second-order agents using matrix theory, graph theory, and LaSalle's invariance principle to derive sufficient conditions which guarantee convergence. The second-order leader–follower pinning consensus for nonlinear multiagent systems was studied in [60] neglecting usual assumptions on the interaction topology such as strong connectedness and having a directed spanning tree. Specifically, the study addressed what kind of agents and how many agents are required to be pinned. Pinning consensus for discrete-time multiagent systems with multiple pinning agents was investigated in [61]. Using Gershgorin disk theorems, sufficient conditions were derived which show that one of root loci of the multiagent system crosses the unit circle earliest along the real axis and then diverges to minus infinity as the pinning gain increases from 0 to infinity, thus the upper bound of the pinning gain that achieves the consensus was obtained. A pinning cluster consensus control problem was discussed for a class of MASs for robotic systems with Lagrangian dynamics in [49] over directed acyclic network topologies.

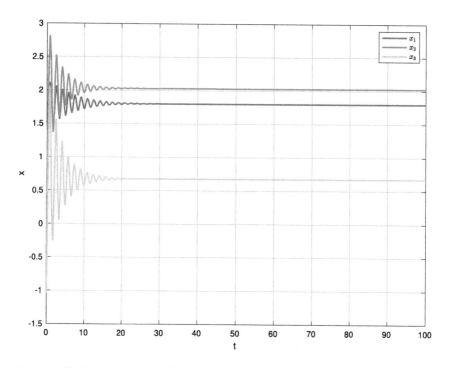

Figure 2.25 Pinning consensus for a network of Chua's oscillators.

2.7.5 Numerical example 2.6

Consider a network of Chua's oscillators [58] (see Fig. 2.25, 2.26, and 2.27):

$$\dot{x}_{i1} = \alpha(x_{i2} - x_{i1} + f(x_{i1})) + c\sum_{j=1}^{N} a_{ij}x_{j1}, \qquad (2.199)$$

$$\dot{x}_{i2} = x_{i1} - x_{i2} + x_{i3}, \qquad (2.200)$$

$$\dot{x}_{i3} = -\beta x_{i2} - \gamma x_{i3}. \qquad (2.201)$$

The system has three equilibrium states $x^0 = [0\ 0\ 0]$, $x^1 = [1.8586\ 0.0048\ -1.8539]$, $x^2 = [-1.8586\ -0.0048\ 1.8539]$. The system parameters are $\alpha = 10$, $\beta = 15$, $\gamma = 0.0385$, $a = -1.27$, and $b = -0.68$. It is required to stabilize the states of the system at x^1. The control protocol u_i is designed as

$$u_i = \begin{cases} cd(x_1^+ - x_{i1}), & i = i_1, \ldots, i_l, \\ 0, & \text{otherwise.} \end{cases} \qquad (2.202)$$

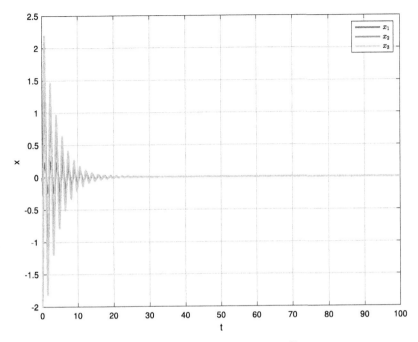

Figure 2.26 Pinning consensus for a network of Chua's oscillators.

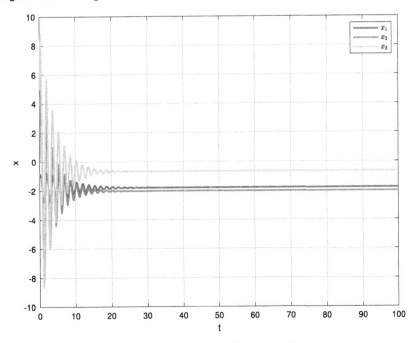

Figure 2.27 Pinning consensus for a network of Chua's oscillators.

2.8 Notes

In this chapter, we have provided the main definitions of various consensus problems and algorithms for continuous-time multiagent systems over fixed topologies. These included leaderless, leader–follower, group, bipartite, scaled, and resilient consensus. We introduced different notions of consensus based on some measures and then moved to develop some important conditions for reaching consensus cited in the literature.

Sufficient and necessary conditions for reaching consensus were proposed in [19] in the investigation on forming opinions over antagonistic networks, the conditions described do not depend on global knowledge of coupling weights among the agents. A second-order bipartite consensus protocol was proposed in [20] via impulsive control strategy without the knowledge of relative state information. A fuzzy fault-tolerant control problem was discussed in [25] for an MAS with a physical coupling graph and communication graph represented by a signed network. The study considered actuator bias and unknown nonlinear behaviors. A fixed-time bipartite flocking control problem was investigated for a nonlinear MAS over structurally balanced signed networks in [27].

An interventional bipartite consensus problem was investigated for an MAS described by higher-order dynamics with nonlinear unknown time-varying disturbance in [21], a distributed algorithm was proposed to reach an interventional bipartite consensus using neural networks based adaptive estimators to estimate the nonlinear disturbances. A tracking control problem was solved in [22] for an MAS with higher-order dynamics exhibiting unknown dynamics and hysteresis inputs.

A bipartite output consensus problem has been investigated for a heterogeneous linear MAS in [23] over networks with antagonistic interactions under structurally balanced and directed conditions. Specifically, two fully distributed protocols were proposed based on system states and their observations. A bipartite output consensus problem was investigated for a higher-order linear MAS in [24] considering both competition and cooperation using a novel distributed event-triggering dynamic observer. A finite-time output consensus problem for a heterogeneous linear MAS was considered in [26] via an event-triggering mechanism over directed signed networks.

Consensus in networks with antagonistic interactions and communication delays was investigated in [28] considering linear and nonlinear coupling between agents. A bipartite consensus problem was investigated for a network of robotic systems with inherent parametric uncertainties, input disturbances, and quantized interactions via novel distributed estimator-based control algorithms. A bipartite consensus tracking problem for uncertain nonlinear MAS with actuator fault was investigated in [31] over directed structurally balanced signed graphs using neural network based adaptive control laws to estimate the unknown dynamics.

The chapter was supplemented by numerical examples with simulations to further explain some theoretical concepts.

References

[1] M. Mesbahi, M. Egerstedt, Graph Theoretic Methods in Multiagent Systems, Princeton University Press, 2010.

[2] C. Altafini, Consensus problems on networks with antagonistic interactions, IEEE Trans. Autom. Control 58 (4) (2013).

[3] W. Ren, R.W. Beard, Distributed Consensus in Multi-Vehicle Cooperative Control: Theory and Application, Springer, 2008.

[4] W. Hou, M. Fu, H. Zhang, Z. Wu, Consensus conditions for general second-order multiagent systems with communication delay, Automatica 75 (2017) 293–298.

[5] W. Xie, B. Ma, Average consensus control of nonlinear uncertain multiagent systems, in: Proc. the 36th Chinese Control Conference, 2017, pp. 8299–8303.

[6] X. Niu, Y. Liu, Y. Man, Distributed adaptive consensus of nonlinear multi-agent systems with unknown control coefficients, IFAC-PapersOnLine (2015) 915–920.

[7] X. Zhang, L. Liu, G. Feng, Leader–follower consensus of time-varying nonlinear multi-agent systems, Automatica 52 (2015) 8–14.

[8] H. Su, G. Chen, X. Wang, Z. Lin, Adaptive second-order consensus of networked mobile agents with nonlinear dynamics, Automatica 47 (2) (2015) 368–375.

[9] H. Ren, F. Deng, Mean square consensus of leader-following multi-agent systems with measurement noises and time delays, ISA Trans. 71 (2017) 76–83.

[10] W. Cao, J. Zhang, W. Ren, Leader–follower consensus of linear multi-agent systems with unknown external disturbances, Syst. Control Lett. 82 (2015) 64–70.

[11] W. Guo, Leader-following consensus of the second-order multi-agent systems under directed topology, ISA Trans. 65 (2016) 116–124.

[12] C. Wang, X. Wang, H. Ji, Leader-following consensus for a class of second-order nonlinear multi-agent systems, Syst. Control Lett. 89 (2016) 61–65.

[13] T. Dong, Y. Gong, Leader-following secure consensus for second-order multi-agent systems with nonlinear dynamics and event-triggered control strategy under DoS attack, Neurocomputing (2020), https://doi.org/10.1016/j.neucom.2019.01.113.

[14] J. Yu, L. Wang, Group consensus of multi-agent systems with directed information exchange, Int. J. Syst. Sci. 43 (2) (2012) 334–348.

[15] M. Oyedeji, M. Mahmoud, Couple-group consensus conditions for general first-order multiagent systems with communication delays, Syst. Control Lett. 117 (2018) 37–44.

[16] Y. Shang, Y. Ye, Fixed-time group tracking control with unknown inherent nonlinear dynamics, IEEE Access 5 (2017) 12833–12842.

[17] H. Cai, G. Hu, Consensus-based distributed package-level state-of-charge balancing for grid-connected battery energy storage system, in: Proc. 12th IEEE International Conference on Control and Automation (ICCA), 2016, pp. 365–370.

[18] C. Huang, S. Weng, D. Yue, S. Deng, J. Xie, H. Ge, Distributed cooperative control of energy storage units in microgrid based on multi-agent consensus method, Electr. Power Syst. Res. 147 (2017) 213–223.

[19] B. Hou, Y. Chen, G. Liu, F. Sum, H. Li, Bipartite opinion forming: toward consensus over coopetition networks, Phys. Lett. A 379 (2015) 3001–3007.

[20] Z. Li, W. Wang, Y. Fan, H. Kang, Impulsive bipartite consensus of second-order multiagent systems without relative velocity information, Commun. Nonlinear Sci. Numer. Simul. 80 (2020) 1–9.

[21] Y. Wu, J. Hu, Y. Zhang, Y. Zeng, Interventional consensus for high-order multi-agent systems with unknown disturbances on coopetition networks, Neurocomputing 194 (2016) 126–134.

[22] T. Yu, L. Ma, H. Zhang, Y. Wang, Adaptive bipartite tracking control of nonlinear multiagent systems with unknown hysteresis, IFAC-PapersOnLine 50 (1) (2017) 5516–5521.

[23] Q. Ma, G. Zhou, E. Li, Adaptive bipartite output consensus of heterogeneous linear multi-agent systems with antagonistic interactions, Neurocomputing 373 (2020) 50–55, https://doi.org/10.1016/j.neucom.2019.09.067.

[24] J. Duan, H. Zhang, J. Han, Z. Gao, Bipartite output consensus of heterogeneous linear multi-agent systems by dynamic triggering observer, ISA Trans. 92 (2019) 14–22.

[25] L. Zhao, G. Yang, Fuzzy adaptive fault-tolerant control of multi-agent systems with interactions between physical coupling graph and communication graph, Fuzzy Sets Syst. 385 (2020) 20–36, https://doi.org/10.1016/j.fss.2019.04.005.

[26] J. Duan, H. Zhang, Y. Liang, Y. Cai, Bipartite finite-time output consensus of heterogeneous multi-agent systems by finite-time event-triggered observer, Neurocomputing 365 (2019) 86–93.

[27] Q. Xiao, H. Liu, X. Wang, Y. Huang, A note on the fixed-time bipartite flocking for nonlinear multi-agent systems, Appl. Math. Lett. 99 (2020) 86–93.

[28] J. Lu, X. Guo, T. Huang, Z. Wang, Consensus of signed networked multi-agent systems with nonlinear coupling and communication delays, Appl. Math. Comput. 350 (2019) 153–162.

[29] Q. Deng, J. Wu, T. Han, Q. Yang, X. Cai, Fixed-time bipartite consensus of multiagent systems with disturbances, Physica A 516 (2016) 37–49.

[30] T. Ding, M. Ge, C. Xiong, J.H. Park, Bipartite consensus for networked robotic systems with quantized-data interactions, Inf. Sci. 511 (2019) 22–42.

[31] H. Yang, D. Ye, Adaptive fixed-time bipartite tracking consensus control for unknown nonlinear multi-agent systems: an information classification mechanism, Inf. Sci. 459 (2018) 238–254.

[32] J. Qin, C. Yua, Cluster consensus control of generic linear multi-agent systems under directed topology with acyclic partition, Automatica 49 (2013) 2898–2905.

[33] Y. Chen, J. Lu, F. Han, X. Yu, On the cluster consensus of discrete-time multi-agent systems, Syst. Control Lett. 60 (2011) 517–523.

[34] H. Hu, W. Yu, Q. Xuan, C. Zhang, G. Xie, Group consensus for heterogeneous multiagent systems with parametric uncertainties, Neurocomputing 142 (2014) 383–392.

[35] J. Yu, J. Liu, L. Xiang, J. Zhou, Group consensus in networked mechanical systems with communication delays, in: Proc. IUTAM Symposium on Nonlinear and Delayed Dynamics of Mechatronic Systems, vol. 22, 2017, pp. 107–114.

[36] L. Ji, Y. Zhang, Y. Jiang, Couple-group consensus: a class of delayed heterogeneous multiagent systems in competitive networks, Hindawi Complexity (2018) 1–11.

[37] J. Liu, J. Ji, J. Zhou, L. Xiang, L. Zhao, Adaptive group consensus in uncertain networked Euler–Lagrange systems under directed topology, Nonlinear Dyn. 82 (2015) 1145–1157.

[38] K. Chen, J. Wang, Y. Zhang, F.L. Lewis, Cluster consensus of heterogeneous linear multi-agent systems, IET Control Theory Appl. 12 (11) (2018) 1533–1542.

[39] Y. Han, W. Lu, T. Chen, Cluster consensus in discrete-time networks of multiagents with inter-cluster nonidentical inputs, IEEE Trans. Neural Netw. Learn. Syst. 24 (4) (2013) 566–578.

[40] Y. Han, W. Lu, T. Chen, Achieving cluster consensus in continuous-time networks of multi-agents with inter-cluster non-identical inputs, IEEE Trans. Autom. Control 60 (3) (2015) 793–798.

[41] G. Wen, Y. Yu, Z. Peng, H. Wang, Dynamical group consensus of heterogenous multi-agent systems with input time delays, Neurocomputing 175 (2016) 278–286.

[42] J. Hu, H. Zhu, Adaptive bipartite consensus on coopetition networks, Phys. D: Nonlinear Phenom. 307 (2015) 14–21.

[43] J. Wu, Q. Deng, T. Han, Q. Yang, H. Zhan, Bipartite tracking consensus for multi-agent systems with Lipschitz-type nonlinear dynamics, Phys. A, Stat. Mech. Appl. 525 (2019) 1360–13691.

[44] M. Valcher, P. Misra, Bipartite tracking consensus for multi-agent systems with Lipschitz-type nonlinear dynamics, Syst. Control Lett. 66 (2019) 94–103.

[45] S. Roy, Scaled consensus, Automatica 51 (2015) 259–262.

[46] D. Meng, Y. Jia, Scaled consensus problems on switching networks, IEEE Trans. Autom. Control 61 (6) (2015) 1664–1669.

[47] H.D. Aghbolagh, E. Ebrahimkhani, F. Hashemzadeh, Scaled consensus tracking under constant time delay, IFAC-PapersOnLine 49 (22) (2016) 240–243.

[48] C. Liu, Scaled consensus seeking in multiple non-identical linear autonomous agents, ISA Trans. 71 (2017) 68–75.

[49] Q. Wang, Scaled consensus of multi-agent systems with output saturation, J. Franklin Inst. 354 (14) (2017) 6190–6199.

[50] L. Zhao, Y. Jia, J. Yu, J. Du, $\mathcal{H}_\infty$ sliding mode based scaled consensus control for linear multi-agent systems with disturbances, Appl. Math. Comput. 292 (2017) 375–389.

[51] J. Yu, L. He, Scaled group consensus in agent networks with finite sub-networks under continuous/discrete-time settings, J. Franklin Inst. 355 (2018) 8780–8801.

[52] L. Chen, Y. Wang, X. Lu, T. Zheng, J. Wang, S. Mei, Resilient active power sharing in autonomous microgrids using pinning-consensus-based distributed control, IEEE Trans. Smart Grid 10 (6) (2019) 6802–6811.

[53] Y. Shen, Y. Xu, H. Yao, J. Shi, F. Li, Z. Wen, C. Zhang, W. Liu, Distributed cluster control for multi-microgrids using pinning-based group consensus of multi-agent system, in: Proc. the 5th IEEE Int. Conference on Cloud Computing and Intelligence Systems (CCIS), 2018, pp. 1077–1080.

[54] A. Sakaguchi, T. Ushio, Consensus speed of static pinning consensus control of multi-agent systems, in: Proc. European Control Conference (ECC), 2018, pp. 1423–1428.

[55] Y. Zhou, G. Jiang, F. Xu, Q. Chen, Distributed finite time consensus of second-order multi-agent systems via pinning control, IEEE Access 6 (2018) 45617–45624.

[56] X. Wang, G. Chen, Pinning control of scale-free dynamical networks, Phys. A, Stat. Mech. Appl. 310 (2002) 521–531.

[57] X. Wang, G. Chen, Pinning adaptive synchronization of a general complex dynamical network, Automatica 44 (2008) 996–1003.

[58] F. Chen, Z. Chen, Z. Xiang, L. Liu, Z. Yuan, Reaching a consensus via pinning control, Automatica 45 (2009) 1215–1220.

[59] G. Wen, J. Huang, Z. Peng, Y. Yu, On pinning group consensus for heterogeneous multi-agent system with input saturation, Neurocomputing 207 (2016) 623–629.

[60] Q. Song, J. Cao, W. Yu, Second-order leader-following consensus of nonlinear multi-agent systems via pinning control, Syst. Control Lett. 59 (2010) 553–562.

[61] D. Xu, T. Ushio, On stability of consensus control of discrete-time multi-agent systems by multiple pinning agents, IEEE Control Syst. Lett. 3 (4) (2019) 1038–1043.

[62] A. Sakaguchi, T. Ushio, Dynamic pinning consensus control of multi-agent systems, IEEE Control Syst. Lett. 1 (2) (2017) 340–345.

[63] Z. Wang, H. Li, Cluster consensus in multiple Lagrangian systems under pinning control, IEEE Access 5 (2017) 11291–11297.

CHAPTER 3

Consensus in multiagent systems over time-varying networks

3.1 Introduction

Networks play a crucial role in the behavior of multiagent systems. In reality, networks exhibit some dynamics either intentionally or as a consequence of some external factors such as environmental conditions or a fault scenario. This chapter introduces the concept of a time-varying network and its effect on multiagent systems. In some applications, networks are required to switch between varying topologies usually as a security feature to prevent intrusion. However, most studies in switching network topologies model the switching behavior as random phenomena and provide mathematical analysis to guarantee stability. A review of previous consensus problems is carried out in this chapter considering varying network behavior. This chapter reexamines different consensus problems considering varying network topologies due to both deterministic and random events. Also problems related to time-varying communication delays are reviewed, highlighting promising results. Without loss of generality, we consider an MAS described by the following dynamics:

$$\dot{x}_i = f(x_i, u_i), \quad i \in \mathbb{V}, \tag{3.1}$$

where $\mathbb{V}$ represents a finite set of nodes, x_i and u_i are state and control inputs of the ith agent, respectively, and f is either a linear or nonlinear function. For system (3.1), we make the following definitions of consensus over a time-varying network modeled by a time-varying graph $\mathbb{G}$.

3.1.1 Definitions

In the sequel, we provide some definitions which are basic to the subsequent development.

Definition 3.1 (Time-varying graph). A time-varying graph $\mathbb{G}(t)$ is a graph with time-varying edge set $\mathbb{E}(t)$ and adjacency matrix $\mathbb{A}(t)$. It is not uncommon to find that time-varying graphs may also have a time-varying vertex set $\mathbb{V}(t)$ resulting primarily from loss of nodes or inclusion of new agents into the network.

Advanced Distributed Consensus for Multiagent Systems
https://doi.org/10.1016/B978-0-12-821186-1.00011-8
73

Definition 3.2 (Switching signal). A switching signal $\sigma(t)$ is time-varying function which determines the topology of a time-varying graph $\mathbb{G}(t)$ for all time t, that is, the switching signal determines which topology is activated at each switching time instant.

Definition 3.3 (Markovian switching signal, [9]). The switching signal

$$\sigma(t), \quad t \in \mathbb{R}_{\geq 0}$$

described by the Markovian process $\mathbb{P}(\sigma(t+h))$ with the infinitesimal generator $Q = (q_{ij})$ defined as follows:

$$\mathbb{P}\{(\sigma(t+h)) = j | \sigma(t) = i\} = \begin{cases} q_{ij}h + o(h), \\ 1 + q_{ii}h + o(h), \end{cases} \tag{3.2}$$

where q_{ii} is the transition rate from state i to state j with $q_{ij} \geq 0$ if $i \neq j$, $g_{ii} = -\sum_{j \notin i} q_{ij}$, and $o(h)$ represents an infinitesimal of higher order than h, that is, $\lim_{h \to 0} \dfrac{o(h)}{h} = 0$.

Definition 3.4 (Consensus). The MAS described by $\dot{x} = f(x_i, u_i)$ is said to reach consensus under arbitrary switching signal $\sigma(t)$ if

$$\lim_{t \to \infty} ||x_i(t) - x_j(t)|| = 0 \quad \forall \, i, j \in \mathbb{V} \tag{3.3}$$

holds for any initial condition.

Definition 3.5 (Almost sure consensus). The MAS defined by (3.1) is said to reach almost sure consensus under Markovian switching signal σ_m if

$$Pr\{\lim_{t \to \infty} ||x_i(t) - x_j(t)|| = 0\} = 1 \quad \forall \, i, j \in \mathbb{V} \tag{3.4}$$

holds for any initial condition and any distribution of $\sigma(t)$.

Definition 3.6 (Mean-square consensus). The MAS defined by (3.1) is said to reach mean-square consensus under Markovian switching signal σ_m if

$$\lim_{t \to \infty} \mathbb{E}[||x_i(t) - x_j(t)||] = 0 \quad \forall \, i, j \in \mathbb{V} \tag{3.5}$$

holds for any initial condition and any distribution of $\sigma(t)$.

Definition 3.7 (Union of graphs). Consider the set of graphs

$$\mathbb{G} = \{\mathbb{G}_i, \ldots, \mathbb{G}_s\}, \quad i = 1, \ldots, s,$$

where each graph $\mathbb{G}_i$ has vertex set $\mathbb{V}_i$ and edge set $\mathbb{E}_i$. The union of these graphs is a graph $\mathbb{G}_u = \{\mathbb{V}_u, \mathbb{E}_u\}$, where $\mathbb{V}_u = \bigcup_{i \in \mathbb{S}} V_i$ and $\mathbb{E}_u = \bigcup_{i \in \mathbb{S}} \mathbb{E}_i$.

Definition 3.8 (Jointly-connected graph). The union of graphs $\mathbb{G}_u$ is jointly connected if the union of graphs is a connected graph. Further, a time-varying graph denoted by $\mathbb{G}_{\sigma(t)}$ is jointly-connected across the time-interval $[t, t + T]$, $T > 0$, if the union of graphs $\{\mathbb{G}_{\sigma(s)}^u : s \in [t, t + T]\}$ is jointly connected.

Definition 3.9 (In-degree balanced node/graph, [80]). Consider a graph $\mathbb{G}(\mathbb{V}, \mathbb{E})$ consisting of two subgraphs $\mathbb{G}_1(\mathbb{V}_1, \mathbb{E}_1)$ and $\mathbb{G}_2(\mathbb{V}_2, \mathbb{E}_2)$, with $|\mathbb{V}_1| = n$, $\mathbb{V}_2 = m$, $\mathbb{V} = \mathbb{V}_1 \cup \mathbb{V}_2$, and $\mathbb{E} = \mathbb{E}_1 \cup \mathbb{E}_2$. A node $v_i \in \mathbb{G}_1$ is in-degree balanced to $\mathbb{G}_2$ if $\sum_{j=n+1}^{n+m} a_{ij} = 0$. Likewise, $\mathbb{G}_1$ is in-degree balanced to $\mathbb{G}_2$ if all the nodes in $\mathbb{G}_1$ are in-degree balanced to $\mathbb{G}_2$.

Definition 3.10 (Out-degree balanced node/graph, [80]). For a graph $\mathbb{G}(\mathbb{V}, \mathbb{E})$ consisting of two subgraphs $\mathbb{G}_1(\mathbb{V}_1, \mathbb{E}_1)$ and $\mathbb{G}_2(\mathbb{V}_2, \mathbb{E}_2)$, with $|\mathbb{V}_1| = n$, $\mathbb{V}_2 = m$, $\mathbb{V} = \mathbb{V}_1 \cup \mathbb{V}_2$, and $\mathbb{E} = \mathbb{E}_1 \cup \mathbb{E}_2$, a node $v_i \in \mathbb{G}_1$ is out-degree balanced to $\mathbb{G}_2$ if $\sum_{j=n+1}^{n+m} a_{ji} = 0$. Likewise, $\mathbb{G}_1$ is out-degree balanced to $\mathbb{G}_2$ if all the nodes in $\mathbb{G}_1$ are out-degree balanced to $\mathbb{G}_2$.

Definition 3.11 (In-degree balanced couple, [80]). Consider a graph $\mathbb{G}(\mathbb{V}, \mathbb{E})$ consisting of two subgraphs $\mathbb{G}_1(\mathbb{V}_1, \mathbb{E}_1)$ and $\mathbb{G}_2(\mathbb{V}_2, \mathbb{E}_2)$, with $|\mathbb{V}_1| = n$, $\mathbb{V}_2 = m$, $\mathbb{V} = \mathbb{V}_1 \cup \mathbb{V}_2$, and $\mathbb{E} = \mathbb{E}_1 \cup \mathbb{E}_2$. Subgraphs $\mathbb{G}_1$ and $\mathbb{G}_2$ are said to be in-degree balanced if all the nodes in $\mathbb{G}_1$ are in-degree balanced to $\mathbb{G}_2$, and vice-versa.

Definition 3.12 (Out-degree balanced couple, [80]). Consider a graph $\mathbb{G}(\mathbb{V}, \mathbb{E})$ consisting of two subgraphs $\mathbb{G}_1(\mathbb{V}_1, \mathbb{E}_1)$ and $\mathbb{G}_2(\mathbb{V}_2, \mathbb{E}_2)$, with $|\mathbb{V}_1| = n$, $\mathbb{V}_2 = m$, $\mathbb{V} = \mathbb{V}_1 \cup \mathbb{V}_2$, and $\mathbb{E} = \mathbb{E}_1 \cup \mathbb{E}_2$. Subgraphs $\mathbb{G}_1$ and $\mathbb{G}_2$ are said to be out-degree balanced if all the nodes in $\mathbb{G}_1$ are out-degree balanced to $\mathbb{G}_2$, and vice-versa.

Definition 3.13 (Balanced couple, [80]). Consider a graph $\mathbb{G}(\mathbb{V}, \mathbb{E})$ consisting of two subgraphs $\mathbb{G}_1(\mathbb{V}_1, \mathbb{E}_1)$ and $\mathbb{G}_2(\mathbb{V}_2, \mathbb{E}_2)$, with $|\mathbb{V}_1| = n$, $\mathbb{V}_2 = m$, $\mathbb{V} = \mathbb{V}_1 \cup \mathbb{V}_2$, and $\mathbb{E} = \mathbb{E}_1 \cup \mathbb{E}_2$. Subgraphs $\mathbb{G}_1$ and $\mathbb{G}_2$ are said to be a balanced couple if they are both in-degree and out-degree balanced to each other.

Definition 3.14 (Uniformly balanced couple). If for a time-varying graph $\mathbb{G}(t)$, Definitions 3.11–3.13 hold for all time instances t, then $\mathbb{G}_1$ and $\mathbb{G}_2$ are called uniformly in-degree, out-degree, or balanced couples, respectively.

Definition 3.15 (Double-tree form). Consider a graph $\mathbb{G}(\mathbb{V}, \mathbb{E})$ consisting of two subgraphs $\mathbb{G}_1(\mathbb{V}_1, \mathbb{E}_1)$ and $\mathbb{G}_2(\mathbb{V}_2, \mathbb{E}_2)$, with $|\mathbb{V}_1| = n$, $\mathbb{V}_2 = m$, $\mathbb{V} = \mathbb{V}_1 \cup \mathbb{V}_2$, and $\mathbb{E} = \mathbb{E}_1 \cup \mathbb{E}_2$. Let $T_1 = \{\mathbb{V}(T_1), \mathbb{E}(T_1), \mathbb{A}(T_1)\}$, $T_2 = \{\mathbb{V}(T_2), \mathbb{E}(T_2), \mathbb{A}(T_2)\}$. Then $T = \{T_1, T_2\}$ is a double-tree form associated with $\mathbb{G}$ if T_1, T_2 are trees and $\mathbb{V}(G) = \bigcup_{i=1,2} \mathbb{V}(T_i)$.

3.2 Switching networks

In this section, we consider consensus problems for multiagent systems over networks with switching topologies.

3.2.1 Leaderless consensus problems

Leaderless consensus problems over time-varying networks based on dwell-time conditions have been investigated by earlier researchers; see [1–4]. A necessary and sufficient condition in some of these studies is the presence of a spanning tree in the union of the switching graphs. Consider a network of MAS over a time-varying topology $\mathbb{G}_k$ modeled with first-order integrator dynamics as

$$\dot{x}_i(t) = u_i(t). \tag{3.6}$$

A distributed leaderless consensus algorithm is designed as

$$u_i(t) = \sum_{j=1}^{N} a_{ij}(t)(x_j(t) - x_i(t)). \tag{3.7}$$

In closed-form, protocol (3.7) applied to (3.6) becomes [1]

$$\dot{x}(t) = \mathbb{L}_k x(t), \quad k = s(t). \tag{3.8}$$

Theorem 3.1 ([1]). *Consider the hybrid system (3.8) consisting of the continuous-time agent model dynamics (3.6) and a discrete-time state behavior for the communication topology described by $\Gamma = \{\mathbb{G}\}$ such that any graph $\mathbb{G}$ belonging to the set is a digraph which is strongly connected and balanced, that is,*

$$\Gamma_n = \{\mathbb{G} = (\mathbb{V}, \mathbb{E}, \mathbb{A}) : \mathrm{rank}(\mathbb{L}(\mathbb{G})) = n - 1, \mathbf{1}^T \mathbb{L}(\mathbb{G}) = 0\}, \tag{3.9}$$

then, for any arbitrary switching signal s(t), system (3.8) globally asymptotically solves the average consensus problem.

In [4], a time-varying distributed consensus protocol was designed as

$$u_i(t) = -\sum_{i=1}^{N} \sigma_{ij}(t)a_{ij}(t)[x_i(t) - x_j(t)], \tag{3.10}$$

where $\sigma_{ij} > 0$ is a weighting scale belonging to a finite set.

Theorem 3.2 ([4]). *Consider an infinite-time sequence, $t_1, t_2, \ldots$ on which the interaction topology or weighting factors changes and let $\tau_i = t_{i+1} - t_i \in \mathbb{T}$, $i = 0, 1, 2, \ldots$ If $\mathbb{G}(t_i) \in \bar{G}$ is the interaction graph at time $t = t_i$ and $\sigma_{ij}(t_i) \in \bar{\sigma}$, where $\bar{\sigma}$ is a finite set of arbitrary positive numbers, then system (3.9) achieves consensus asymptotically under protocol (3.10), if there exists an infinite sequence of uniformly bounded, nonoverlapping time intervals $[t_{ij}, t_{ij+l_j})$, $j = 1, 2, \ldots$ starting at $t_{i1} = t_0$, with the property that each interval $[t_{ij}, t_{ij+1})$ is bounded and the union of the graphs across each interval has a spanning tree.*

Consider an MAS described by the following general linear system:

$$\dot{x}_i(t) = Ax_i(t) + Bu_i(t). \tag{3.11}$$

For system (3.11), the following assumption is made:

Assumption 3.2.1. The pair (A, B) is stabilizable and A may have eigenvalues in the open right-half plane.

A leaderless consensus problem for a general linear multiagent system was investigated in [5] for the MAS (3.11) under the following protocol:

$$\dot{x}_i(t) = Ax_i(t) + Bu_i(t). \tag{3.12}$$

Theorem 3.3. *Consider the MAS modeled by (3.11), satisfying Assumption 3.2.1, having a time-varying communication topology $\mathbb{G}(t)$, which is weakly connected and balanced across over uniformly bounded time intervals $[t_k, t_k + 1)$, $k \in \mathbb{T}$, with $s_{k+1} - s_k \leq \tau$ for some $\tau > 0$. The MAS (3.11), under protocol (3.12) solves the leaderless consensus problem provided there exist $P > 0$ and $\epsilon > 0$ satisfying the LMIs (3.13) and guaranteeing the existence of the feedback gain $K = B^T P$ such that*

$$PA + A^T P - 2\mu PBB^T P < -\mu I, \tag{3.13}$$

$$\mu > M\lfloor T/\tau \rfloor[(\theta + 1)^2 - 1], \tag{3.14}$$

where

$$\mu < \min\{\lambda_{k,i} | k \in \tau, i = 2, \ldots, N\},$$
$$\theta = \bar{A}\varepsilon \, T e^{\bar{A}\epsilon T}(1 - e^{\lfloor T/\tau \rfloor \bar{A}\epsilon T}/(1 - e^{\bar{A}\epsilon T})),$$
$$M = \|I_N \otimes (PA + A^T P) - (L_{\sigma(s) + L_{\sigma(s)}^T}) \otimes BB^T\|.$$

Definition 3.16 (Average dwell time). Let $N_\sigma(t, \tau)$ be the number of discontinuities of a switching signal $\sigma(t)$ on the time-interval (τ, t), then the switching signal $\sigma(t)$ is said to have average dwell time $\tau_a > 0$ if for each $0 \le \tau \le t$,

$$N_\tau(t, \tau) \le N_0 + \frac{t - \tau}{\tau_a}, \quad N_0 > 0. \tag{3.15}$$

A leaderless output consensus problem was investigated in [6] for the MAS with dynamics

$$x_i(t) = Ax_i(t) + Bu_i(t), \quad i \in \mathbb{V}, \tag{3.16}$$
$$y_i(t) = Cx_i(t),$$

with relative output errors designed as

$$z_i(t) = \sum_{i=1}^{N} a_{ij}(t)(y_j(t) - y_i(t)), \quad i \in \mathbb{V}, \tag{3.17}$$

and a dynamic compensator designed as:

$$\dot{\zeta}_i = A_c \zeta_i + B_c z_i, \tag{3.18}$$
$$u_i = C_c \zeta_i + D_c z_i.$$

Applying compensator (3.18) to (3.16), we arrive at

$$\dot{\bar{x}} = [I_N \otimes \bar{A} - \mathbb{L}_{\sigma(t)} \otimes \bar{B}\bar{C}]\bar{x}, \tag{3.19}$$
$$\bar{A} = \begin{bmatrix} A & BC_c \\ 0 & A_c \end{bmatrix}, \quad \bar{B} = \begin{bmatrix} BD_c \\ B_c \end{bmatrix}, \quad \bar{C} = \begin{bmatrix} C & 0 \end{bmatrix}.$$

Assumption 3.2.2 ([6]). All the eigenvalues of A lie in the closed left–half complex plane and at least one eigenvalue lies on the imaginary axis.

Assumption 3.2.3 ([6]). The switching signal $\sigma(t)$ takes its value on a finite index set $\mathbb{P} = \{1, \ldots, p\}$, that is, $\sigma(t) \in \mathbb{P}$ for $t > 0$ and the corresponding set $\{\mathbb{G}_1, \ldots, \mathbb{G}_p\}$ of digraphs contains r connected digraphs for $1 \leq r \leq p$.

Theorem 3.4 ([6]). *For the MAS described by the linear dynamics in* (3.16), *under Assumptions* 3.2.2–3.2.3, *the MAS* (3.16) *solves the leaderless consensus problem with a convergence rate δ if there exists an average dwell time $\tau_a > 0$ and a time $\tau^* \geq 0$ such that*

$$\tau_a(\sigma^* - \sigma) - a > 0, \tag{3.20}$$

$$\tau^c(t, 0) \geq \frac{\tau_a(\delta^d + \delta) + a}{\tau_a(\delta^c - \delta) - a} \geq \tau^d(t, 0) \quad \forall t \geq \tau^*, \tag{3.21}$$

where $0 < \delta < \delta^c$, while $\tau^c(t, 0)$ and $\tau^d(t, 0)$ are the total activation times of connected and disconnected digraphs $[0, t)$, respectively.

Leaderless consensus was investigated for an MAS with nonlinear dynamics (3.22) in [7]

$$\dot{x}_i(t) = Ax_i(t) + Bu_i(t) + Df(x_i(t), t), \quad i = 1, 2, \ldots, N, \tag{3.22}$$

where $x_i(t)$ and $u_i(t)$ represent state and control inputs; A, B and D are constant system matrices; $f : \mathbb{R}^n \times [0, \infty > \rightarrow \mathbb{R}^m$ is a continuously differentiable vector-valued function representing the nonlinearities which satisfies the Lipschitz condition

$$\|f(x(t), t) - f(y(t), t)\| \leq \rho \|x(t) - y(t)\| \quad \forall x, y \in \mathbb{R}^n, t \geq 0, \tag{3.23}$$

where $\rho > 0$ is a constant scalar. The distributed consensus algorithm was proposed for the system as follows:

$$u_i(t) = cK \sum_{j=1}^{N} a_{ij}^{\sigma(t)}(t)(x_j(t) - x_i(t)), \tag{3.24}$$

where K is a feedback matrix to be designed and c is a coupling strength to be chosen. System (3.22) under protocol (3.24) becomes

$$\dot{x}(t) = [I_n \otimes A - c\mathbb{L}_{\sigma(t)} \otimes BK]x(t) + (I_n \otimes D)F(x, t), \tag{3.25}$$

where

$$x(t) = [x_1(t)^t, x_2(t)^t, \ldots, x_n(t)^t],$$

$$F(x, t) = [f_1(x, t)^t, f_2(x, t)^t, \ldots, f_n(x, t)^t],$$

and $\mathbb{L}_{\sigma(t)}$ represents the switched Laplacian matrix describing the topology at the switching time instant.

Assumption 3.2.4 ([7]). Each graph $G_{\sigma(t)}$ at the switching instant t contains a directed spanning tree.

Lemma 3.1 ([8]). *Given a Laplacian matrix $\mathbb{L} \in \mathfrak{R}^{N \times N}$ of a graph $\mathbb{G}$ and a full row rank matrix $\mathbf{E} \in \mathbb{R}^{N-1 \times N}$ defined as*

$$\mathbf{E} = \begin{bmatrix} 1 & -1 & 0 & \ldots & 0 \\ 0 & -1 & -1 & \ldots & 0 \\ \vdots & \vdots & \ddots & \ddots & \vdots \\ 1 & -1 & 0 & \ldots & 0 \end{bmatrix}, \tag{3.26}$$

there exist a matrix $M \in \mathbb{R}^{N \times (N-1)}$ such that $\mathbb{L} = M\mathbf{E}$. If the graph has a directed spanning tree, M is a full column rank and $\mathrm{Re}(\lambda(M\mathbf{E})) > 0$.

Under Assumption 3.2.4, using Lemma 3.1, [7] concludes that, given a time-varying Laplacian $\mathbb{L}_{\sigma(t)}$, $\sigma(t) \in \mathbb{P}$, there exists a corresponding full column rank matrix $M_{\sigma(t)}$ such that $\mathbb{L}_{\sigma(t)} = M_{\sigma(t)}\mathbf{E}$ with $\mathbf{E}$ defined as in (3.26) and the eigenvalues of $M_{\sigma(t)}\mathbf{E}$ have positive real parts.

Lemma 3.2 ([7]). *For the matrix $\mathbf{E}M^{(i)}$, $i \in \mathbb{P}$ satisfying Assumption 3.2.4, there exist a positive definite matrix $Q^{(i)}$ and a common positive scalar α_0 for $i \in \mathbb{P}$, such that the following condition is satisfied:*

$$(\mathbf{E}M^{(i)})^t Q^{(i)} + Q^{(i)}\mathbf{E}M^{(i)} > \alpha_0 Q^{(i)}. \tag{3.27}$$

These take us to

Theorem 3.5 ([7]). *The MAS described by (3.25) solves the leaderless consensus problem under restricted switching topologies if there exist real scalars $c > 0$ and $\beta_0 > 0$, and positive definite matrix $P > 0$ such that the following LMI is feasible:*

$$\begin{pmatrix} A^T + PA - C\alpha_0 PBB^T P + \beta_0 P & PD & I_n \\ \bullet & -\dfrac{1}{\phi_1}I_n & 0 \\ \bullet & \bullet & -\dfrac{\phi_2}{\rho^2}I_n \end{pmatrix} > 0, \tag{3.28}$$

where $\phi_1 = \max_{i\in\mathbb{P}}\{\lambda(Q^{(i)})\}$, $\phi_2 = \min_{i\in\mathbb{P}}\{\lambda(Q^{(i)})\}$. The feedback matrix K is designed as $K = B^T P$, and further, the average dwell time satisfies

$$\tau_a > \tau_a^* = \frac{\ln h_0}{\beta_0},$$

$$h_0 = \frac{\phi_1}{\phi_2}. \qquad (3.29)$$

Assumption 3.2.5. Each graph $\mathbb{G}_{\sigma(t)} \in \mathbb{G}$ is strongly connected and balanced.

Lemma 3.3. *Suppose that the graph $\mathbb{G}_{\sigma(t)}$ is strongly connected and balanced, then the matrix $\mathbf{E}(\mathbb{L} + \mathbb{L}^T)\mathbf{E}$ is positive definite.*

Theorem 3.6 ([7]). *The MAS described by (3.25) solves the leaderless consensus problem under arbitrary switching topologies provided there exist real scalars $c > 0$ and $\beta_0 > 0$, and positive definite matrix $P > 0$ such that the following LMI is feasible:*

$$\begin{bmatrix} A^t P + PA - \dfrac{c\lambda_0}{\tilde\phi_1}PBB^t P & PD & I_n \\[2mm] \bullet & -\dfrac{1}{\tilde\phi_2}I_n & 0 \\[2mm] \bullet & \bullet & -\dfrac{1}{\rho^2\tilde\phi_1}I_n \end{bmatrix} > 0, \qquad (3.30)$$

where $\tilde\phi_1 = \max_{i\in\mathbb{P}}\{\lambda(EE^t)\}$, $\tilde\phi_2 = \min_{i\in\mathbb{P}}\{\lambda(EE^T)\}$, $\lambda_0 = \min_{i\in\mathbb{P}}\{\lambda(E((L^{(i)})^T + L^{(i)})E^T)\}$. The feedback matrix K is designed as $K = B^T P$, and further, the average dwell time satisfies

$$\tau_a > \tau_a^* = \frac{\ln h_0}{\beta_0},$$

$$h_0 = \frac{\phi_1}{\phi_2}. \qquad (3.31)$$

3.2.2 Numerical example 3.1

Consider the MAS (2.9) consisting of three agents described by the graphs in Fig. 3.1, (a) and (b), with the corresponding Laplacian matrices (3.32) and (3.33), respectively, representing switching between two distinct but stable topologies:

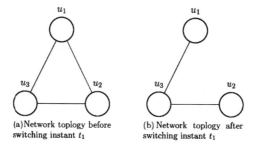

(a) Network toplogy before switching instant t_1

(b) Network toplogy after switching instant t_1

Figure 3.1 Graphs on $n = 3$ vertices.

$$\mathbb{L} = \begin{bmatrix} 4 & -3 & -1 & 1 & -1 \\ -3 & 3 & 0 & -1 & 1 \\ -1 & 0 & 1 & 0 & 0 \\ 1 & -1 & 0 & 3 & -3 \\ -1 & 1 & 0 & -3 & 3 \end{bmatrix}, \tag{3.32}$$

$$\mathbb{L} = \begin{bmatrix} 4 & -3 & -1 & 1 & -1 \\ -3 & 3 & 0 & -1 & 1 \\ -1 & 0 & 1 & 0 & 0 \\ 1 & -1 & 0 & 3 & -3 \\ -1 & 1 & 0 & -3 & 3 \end{bmatrix}. \tag{3.33}$$

3.2.3 Leader–follower consensus

A leader–follower consensus problem was investigated for MAS with double-integrator dynamics in [63] over networks with time-varying topologies (see Fig. 3.2). Let the follower agents be represented by

$$\dot{x}_i(t) = v_i(t),$$
$$\dot{v}_i(t) = u_i(t), \tag{3.34}$$

and the leader agents are defined by the following dynamics

$$\dot{x}_0(t) = v_0(t),$$
$$\dot{v}_0(t) = 0. \tag{3.35}$$

The consensus objective necessitates that the leader and follower position and velocity states converge as follows:

$$\lim_{t \to \infty} |x_i(t) - x_0(t)| = 0, \tag{3.36}$$

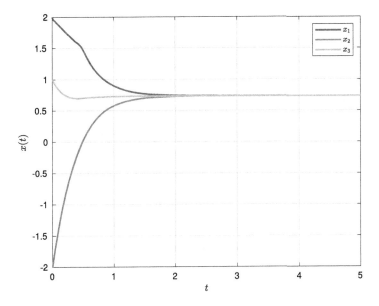

Figure 3.2 Consensus under switching topologies.

$$\lim_{t \to \infty} |v_i(t) - v_0(t)| = 0. \qquad (3.37)$$

The following distributed consensus protocol was proposed by [63]:

$$u_i(t) = -\alpha \left[\sum_{j \in \mathbb{N}^i} a_{ij}(t)(x_i(t) - x_j(t)) + b_i(t)(x_i(t) - x_0(t)) \right]$$
$$- \beta(v_i(t) - v_0(t)), \qquad (3.38)$$

under the following

Assumption 3.2.6 ([63]).

- The velocity communication topology between the leader and follower agents is connected and fixed. There is no velocity information exchange among the follower agents and the constant velocity of the leader is known for any follower agent.
- The position communication topologies of agents are time-varying.
- The position communication topologies between a fixed time interval $[t_r, t_{r+1}]$, $r = 1, 2, \dots$ are jointly connected.

Theorem 3.7 ([63]). *Consider an MAS network with leader and follower agents defined in (3.35) and (3.34). Under Assumption 3.2.6 and if the following in-*

equalities are fulfilled for any time $t \in (0, +\infty)$:

$$\lambda_n(\mathbb{L}_\sigma + \mathbb{B}_\sigma) < \frac{\beta^2}{\alpha}, \tag{3.39}$$

$$\lambda_n(\mathbb{L}_\sigma) < \beta, \alpha < \beta, \tag{3.40}$$

protocol (3.38) solves the second-order leader–follower consensus problem.

Corollary 3.1 ([63]). *Suppose the MAS network consists of fixed leader agents (3.41) and dynamic follower agents (3.34):*

$$x_0 = c,$$
$$v_0 = 0. \tag{3.41}$$

If Assumption 3.2.6 holds and the following inequalities are fulfilled:

$$\lambda_n(\mathbb{L}_\sigma + \mathbb{B}_\sigma) < \frac{\beta^2}{\alpha}, \tag{3.42}$$

$$\lambda_n(\mathbb{L}_\sigma) < \beta, \alpha < \beta, \tag{3.43}$$

then, under the following protocol

$$u_i(t) = -\alpha \left[\sum_{j \in \mathbb{N}^i} a_{ij}(t)(x_i(t) - x_j(t)) + b_i(t)(x_i(t) - x_0(t)) \right]$$
$$- \beta(v_i(t)), \tag{3.44}$$

the MAS (3.41), (3.34) solves the second-order leader–follower consensus problem:

$$\lim_{t \to \infty} x_i(t) = c,$$
$$\lim_{t \to \infty} v_i(t) = 0. \tag{3.45}$$

Consider the following MAS with the leader dynamics (3.46) and follower dynamics (3.47):

$$\dot{x}_0(t) = Ax_0(t), \tag{3.46}$$
$$\dot{x}_i(t) = Ax_i(t) + Bu_i(t). \tag{3.47}$$

Assumption 3.2.7. The eigenvalues of matrix A have positive real parts.

Assumption 3.2.8. Both the leader and follower agents have the same dynamics.

Consider the following distributed leader–follower consensus algorithm over a time-varying network [59]:

$$u_i(t) = K \sum_{j \in \mathbb{N}_i(t)} (x_j(t) - x_i(t)) + Kd_i(t)(x_0(t) - x_i(t)), \quad \forall i \in \mathbb{V}. \qquad (3.48)$$

Theorem 3.8. *The MAS described by (3.46) and (3.47) solves the leader–follower consensus problem under protocol (3.48) provided that Assumptions 3.2.2–3.2.3 are satisfied and there exists P such that:*

$$PA + A^T P - 2\delta_{\min} PBB^T P + \delta_{\min} I < 0, \qquad (3.49)$$

$$PA + A^T P \leq 0, \qquad (3.50)$$

which allows the feedback matrix to be designed as $K = B^T P$ and $\lambda_{\min} = \min\{\lambda_p : p \in \mathbb{P}\}$, $\lambda_p := \lambda_{\min}(H_p)$, and $H_p = L_p + D_p$, $p \in \mathbb{P}$.

3.2.4 Group consensus

Consider an MAS defined by the following single-integrator dynamics:

$$\dot{x}_i = u_i(t). \qquad (3.51)$$

A time-varying couple-group distributed consensus protocol is given as

$$u_i(t) = \begin{cases} \sum_{j \in \mathbb{N}_i^!} a_{ij}(t)[x_j(t) - x_i(t)] + \sum_{j \in \mathbb{N}_i^!} a_{ij}(t)x_j(t) & \forall i \in \chi_1, \\ \sum_{j \in \mathbb{N}^i} a_{ij}(t)[x_j(t) - x_i(t)] + \sum_{j \in \mathbb{N}_i^!} a_{ij}(t)x_j(t) & \forall i \in \chi_2. \end{cases} \qquad (3.52)$$

Applying (3.52) to (3.51), we get

$$\dot{x}_i(t) = \mathbb{L}(\mathbb{G}(t))x_i(t). \qquad (3.53)$$

Some lemmas and theorems were proposed in [80] for reaching couple-group consensus over networks with switching topologies using definitions based on double-tree form transformations.

Definition 3.17 (Double-tree transform, [80]). We say that $T_1 = (\mathbb{V}(T_1), \mathbb{E}(T_1), \mathbb{A}(T_1))$ and $T_2 = (\mathbb{V}(T_1), \mathbb{E}(T_1), \mathbb{A}(T_1))$ denote two trees in a double-tree form $T = (T_1, T_2)$ associated with $\mathbb{G}$ if
- for $i = 1, 2$, T_i is a tree and
- $\mathbb{V}(T_1) \cup \mathbb{V}(T_2) = \mathbb{V}$.

However, the double-tree form T associated with $\mathbb{G}$ is not necessarily a tree of $\mathbb{G}$, and is not a spanning tree of $\mathbb{G}$. Based on the double-tree form T, it is possible to derive a double-tree form transformation γ_T, defined as

$$\gamma_T = [\gamma_{ij}(t)], \quad i,j \in \mathbb{V}, \tag{3.54}$$

$$\gamma_{ij}(t) = x_i(t) - x_j(t). \tag{3.55}$$

Lemma 3.4 ([80]). *Protocol (3.52) solves the group consensus problem if and only if $\lim_{t\to\infty} \gamma_T(t) = 0$.*

Lemma 3.5 ([80]). *Given a double-tree transformation L_T, system (3.53) can be transformed to a reduced order form*

$$\dot{\gamma}_T(t) = \phi(T_{r(t)}, \mathbb{G}_{r(t)})\gamma_T(t), \tag{3.56}$$

where $r(t)$ is a switching signal.

Lemma 3.6 ([80]). *The distributed consensus protocol (3.52) solves the group consensus problem if the system*

$$\dot{\gamma}_T(t) = \phi(T_{r(t)}, \mathbb{G}_{r(t)})\gamma_T(t) \tag{3.57}$$

is stable under any arbitrary switching signal, $r(t)$.

Lemma 3.7 ([80]). *The distributed consensus protocol (3.52) solves the group consensus problem asymptotically if there exist matrices such that*

$$(B\mathbb{L}_i C)^t P + P B \mathbb{L}_i C > 0 \quad \forall i = 1, \ldots, p. \tag{3.58}$$

3.2.5 Numerical example 3.2

In this numerical example, we consider an MAS described by the dynamics (3.51)–(3.52), and represented by the topology in Fig. 2.3. Here, we are interested in simulating the behavior of the MAS subject to some distributed attacks in the network, which are described by a switching behavior in the network topology. We assume an MAS modeled by the graph in Fig. 2.3 with the following Laplacian (3.59) and adjacency matrix (3.60):

$$\mathcal{L} = \begin{bmatrix} 4 & -3 & -1 & 1 & -1 \\ -3 & 3 & 0 & -1 & 1 \\ -1 & 0 & 1 & 0 & 0 \\ 1 & -1 & 0 & 3 & -3 \\ -1 & 1 & 0 & -3 & 3 \end{bmatrix}, \tag{3.59}$$

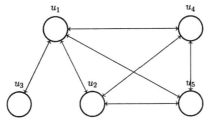

Figure 3.3 Graph topology depicting the interconnections among agents.

$$\mathcal{D} = \begin{bmatrix} 0 & 3 & 1 & -1 & 1 \\ 3 & 0 & 0 & 1 & -1 \\ 1 & 0 & 0 & 0 & 0 \\ -1 & 1 & 0 & 0 & 3 \\ 1 & -1 & 0 & 3 & 0 \end{bmatrix}. \tag{3.60}$$

The interaction graph described by Fig. 3.3 consists of subgroups $\mathcal{G}_1$ and $\mathcal{G}_2$. A simple check reveals that the graph Laplacian (3.59) satisfies the conditions cited in Lemma 3.7. The partitioning of the Laplacian matrix is given by:

$$\mathcal{L}_{11} = \begin{bmatrix} 4 & -3 & -1 \\ -3 & 3 & 0 \\ -1 & 0 & 1 \end{bmatrix}, \quad \mathcal{L}_{22} = \begin{bmatrix} 3 & -3 \\ -3 & 3 \end{bmatrix}, \tag{3.61}$$

$$\mathcal{L}_{12} = \begin{bmatrix} 1 & -1 \\ -1 & 1 \\ 0 & 0 \end{bmatrix}, \tag{3.62}$$

$\mu_i = 0.0000, 1.3542, 6.6458$, $\mu_j = 0, 6$, and $\mu_{\max}(\mathcal{L}_{12}\mathcal{L}_{21}) = 4$.

The following consensus protocol is proposed:

$$u_i(t) = \begin{cases} \alpha_1 \displaystyle\sum_{\substack{j=1 \\ j\neq i \\ j\notin\Gamma(t)}}^{n} a_{ij}(x_j(t) - x_i(t)) + \beta_1 \displaystyle\sum_{\substack{j=n+1 \\ j\neq i \\ j\notin\Gamma(t)}}^{n+m} a_{ij}x_j(t) & \forall i \in \mathcal{G}_1, \\[2em] \alpha_2 \displaystyle\sum_{\substack{j=n+1 \\ j\neq i \\ j\notin\Gamma(t)}}^{n+m} a_{ij}(x_j(t) - x_i(t)) + \beta_2 \displaystyle\sum_{\substack{j=1 \\ j\neq i \\ j\notin\Gamma(t)}}^{n} a_{ij}x_j(t) & \forall i \in \mathcal{G}_2, \end{cases} \tag{3.63}$$

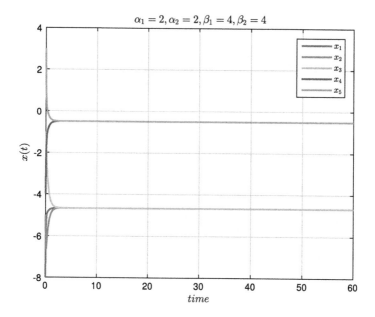

Figure 3.4 Nominal case with coupling gains $\alpha_1 = \alpha_2 = 2, \beta_1 = \beta_2 = 4$.

under the following constraints on the coupling gains:

$$b^2\gamma_b^2 + 4a^2 - 4ab\gamma_a + 4b^2(\alpha_1\alpha_2\gamma_c - \beta_1\beta_2\mu_l) > 0, \tag{3.64}$$

$$2a - b\gamma_a \neq 0. \tag{3.65}$$

We assume the simple case when we have a multiagent system with single integrator dynamics, which corresponds to $a = 0$, $b = 1$. In Figs. 3.4 and 3.5, we present nominal cases without the influence of an attacker. Fig. 3.4 shows the case when the coupling strengths within and between subgroups are $\alpha_1 = \alpha_2 = 2$ and $\beta_1 = \beta_2 = 4$, while Fig. 3.5 represents the case with $\alpha_1 = \alpha_2 = 1$ and $\beta_1 = \beta_2 = 1$. Take note that in both instances, the MAS achieves group consensus with respect to the conditions in (3.64) and as shown by simulation plots.

Now, we consider the case where subgroup 1 is under attack. Specifically, we assume that the intruder launches a DoS attack between agents 1 and 2. As such, the graph Laplacian for the attack modes and the entire network is given by:

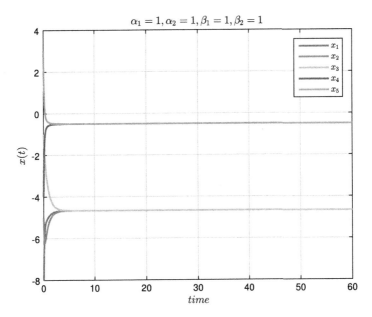

$$\alpha_1 = 1, \alpha_2 = 1, \beta_1 = 1, \beta_2 = 1$$

Figure 3.5 Nominal case with coupling gains $\alpha_1 = \alpha_2 = 1$, $\beta_1 = \beta_2 = 1$.

$$\mathcal{L} = \begin{bmatrix} 3 & 0 & -1 & 1 & -1 \\ 0 & 2 & 0 & -1 & 1 \\ -1 & 0 & 1 & 0 & 0 \\ 1 & -1 & 0 & 3 & -3 \\ -1 & 1 & 0 & -3 & 3 \end{bmatrix}, \tag{3.66}$$

$$\mathcal{L}_{11}^t = \begin{bmatrix} 1 & 0 & 0 \\ 0 & 0 & 0 \\ 0 & 0 & 0 \end{bmatrix}. \tag{3.67}$$

The eigenvalues of the Laplacian matrices representing the entire network after the influence of the attacker and the attack mode are given as follows. In Fig. 3.6, we assume that the coupling strengths are $\alpha_1 = \alpha_2 = 2$ and $\beta_1 = \beta_2 = 4$. We can infer from the foregoing conditions and Fig. 3.6 that consensus is not achieved. Next we modify the coupling strengths to $\alpha_1 = \alpha_2 = 1$ and $\beta_1 = \beta_2 = 1$ and can confirm by Fig. 3.7 and established conditions in (3.68) that group consensus is achieved.

$$b^2(\gamma_2^*)^2 + 4a^2 - 4ab\gamma_2 + 4b^2[\alpha_1\alpha_2[\eta_i \mathrm{Re}(\mu_j) - \eta_i^* \mathrm{Im}(\mu_j)] - \beta_1\beta_2\mu_i] > 0, \tag{3.68}$$

$$2a - b\gamma_2 \neq 0, \tag{3.69}$$

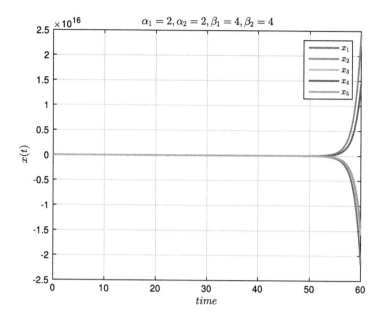

Figure 3.6 Attack in subgroup 1 with coupling gains $\alpha_1 = \alpha_2 = 2$, $\beta_1 = \beta_2 = 4$.

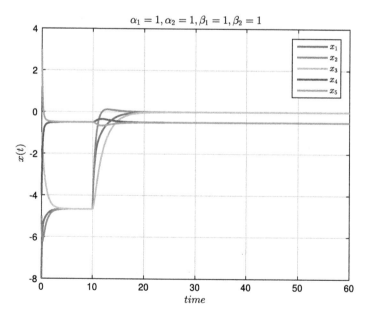

Figure 3.7 Attack in subgroup 1 with coupling gains $\alpha_1 = \alpha_2 = 1$, $\beta_1 = \beta_2 = 1$.

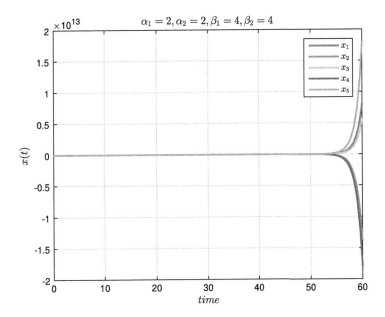

Figure 3.8 Attack in subgroup 2 with coupling gains $\alpha_1 = \alpha_2 = 2$, $\beta_1 = \beta_2 = 4$.

where $\gamma_2 = \alpha_1 \eta_i + \alpha_2 \mathrm{Re}(\mu_j)$, $\gamma_2^* = \alpha_1 \eta_i^* + \alpha_2 \mathrm{Im}(\mu_j)$, $\eta_i = \mathrm{Re}(\mu_i) - \mathrm{Re}(\mu_i^t)$, $\eta_i^* = \mathrm{Im}(\mu_i) - \mathrm{Im}(\mu_i^t)$, μ_l is the maximum eigenvalue of $\hat{\mathcal{L}}_{12}\hat{\mathcal{L}}_{21}$.

Secondly, we consider the case where the attacker is operating within subgroup 2 interrupting the communication link between agents 4 and 5. Again for this case, we compute the Laplacian of the network and attack modes after the influence of the attacker as given by (3.72)–(3.73). Again, we first assume the coupling strengths $\alpha_1 = \alpha_2 = 2$ and $\beta_1 = \beta_2 = 4$, and we can observe that the conditions in (3.70) are violated and as shown in Fig. 3.8, group consensus is not achieved and the system becomes unstable. Again, we readjust the coupling strengths to $\alpha_1 = \alpha_2 = 1$ and $\beta_1 = \beta_2 = 1$ and can observe in Fig. 3.9 that although the link between agents 4 and 5 is interrupted, group consensus is achieved:

$$b^2(\gamma_3^*)^2 + 4a^2 - 4ab\gamma_3 + 4b^2[\alpha_1\alpha_2[\mathrm{Re}(\mu_i)\eta_j - \mathrm{Im}(\mu_i)\eta_j^*] - \beta_1\beta_2\mu_l] > 0,$$

$$(3.70)$$

$$2a - b\gamma_3 \neq 0,$$

$$(3.71)$$

where $\gamma_3 = \alpha_1\mathrm{Re}(\mu_i) + \alpha_2\eta_j$, $\gamma_3^* = \alpha_1\mathrm{Im}(\mu_i) + \alpha_2\eta_j^*$, $\eta_i = \mathrm{Re}(\mu_i) - \mathrm{Re}(\mu_i^t)$, $\eta_i^* = \mathrm{Im}(\mu_i) - \mathrm{Im}(\mu_i^t)$, $\eta_j = \mathrm{Re}(\mu_j) - \mathrm{Re}(\mu_j^t)$, $\eta_j^* = \mathrm{Im}(\mu_j) - \mathrm{Im}(\mu_j^t)$, μ_l is

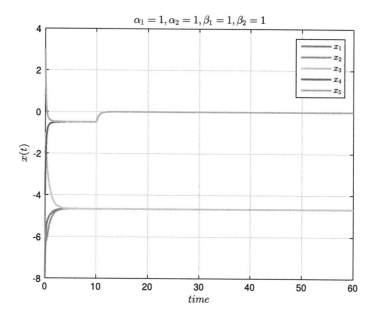

Figure 3.9 Attack in subgroup 2 with coupling gains $\alpha_1 = \alpha_2 = 1$, $\beta_1 = \beta_2 = 1$.

the maximum eigenvalue of $\hat{\mathcal{L}}_{12}\hat{\mathcal{L}}_{21}$,

$$\mathcal{L} = \begin{bmatrix} 4 & -3 & -1 & 1 & -1 \\ -3 & 3 & 0 & -1 & 1 \\ -1 & 0 & 1 & 0 & 0 \\ 1 & -1 & 0 & 2 & 0 \\ -1 & 1 & -1 & 0 & 2 \end{bmatrix}, \tag{3.72}$$

$$\mathcal{L}'_{22} = \begin{bmatrix} 1 & -3 \\ -3 & 1 \end{bmatrix}. \tag{3.73}$$

Finally, we examine the case when the attacker is able to influence both subgroups. In this scenario, the communication links between agents 1 and 2 and agents 4 and 5 are affected. The Laplacian matrix after the attack is given by (3.74). Yet again, we first assume the coupling strengths $\alpha_1 = \alpha_2 = 2$ and $\beta_1 = \beta_2 = 4$ and can observe that the conditions in Theorem 3.8 are violated, and as shown in Fig. 3.10, group consensus is not achieved and the system becomes unstable. However, it is interesting to note that when we adjusted the coupling strengths to $\alpha_1 = \alpha_2 = 1$ and $\beta_1 = \beta_2 = 1$, the agents were able to reorganize themselves after the influence of the attack. As shown in Fig. 3.11 before the attack, we can observe that agents

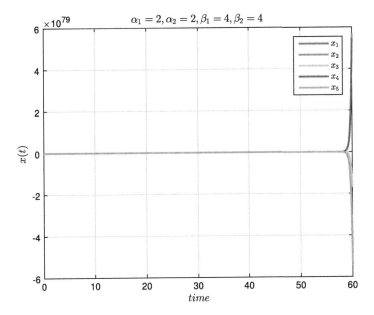

Figure 3.10 Attack in both subgroups with coupling gains $\alpha_1 = \alpha_2 = 2, \beta_1 = \beta_2 = 4$.

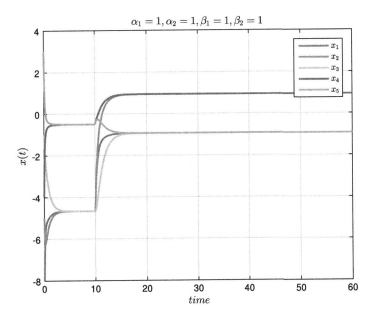

Figure 3.11 Attack in both subgroups with coupling gains $\alpha_1 = \alpha_2 = 1, \beta_1 = \beta_2 = 1$.

1, 2, 3 form a single consensus, as do agents 4 and 5. But after the loss of communication links between agents 1 and 2 and agents 4 and 5, we can observe that agents 2 and 4, and agents 1, 3 and 5 form a new consensus:

$$
\mathcal{L} = \begin{bmatrix}
3 & 0 & -1 & 1 & -1 \\
0 & 2 & 0 & -1 & 1 \\
-1 & 0 & 1 & 0 & 0 \\
1 & -1 & 0 & 2 & 0 \\
-1 & 1 & 0 & 0 & 2
\end{bmatrix}. \tag{3.74}
$$

3.2.6 Bipartite consensus

Consider the following distributed bipartite consensus protocol:

$$
\dot{x}_i(t) = \sum_{j \in \mathbb{N}_i(t)} a_{ij}(t)[x_j(t) - \operatorname{sgn}(a_{ij}(t))x_i(t)], \quad \forall i \in \mathbb{V}, \tag{3.75}
$$

$$
\dot{x}(t) = -\mathbb{L}(t)x(t), \tag{3.76}
$$

where $\mathbb{L}(t) = [l_{ij}(t)] \in \mathbb{R}^{n \times n}$ and $l_{ij} = \begin{cases} \sum\limits_{k \in \mathbb{N}_i(t)} |a_{ik}(t)|, & j = i, \\ -a_{ij}(t), & j \neq i. \end{cases}$

To proceed further, consider the following MAS with nonlinear dynamics over switching graphs described by [90]

$$
\dot{x}_i(t) = f(x_i(t)) + \sum_{j \in \mathbb{N}_i(t)} a_{ij}^{\sigma(t)}[x_j(t) - \operatorname{sgn}(a_{ij}^{\sigma(t)})x_i(t)], \quad \forall i \in \mathbb{V}, \tag{3.77}
$$

where $\sigma : [0, \infty) \to \mathbb{S}$ is piecewise constant and right continuous function whose switching instants $\{t_i : i = 0, 1, \ldots\}$, satisfy $t_{k+1} - t_k \geq \tau > 0, \forall k \geq 0$. According to [90], the following assumptions are necessary to derive important results for pinning bipartite consensus:

Assumption 3.2.9. The nonlinear function $f(x)$ satisfies the following one-sided Lipschitz condition for some $x, y \in \mathbb{R}$:

$$
(x - y)(f(x) - f(y)) \leq m(x - y)^2, \quad m > 0. \tag{3.78}
$$

Assumption 3.2.10. All the signed graphs $\mathbb{G}_k, k \in S$ are connected and the vertices of each possible topology can be partitioned into two subsets $\mathbb{V}_k^1$ and $\mathbb{V}_k^2$ such that $a_{ij}^k > 0 \,\forall\, i, j \in \mathbb{V}_p, p = 1, 2$ and $a_{ij}^k \leq 0$, noting that $\{(i, j)| i \in \mathbb{V}_k^1, j \in \mathbb{V}_k^2, i \neq j\}$.

This leads us to the following result:

Theorem 3.9. *The MAS described by the nonlinear dynamics (3.77) achieves pinning bipartite consensus under Assumptions 3.2.9–3.2.10 if there exists $\epsilon > 0$ such that $f(x) > 0 \ \forall x \in (0, \epsilon)$ and $f(x) < 0 \ \forall x \in (-\epsilon, 0)$, $s(0) \neq 0$, and there exist $\beta_i^k, i = 1, 2, \ldots, l$ such that*

$$mI_n - c\phi L^k \phi + cD^k < 0, \quad k = 1, 2, \ldots, p, \tag{3.79}$$

where $D^k = \text{diag}\{\beta_1^k, \beta_2^k, \ldots, \beta_l^k, 0, \ldots, 0\}$.

Next we direct attention to another type of consensus profile.

3.2.7 Scaled consensus

Scaled consensus over switching networks was introduced for an MAS with single-integrator dynamics over directed networks and studied in [94]. Consider an MAS described by the following single-integrator dynamics:

$$\dot{x}_i(t) = u_i(t). \tag{3.80}$$

A distributed scaled-consensus protocol over a time-varying topology is proposed as

$$u_i(t) = -\sum_{j=1}^{N} a_{ij}(t)[\omega_{ij}(t)x_j(t) - x_i(t))], \tag{3.81}$$

where $\omega_{ij}(t)$ is a weighted scalar ratio between $x_j(t)$ and $x_i(t)$ defined as

$$\omega_{ij}(t) = \frac{\omega_i(t)}{\omega_j(t)}. \tag{3.82}$$

Definition 3.18 ([94]). Given any positive constants $\alpha > 0$ and $\beta > 0$, an initial time $t_0 \geq 0$ and initial state $\{x_i(t_0), i \in \mathbb{V}\}$, we say that system (3.80) under protocol (3.81) solves the scaled consensus problem exponentially if the following condition is satisfied:

$$||x_i(t) - \omega_i(t)x_f|| \leq \alpha e^{-\beta(t-t_0)} \quad \forall t \geq t_0, \tag{3.83}$$

where $x_f \in \mathbb{R}^n$ is a constant state reference vector and ω_i is a time varying weight scale.

Theorem 3.10. *System* (3.80) *under consensus protocol* (3.81) *solves the scaled consensus problem exponentially provided the following Assumptions 3.2.11, 3.2.12, and 3.2.13 hold.*

Assumption 3.2.11 ([94]). For any initial time t_0, there exist an infinite time sequence $\{t_k : k \in \mathbb{Z}_+\}$ satisfying $t_k \to \infty$ as $k \to \infty$ with a uniformly bounded dwell time $\tau_k = t_{k+1} - t_k \geq \underline{\tau} > 0$ such that $\mathbb{G}(t) = \mathbb{G}_k \forall t \in [t_k, t_{k+1})$ and $\mathbb{G}_k \in \mathbb{G}_s \triangleq \{G_{s1}, G_{s2}, \ldots, G_{sM}\}$ for some finite positive integer M.

Assumption 3.2.12 ([94]). The time sequence $\{t_k : k \in \mathbb{Z}\}$ in Assumption 5.3.1 is such that $\omega_i = \omega_{i,k} \neq 0 \, \forall \, t \in [t_k, t_{k+1})$, $i \in \mathbb{V}$.

Although, the scales can be positive or negative, the following bounds are defined on the time-varying weight scales $\omega_i(t)$:

$$\underline{b}_{\omega_i} \leq \omega_{i,k} \leq \bar{\omega}_i \quad \forall k, \tag{3.84}$$

$$\underline{b}_{\omega_i} \bar{\omega}_i > 0 \quad i \in \mathbb{V}, \tag{3.85}$$

$$\bar{b}_\omega = \max_{i \in \mathbb{V}} \{ |\underline{b}_{\omega_i}|, |\bar{b}_{\omega_i}| \} < \infty, \tag{3.86}$$

$$\underline{b}_\omega = \min_{i \in \mathbb{V}} \{ |\underline{b}_{\omega_i}|, |\bar{b}_{\omega_i}| \} < \infty. \tag{3.87}$$

Assumption 3.2.13 ([94]). There exist a infinite sequence $\{k_j : k_0 = 0, 0 < k_{j+1} - k_j \leq h, \forall j \in \mathbb{Z}_+\}$ for some positive h such that the collection of digraphs

$$\{\mathbb{G}_{kj}, \mathbb{G}_{kj+1}, \ldots, \mathbb{G}_{kj+1-1}\}$$

jointly have a spanning tree for all $j \in \mathbb{Z}_+$.

3.2.8 Related studies

In [1], some analysis for this type of hybrid dynamical systems were presented for switching networks (and time delays) considering single-integrator models. In particular, the study investigates independently consensus conditions on directed networks with switching topologies and undirected networks with time-delays. A necessary condition for reaching consensus is that the union of the possible graph topologies over all the switching time instants has a rooted out branching for directed networks and in undirected networks is connected. The authors of [36] studied consensus for a second-order MAS with time-delays over fixed and switching topologies using eigenvalue analysis and Lyapunov–Razumikhin functions.

A leader–follower consensus was proposed for an MAS with hetero-geneous second-order time-varying nonlinear dynamics under directed switching topology in [78]. Using topology-dependent Lyapunov functions and appropriate choice of time-varying regulation factor and coupling strength, exponential consensus tracking was achieved. A leader–follower consensus for second-order multiagent system under the influence of coupling time delay was investigated in [61]. For each agent, a decentralized control law was constructed to track the leader dynamics under fixed and switching topologies.

Decentralized finite-time adaptive consensus was studied in [21] for an MAS modeled with first-order dynamics over fixed and switching topologies satisfying detail balanced conditions. Consensus of higher-order MAS with switching topologies and time-delay networks was studied in [22] using nearest-neighbor rule without requiring the communication topology to have spanning trees. Asynchronous consensus was studied in [10] for an MAS over networks with switching topologies and time-delays using nonnegative matrix and graph theories. Mean-square average consensus was studied in [19] for an MAS over time-varying topologies and stochastic communication noises. The same paper discussed an average consensus problem for a discrete-time MAS with limited communication data-rate over switching networks under the assumption that the graph is jointly connected. A compensation method was used to design a distributed encoding–decoding scheme based on difference quantization with dynamic scaling. A weak condition for global consensus was discussed in [17] for a continuous-time MAS over undirected switching topologies without requiring persistent excitation of switching signals.

Finite-time consensus of switched multiagent systems consisting of discrete and continuous-time subsystems was studied in [15] based on a switching control method. Consensus in continuous-time switched nonlinear systems was studied [16] where two theorems were proved based on local quadratic functions under weaker graph topology assumptions. Asymptotic consensus over a class of switching graphs was investigated in [13], relaxing some reciprocity and connectivity assumptions by using theories on switched systems to reach consensus. Consensus of switched multiagent systems was studied in [14] for an MAS composed of discrete and continuous time subsystems using Lyapunov and graph theory under arbitrary switching over undirected connected, directed, and switching topologies.

Distributed H_∞ leader–follower consensus tracking control for an MAS with linear dynamics Model VIIA over directed switching topologies was proposed in [68] without the assumption of zero input for the leader agent. The proposed approach employs multiple Lyapunov functions and algebraic graph theory to derive LMI conditions for stability. Leader–follower consensus for an MAS with switching jointly reachable interconnection and noncoupling time delays was studied in [70] where the leader was considered globally-unreachable. A leader–follower consensus analysis inspired by hierarchy and coordination in pigeons considering switching topologies was carried out in [77] with the objective of investigating the effect of hierarchical topologies and self-loops on convergence properties of the leader–follower MAS. The leader–follower consensus for an MAS with double-integrator dynamics over switching jointly reachable topologies subject to noncoupling time-delays was proposed in [70]. The authors of [69] studied leader–follower consensus for a chaotic MAS over directed intermittent networks with switching topologies and unknown time-delays using graph-dependent Lyapunov analysis.

Leader–follower consensus for a second-order multi-agent system under the influence of coupling time delay was investigated in [61]. For each agent, a decentralized control law was constructed to track the leader dynamics under fixed and switching topologies. Distributed cooperative control was studied in [65] for quadcopters based on a leader-following strategy considering the effect of packet dropouts. The effect of dropout in the network was modeled using Bernoulli distribution while sufficient conditions for stability were derived using Lyapunov-based analysis and linear matrix inequalities (LMIs). Distributed H_∞ leader–follower consensus tracking control for an MAS with linear dynamics over directed switching topologies was proposed in [68] without the assumption of zero input for the leader agent. The proposed approach employs multiple Lyapunov functions and algebraic graph theory to derive LMI conditions for stability. Leader–follower consensus was proposed for an MAS with heterogeneous second-order time-varying nonlinear dynamics under directed switching topology in [78]. Using topology-dependent Lyapunov functions and appropriate choice of time-varying regulation factor and coupling strength, exponential consensus tracking was achieved.

The authors of [36] studied consensus for a second-order MAS with time-delays over fixed and switching topologies using eigenvalue analysis and Lyapunov–Razumikhin functions. The approach relied on double-tree transformations for converting agent dynamics into reduced-order systems.

Group consensus for an MAS with communication delays was investigated in [81] using nonnegative matrix and graph-theoretic analysis. The discussion considered networks with both fixed and switching topologies. Group consensus was studied in [80] for an MAS over networks with finite arbitrary switching topologies and communication delays. Convergence analysis was also investigated with time-varying topologies. Group consensus was studied in [80] for an MAS over networks with finite arbitrary switching topologies and communication delays. A group-consensus problem for a second-order discrete time MAS was investigated in [83] over switching networks with time-varying delays using nonnegative matrix and graph theories. Group consensus for networked mechanical systems over directed acyclic topology under the influence of communication delays was studied in [84] based on neural network based adaptive control scheme. The approach relied on double-tree transformations for converting agent dynamics into reduced-order systems. Group consensus for an MAS with communication delays was investigated in [81] using nonnegative matrix and graph-theoretic analysis. The discussion considered networks with both fixed and switching topologies.

Distributed H_∞ consensus filtering was proposed for an MAS with time-delay over network with switching topology and packet dropouts in [55] to guarantee robustness of the error system to random and time-varying delays, varying channel-induced packet losses, switching network topologies and minimization of estimation deviations in H_∞ sense. A nonfragile control scheme was proposed in [44] for finite-time consensus of an MAS with time-varying input delay over switching network topology. Using the famous Lyapunov–Krasovskii function, stability and graph theories, sufficient conditions were proposed for the solvability of finite-time consensus.

Consensus conditions for a class of nonlinear multiagent systems with time-varying delays were established in [38] using Lyapunov–Razumikhin function under both fixed and directed topologies. A requirement for consensus according to the authors is that under fixed topologies, the directed graph should contain a spanning tree, while in the case of switching topologies, the union of the subset of persistent graphs should contain a spanning tree. Necessary and sufficient conditions were given for discrete-time networked MAS with directed topology and diverse time-varying topologies in [43] based on networked predictive control schemes.

In [91], an adaptive protocol was proposed to achieve bipartite consensus tracking over networks which allow for cooperation and competition modeled by signed graphs. Further, the considered second-order integrator dynamics includes unknown disturbances represented by linear parametrized models. Convergence analysis were also investigated with time-varying topologies.

Bipartite consensus for second-order discrete-time multiagent systems over asynchronous networks with switching communication topologies was investigated in [92]. An important assumption in this analysis is that the communication graph is structurally balanced with the union of possible topologies having a spanning tree.

Asymptotic consensus over a class of switching graphs was investigated in [13], relaxing some reciprocity and connectivity assumptions by using theories on switched systems to reach consensus. Consensus of switched multiagent systems was studied in [14] for an MAS composed of discrete- and continuous-time subsystems using Lyapunov and graph theories under arbitrary switching over undirected connected, directed, and switching topologies. Finite-time consensus of switched multiagent systems consisting of discrete- and continuous-time subsystems was studied in [15] based on a switching control method. Consensus in continuous-time switched nonlinear systems was studied [16] where two theorems were proposed based on local quadratic functions under weaker graph topology assumptions. A weak condition for global consensus was discussed in [17] for a continuous-time MAS over undirected switching topologies without requiring persistent excitation of switching signals.

Decentralized finite-time adaptive consensus was studied in [21] for an MAS modeled with first-order dynamics over fixed and switching topologies satisfying detail balanced conditions. Consensus of higher-order MAS with switching topologies and time-delay networks was studied in [22] using nearest-neighbor rule without requiring the communication topology to have spanning trees. The authors of [19] discussed an average consensus problem for a discrete-time MAS with limited communication data-rate over switching networks under the assumption that the graph is jointly connected. A compensation method was used to design a distributed encoding–decoding scheme based on difference quantization with dynamic scaling.

Multiagent synchronization problem was studied in [18] for Euler–Lagrange systems with parametric uncertainties under unidirectional dynamically changing network topologies via adaptive control techniques based on the assumption that the union graph has a directed spanning tree frequently.

3.3 Random (Markovian) networks

Here, we consider consensus problems over randomly (or stochastically) switching networks. In some studies, the network topology is described as using a Markovian process (3.2). Here, we consider a communication graph modeled by a Markovian process. That is, the switching process $\{\theta(t), t \geq 0\}$ is driven by a time-homogeneous Markov chain, and for each state of M, there is a possible topology structure. The Markov process is defined by the triple (Ω, F, P), where Ω is the sample space, F is the σ-algebra of subsets of Ω, and P is a probability measure on F. The infinitesimal generator of the Markov process $\{\theta(t), t \geq 0\}$ is $\lambda = (\lambda_{ij})$ defined probabilistically as

$$P\{\theta(t+h) = j | \theta(t) = i\} = \begin{cases} \lambda_{ij} h + o(h), & \text{if } i \neq j, \\ 1 + \lambda_{ii} + o(h), & \text{otherwise,} \end{cases} \tag{3.88}$$

where λ_{ij} is the transition rate from state i to state j with $\lambda_{ij} \geq 0$ if $i \neq j$.

3.3.1 Leaderless consensus

We start with the following definition

Definition 3.19. The MAS defined by the continuous-time dynamics $\dot{x}_i(t) = f(x_i(t), u_i(t))$ solves the mean-square scaled consensus problem as defined by (3.105), if there exists a time-varying consensus protocol gain such that the ratios among the state variable reach specified constants asymptotically in the mean-square sense, i.e.,

$$\lim_{t \to \infty} \mathbb{E}[|x_i(t) - x_j(t)|] = 0 \ \forall i, j \in \mathbb{V}, \tag{3.89}$$

where $\alpha_i \neq 0$, and condition (3.89) holds for any initial data $x_0(t_0)$ and initial distribution of $\pi(t_0)$ and $\lambda_{ij}(t_0)$, where the expectation $\mathbb{E}(\cdot)$ is taken under measure P.

In [9], the leaderless consensus problem was considered for an MAS over Markovian switching topologies. Consider the MAS with the following general linear dynamics:

$$\dot{x}_i(t) = Ax_i(t) + Bu_i(t), \tag{3.90}$$

where $x_i(t) \in \mathbb{R}^n$ and $u_i \in \mathbb{R}^m$ are represent agent state and input, respectively. In [9], the following consensus protocol is proposed:

$$u_i(t) = K \sum_{j \in V} a_{ij}(t)[x_j(t) - x_i(t)]. \tag{3.91}$$

Assumption 3.3.1 ([9]). The continuous-time Markov process with transition rate matrix Q is ergodic.

Assumption 3.3.2 ([9]). All the possible graphs $\mathbb{G}_i$, $i \in \mathbb{S}$, are detailed balanced.

Assumption 3.3.3 ([9]). A is not Hurwitz.

Theorem 3.11 ([9]). *Suppose Assumptions 3.3.1–3.3.3 are satisfied. Then the MAS described by (3.90) under protocol (3.91) solves the leaderless consensus problem over switching topology $\mathbb{G}_{\sigma(t)}$ with the Markovian switching signal $\sigma(t)$ provided that the following conditions are satisfied:*

- *$\mathbb{G}_u = \bigcup_{i \in \mathbb{S}} \mathbb{G}_i$, the union of all possible graphs, contains a spanning tree.*
- *The pair $\{A, B\}$ is stabilizable, and*
- *The state feedback matrix is defined as $K = \tau B^t P$, where P is a solution of*

$$PA + A^t P - 2PBB^t P + \alpha P < 0. \tag{3.92}$$

The solution of (3.92) exists provided that the pair (A, B) is controllable and α dictates the speed of convergence, $\tau \geq \underline{\pi}^c \lambda(\frac{1}{2}(\mathbb{L} + \mathbb{L}^t))$, $\underline{\pi}^c = \min_{i \in \mathbb{S}} \{\pi_i^c\}$, where π_i^c belongs to an invariant distribution $\pi^c = \{\pi_1^c, \ldots, \pi_s^c\}$ which describes the state space of all possible network topologies which ensure that each graph state is reachable from any graph state in the set.

3.3.2 Leader–follower consensus

Here, we discuss relevant research findings on the leader–follower consensus problems over stochastically switching networks defined by Markovian probabilistic models.

Definition 3.20. The MAS consisting of N agents over a time-varying graph $\mathbb{G}(\theta(t))$ with leader agents defined by $\dot{x}_0(t) = f(x_0(t))$ and follower agents defined by $\dot{x}_i(t) = f(x_i(t), u_i(t))$ solves the mean-square leader–follower consensus problem if there exists a time-varying consensus protocol gain such that

$$\lim_{t \to \infty} \mathbb{E}[|x_i(t) - x_0(t)|] = 0 \quad \forall i, j \in \mathbb{V} \tag{3.93}$$

holds for any initial conditions $x_0(t_0)$ and initial distribution of $\pi(t_0)$ and $\lambda_{ij}(t_0)$, where the expectation $\mathbb{E}(\cdot)$ is taken under measure P.

3.3.3 Group consensus

Here, we discuss relevant research findings on group consensus problems over stochastically switching networks defined by Markovian probabilistic models.

Definition 3.21. The MAS consisting of $N + M$ agents defined by the continuous-time dynamics $\dot{x}_i(t) = f(x_i(t), u_i(t))$ solves the mean–square couple group consensus problem if there exists a time-varying consensus protocol gain such that

$$\lim_{t \to \infty} \mathbb{E}[|\alpha_i x_i(t) - \alpha_j x_j(t)|] = 0 \quad \forall i, j = 1, 2, \ldots, N, \tag{3.94}$$

$$\lim_{t \to \infty} \mathbb{E}[|\alpha_i x_i(t) - \alpha_j x_j(t)|] = 0 \quad \forall i, j = N + 1, N + 2, \ldots, N + M, \tag{3.95}$$

holds for any initial data $x_0(t_0)$ and initial distribution of $\pi(t_0)$ and $\lambda_{ij}(t_0)$, where the expectation $\mathbb{E}(\cdot)$ is taken under measure P.

In [88], the couple-group consensus problem was studied for an MAS over stochastically switching topologies, that is, an MAS was modeled by

$$\dot{x}_i(t) = A x_i(t) + B u_i(t), \quad i = 1, 2, \ldots, N + M. \tag{3.96}$$

For the MAS (3.96), the following consensus protocol was proposed:

$$u_i(t) = \begin{cases} K\left(\sum_{j \in \mathbb{N}_i^1(t)} a_{ij}(t)(x_j(t) - x_i(t)) + \sum_{j \in \mathbb{N}_i^2(t)} a_{ij}(t)x_j(t) \right) \\ \forall \; i, j = 1, 2, \ldots, N, \\ K\left(\sum_{j \in \mathbb{N}_i^2(t)} a_{ij}(t)(x_j(t) - x_i(t)) + \sum_{j \in \mathbb{N}_i^1(t)} a_{ij}(t)x_j(t) \right) \\ \forall i, j = N + 1, N + 2, \ldots, N + M, \end{cases} \tag{3.97}$$

where K is a feedback matrix to be designed. For the MAS (3.96) under protocol (3.97), the following assumptions were made:

Assumption 3.3.4 ([88]).

- The Markov process is ergodic and

$$\sum_{j=N+1}^{N+M} a_{ij}^{(k)} = \alpha_k \quad \forall i = 1, \ldots, N, k \in S, \tag{3.98}$$

$$\sum_{j=1}^{N} a_{ij}^{(k)} = \beta_k \quad \forall i = N + 1, \ldots, N + M, k \in S, \tag{3.99}$$

where $\alpha^{(k)}$ and $\beta^{(k)}$ are constants.

- Both subgraphs $\mathbb{G}_1^s$ and $\mathbb{G}_2^s$ are balanced and contain spanning trees.

Theorem 3.12 ([88]). *If Assumption 3.3.4 is fulfilled and the pair (A, B) is stabilizable, the MAS (3.96) achieves couple group consensus with feedback gain $K = \sigma B^T P_s$, where the coefficient σ satisfies*

$$\sigma \geq \rho \min_{k \in S}(\{\pi_k\})^{-1}, \tag{3.100}$$

$$0 < \rho < \min\{\lambda_2(\hat{L}_1^S), \lambda_2(\hat{L}_2^S)\}, \tag{3.101}$$

the following inequalities are fulfilled:

$$\text{diag}(\lambda_2(\hat{L}_2^S) - \rho, \lambda_M(\hat{L}_2^s) - \rho) \geq \Xi^t \text{diag}^{-1}(\lambda_2(\hat{L}_2^S) - \rho, \lambda_N(\hat{L}_2^s) - \rho) \Xi^t, \tag{3.102}$$

and P satisfies the following LMI:

$$PA + A^t P - 2PBB^t P + \tau P < 0, \tag{3.103}$$

where τ dictates speed of convergence, and

$$\Xi = U_1^t(W_1^S + (W_2^S)^t)U_2 \in \mathbb{R}^{(N-1)\times(M-1)}.$$

3.3.4 Bipartite consensus

Similarly, we start with a basic definition

Definition 3.22. The MAS defined by the continuous-time dynamics $\dot{x}_i(t) = f(x_i(t), u_i(t))$ solves the mean-square scaled consensus problem as defined by (3.104), if there exists a time-varying consensus protocol gain such that the ratios among the state variable reach specified constants asymptotically in the mean-square sense, i.e.,

$$\lim_{t \to \infty} \mathbb{E}[|x_i(t) - s_i x_j(t)|] = 0 \quad \forall i, j \in \mathbb{V}, \tag{3.104}$$

where $s_i = 1 \ \forall j \in \mathbb{V}_1$, $s_{ij} = -1 \ \forall j \in \mathbb{V}_2$ and condition (3.105) holds for any initial data $x_0(t_0)$ and initial distribution of $\pi(t_0)$ and $\lambda_{ij}(t_0)$, where the expectation $\mathbb{E}(\cdot)$ is taken under measure P.

3.3.5 Scaled consensus

Again, we give the following definition:

Definition 3.23. The MAS defined by the continuous-time dynamics $\dot{x}_i(t) = f(x_i(t), u_i(t))$ solves the mean-square scaled consensus problem as defined by (3.105), if there exists a time-varying consensus protocol gain such that the ratios among the state variable reach specified constants asymptotically in the mean-square sense

$$\lim_{t \to \infty} \mathbb{E}[|\alpha_i x_i(t) - \alpha_j x_j(t)|] = 0 \quad \forall i, j \in \mathbb{V}, \tag{3.105}$$

where $\alpha_i \neq 0$, and condition (3.105) holds for any initial data $x_0(t_0)$ and initial distribution of $\pi(t_0)$ and $\lambda_{ij}(t_0)$, where the expectation $\mathbb{E}(\cdot)$ is taken under measure P.

Scaled consensus problem was investigated for an MAS with single-integrator (3.106) dynamics in [95] over networks exhibiting Markovian switching behavior:

$$\dot{x}_i(t) = u_i(t), \tag{3.106}$$

$$\dot{u}_i(t) = c(t)\mathrm{sgn}(\alpha_i) \sum_{j \in \mathbb{N}^i} a_{ij}(\theta(t))[\alpha_j x_j(t) + \sigma_{ij}\rho_{ji} - \alpha_i x_i(t)], \tag{3.107}$$

where $\sigma_{ij} \in \mathbb{R}$ represent noise intensity and $\rho_{ij} \in \mathbb{R}$ is a 1-dimensional measurement noise which satisfies $\int_0^t \rho_{ji}(s)ds = w_{ji}(t)$.

3.3.6 Related studies

Couple-group consensus for a second-order discrete MAS over fixed and stochastically switching topologies was investigated in [87]. Using linear transformations on the error system, necessary and sufficient conditions were reported for mean-square couple-group consensus. Couple-group consensus for an MAS was studied by [88] under Markovian switching topologies. Under the assumption of stabilizability of the agents, it was proved that couple-group consensus can be reached under some mild algebraic and topological conditions. The L^1 group consensus was investigated by [89] for a discrete-time MAS with switching topologies and external stochastic inputs based on ergodicity and matrix theory. Mean-square average consensus was studied in [19] for an MAS over time-varying topologies and stochastic communication noises.

A leader–follower consensus problem was solved in [76] for a swarm-based multi-UUV system under multiindependent switching topologies and time-delays. The switching network was modeled as a Markovian switching swarm system, and Lyapunov–Krasovskii functionals were employed to derive sufficient conditions for reaching a consensus. Additionally, local state feedback control loops were designed to ensure agents are stochastically stable.

An LMI-based approach was proposed in [11] for consensus analysis of an MAS with time-delays and uncertain switching topologies modeled by Markovian jumps with uncertain rates of transitions.

A consensus stabilization problem was investigated in [54] for a stochastic MAS with Markovian switching topology, noises, and time delay. Due to difficulties related to Lyapunov-based approaches for proving convergence over switching digraphs, an ergodicity-based approach was proposed which does not require the double stochasticity condition as in the case of quadratic Lyapunov-based approaches.

3.4 Time-varying delay networks

In the sequel, we provide results pertaining to consensus problems and related results over time-varying networks.

3.4.1 Leaderless consensus

In [58], the consensus problem for an MAS exhibiting linear dynamics with time-varying self-delay was examined:

$$x_i(t) = Ax_i(t - \tau(t)) + Bu_i(t), \tag{3.108}$$

where $\tau(t)$ represent time-varying self-delay. The distributed control protocol was designed as

$$u_i(t) = K \sum_{v_j \in \mathbb{V}} (a_{ij}(t) + \Delta a_{ij}(t))[x_j(t - \tau(t)) - x_i(t - \tau(t))], \tag{3.109}$$

where $K \in \mathbb{R}^{m \times n}$ is a common consensus gain matrix and Δa_{ij} represent topology uncertainties. The switching network is governed by a probabilistic Markovian process which is assumed to be unrelated to the perturbations of $\Delta a_{ij}(t)$. The consensus protocol was designed such that the MAS reaches almost sure consensus under the following assumptions:

- The family of communication graphs $\mathbb{G}_1, \mathbb{G}_2, \ldots, \mathbb{G}_n$ is balanced.
- The Markov process $\theta(t)$ is ergodic.
- $0 \le \tau(t) \le h$ and $\dot{\tau}(t) \le \kappa \le 1$ for $t \ge 0$, where $h, \kappa \ge 0$ which implies that the delay $\tau(t)$ is differentiable.

3.4.2 Leader–follower consensus

A leader–follower consensus problem was investigated in [63] for an MAS over switching networks with time-varying delays. For the MAS with leader and following agents in (3.34), the following consensus algorithm was proposed:

$$u_i(t) = -\alpha \left[\sum_{j \in \mathbb{N}^i} a_{ij}(t)(x_i(t - \tau(t)) - x_j(t - \tau(t))) \right.$$

$$\left. + \left[\sum_{j \in \mathbb{N}^i} b_i(t)(x_i(t - \tau(t)) - x_0(t - \tau(t))) \right] - \beta(v_i(t) - v_0(t)). \quad (3.110)$$

Assumption 3.4.1. The time-varying communication delay $\tau(t)$ in (3.110) satisfy $0 \le \tau(t) < h$, $t \ge 0$, $h > 0$.

Theorem 3.13 ([63]). *Consider the leader–follower MAS with double-integrator dynamics (3.34)–(3.35) under protocol (3.110). Suppose Assumptions 3.3.4 and 3.4.1 are fulfilled simultaneously and the following inequality holds:*

$$\lambda_n(\mathbb{L}_\sigma + \mathbb{B}_\sigma) < \min \left\{ \frac{2}{\alpha(h + \frac{1}{\Delta})}, \frac{1 + \sqrt{1 + \Delta h}}{\alpha h} \right\}, \quad (3.111)$$

where $\Delta = 2\beta - h - 2 > 0$, then the MAS (3.34)–(3.35) solves the second-order consensus problem.

The leader–follower consensus problem over time-varying delay networks was studied in [79] for an MAS with general linear dynamics:

$$x_0(t) = Ax_0(t),$$
$$x_i(t) = Ax_i(t) + Bu_i(t, \tau_{ij}), \quad (3.112)$$

where $x_0(t)$ and $x_i(t)$ represent the states of leader and follower agents, respectively, and $u_i(t, \tau_{ij})$ is a time-delay control input to be designed such that the following objective is realized:

$$\lim_{t \to \infty} |x_i(t) - x_0(t)| = 0. \quad (3.113)$$

The following adaptive distributed control protocol was proposed to solve the problem:

$$u_i(t) = -\sum_{j=0}^{N} a_{ij} \kappa_{ij}^T(t)[x_i(t - \tau_{ij}(t)) - x_j(t - \tau_{ij}(t))], \quad (3.114)$$

where $\kappa_{ij}^t(t)$ are adaptive control gains designed as follows:

$$\dot{\kappa}_{ij,k}^t(t) = \zeta_{ij,k}|x_{i,k}(t - \tau_{ij}(t)) - x_{j,k}(t - \tau_{ij}(t))|^2, \qquad (3.115)$$

with $x_{i,k}$ being the kth component of the state vector x_i, $k = 1, \ldots, n$, and $\zeta_{ij,k} \in \mathbb{R}^+$ being positive constants.

3.4.3 Bipartite consensus

Bipartite consensus for an MAS modeled by double-integrator dynamics with time-varying network delays was investigated in [93] under structural balance and gauge transformation conditions. Consider the following MAS with double-integrator dynamics:

$$\dot{x}_i(t) = v_i(t),$$
$$\dot{v}_i(t) = u_i(t). \qquad (3.116)$$

For system (3.116), the following distributed bipartite consensus protocol was proposed in [93]:

$$u_i(t) = -\gamma_1 v_i(t) - \sum_{j \in \mathbb{N}_i} \beta_1 |a_{ij}|[v_i(t - \tau_{ij}(t)) - \text{sgn}(a_{ij})v_j(t - \tau_{ij}(t))]$$
$$+ \beta_2 |a_{ij}|[x_i(t - \tau_{ij}(t)) - \text{sgn}(a_{ij})x_j(t - \tau_{ij}(t))] \qquad (3.117)$$

where γ_1, β_1, β_2 are positive gains, and $0 \leq \tau_{ij} \leq h$, $h > 0$, with $\tau_{ij}(t)$ representing the unknown time-varying delay.

Assumption 3.4.2. The MAS is described by a graph $\mathbb{G}$ which is structurally balanced, that is, there exists a partition into $\mathbb{V}_1$ and $\mathbb{V}_2$ such that $\mathbb{V}_1 \cap \mathbb{V}_2 = \emptyset$ and $\mathbb{V}_1 \cup \mathbb{V}_2 = \mathbb{V}$, with the following adjacency relationships: $a_{ij} \geq 0 \, \forall i, j \in \mathbb{V}_a$ ($a \in \{1, 2\}$) and $a_{ij} \leq 0 \, \forall i \in \mathbb{V}_a, j \in \mathbb{V}_b$, $a \neq b$ ($a, b \in \{1, 2\}$).

Let $\mathbb{P} = \{\mathbf{P}|\mathbf{P} = \text{diag}(v_1, \ldots, v_2, \ldots, v_n), \; v_i \in \{+1, -1\}\}$ be employed in the similarity transformation

$$\zeta = \mathbf{P}x, \quad \eta = \mathbf{P}v. \qquad (3.118)$$

Then the MAS (3.116), under (3.117) when $\tau_{ij}(t) = 0 \, \forall i, j$, transforms to:

$$\dot{\zeta}(t) = \eta,$$
$$\dot{\eta}(t) = -\gamma_1 \eta - \beta_1 \mathbb{L}_p \eta - \beta_2 \mathbb{L}_p \zeta, \qquad (3.119)$$

where $\mathbb{L}_p = \mathbf{P}\mathbb{L}\mathbf{P}$.

Theorem 3.14 ([93]). *The MAS (3.116) over a communication graph $\mathbb{G}$ under control protocol (3.117) achieves bipartite consensus asymptotically if the associated graph $\mathbb{G}$ is structurally balanced and there exist positive definite matrices ζ, n_s, $R_s \in \mathbb{R}^{(2n-1)\times(2n-1)}$ such that the following inequality is satisfied:*

$$
\begin{bmatrix} \tilde{\phi}^t \zeta + \zeta \tilde{\phi} + \sum_{s=1}^{m} \zeta_s - \sum_{s=1}^{m} R_s & \zeta \tilde{\mathbb{L}} + \tilde{R} \\ \tilde{\mathbb{L}}^t + \tilde{\mathbb{L}}^t & -(1-\rho)\hat{\eta} - \hat{R} \end{bmatrix}
$$
$$
+ k^2 \begin{bmatrix} \tilde{\phi}^t \\ \tilde{\mathbb{L}}^t \end{bmatrix} \left(\sum_{s=1}^{m} R_s \right) \begin{bmatrix} \tilde{\phi}^t & \tilde{\mathbb{L}} \end{bmatrix} < 0, \tag{3.120}
$$

where

$$
\hat{\eta} = \mathrm{diag}\{\eta_1, \eta_2, \ldots, \eta_s\},
$$
$$
\hat{R} = \mathrm{diag}\{R_1, R_2, \ldots, R_s\},
$$
$$
\tilde{R} = \mathrm{diag}\{\tilde{R}_1, \tilde{R}_2, \ldots, \tilde{R}_s\},
$$
$$
\tilde{\mathbb{L}} = \mathrm{diag}\{\tilde{\mathbb{L}}_1, \tilde{\mathbb{L}}_2, \ldots, \tilde{\mathbb{L}}_m\}.
$$

3.5 Notes

This chapter has presented a systematic journey into the different problems of consensus in multiagent systems over time-varying networks. We complement this journey by reviewing additional and related works.

The area of consensus over random graphs currently remains an open-area of research with numerous problems to consider. Necessary and sufficient conditions were derived for stochastic discrete-time dynamical systems over random graphs in [56] using ergodicity and probabilistic arguments. In [57], consensus problems for a class of MASs with dynamics of Model VIIA were studied under time-varying stochastic network topologies. Using the concept of super-martingales to show that the probability of network connectivity is not zero, the states of the agents reach consensus over the proposed algorithms without requiring to know the set of feasible network topologies, thereby reducing computational costs.

The leader–follower consensus for a class of discrete-time multiagent systems with large-delay sequences under directed topology was studied in [60]. Some novel Lyapunov functionals were proposed to analyze consensus problems under two different types of time-varying delays. Sufficient conditions for ensuring leader–follower consensus were established using LMIs.

A containment control problem was solved in [68] for a leader–follower MAS over dynamically switching network topologies with time-varying delays using Lyapunov–Krasovskii functionals. Distributed algorithms were proposed for an MAS with first- and second-order dynamics.

A group-consensus problem for a second-order discrete time MAS was investigated in [83] over switching networks with time-varying delays using nonnegative matrix and graph theories. A time-delay based PD-like consensus protocol was proposed for an MAS over fast arbitrary switching networks in [26] using a system transformation to convert the problem into a stability problem of a switched delay system. Consensus for a network of MASs with asynchronous update times has been studied by some researchers [10]. The results discussed in this paper consider a combination of asynchronous update of the state dynamics of each agent, time-varying network topologies and time-delays. An LMI-based approach was proposed in [11] for consensus analysis of an MAS with dynamics with time-delays and uncertain switching topologies modeled by Markovian jumps with uncertain rates of transitions.

Average consensus for a nonlinear MAS modeled by single integrator dynamics over switched networks with heterogeneous time-delays was investigated in [12].

Distributed $\mathbb{H}_\infty$ consensus filtering was proposed for an MAS with time-delay over a network with switching topology and packet dropouts in [55] to guarantee robustness of the error system to random and time-varying delays, varying channel induced packet losses, switching network topologies, and minimization of estimation deviations in the H_∞ sense.

Synchronous hybrid event and time-based consensus in an MAS with time delays was studied in [28] for feedback with common time delay, multiple time-invariant delays, and multiple time varying delays using algebraic graph theory.

Adaptive consensus for a nonlinear MAS over time-delay networks was studied in [47] using radial basis neural network functions for handling uncertain nonlinear dynamics, while Lyapunov–Krasovskii functionals were used to handle uncertain time delays.

References

[1] R. Olfati-Saber, R.M. Murray, Consensus problems in networks of agents with switching topology and time-delays, IEEE Trans. Autom. Control 49 (9) (2004) 1520–1533.
[2] A. Jadbabaie, J. Lin, A.S. Morse, Coordination of groups of mobile autonomous agents using nearest neighbor rules, IEEE Trans. Autom. Control 48 (6) (2003) 988–1001.

[3] L. Moreau, Stability of continuous-time distributed consensus algorithms, in: Proc. the 43rd IEEE Conference on Decision and Control, vol. 4, 2004, pp. 3998–4003.

[4] W. Ren, J. Lin, R.W. Beard, Consensus seeking in multiagent systems under dynamically changing interaction topologies, IEEE Trans. Autom. Control 50 (5) (2005) 655–661.

[5] X. Zhang, L. Chen, Y. Chen, Consensus analysis of multi-agent systems with general linear dynamics and switching topologies by non-monotonically decreasing Lyapunov function, Int. J. Syst. Sci. 7 (1) (2019) 179–188.

[6] D. Vengertsev, H. Kim, J. Seo, H. Shim, Consensus of output-coupled high-order linear multi-agent systems under deterministic and Markovian switching networks, Int. J. Syst. Sci. 46 (10) (2015) 1790–1799.

[7] W. Liu, S. Zhou, Y. Qi, X. Wu, Leaderless consensus of multi-agent systems with Lipschitz nonlinear dynamics and switching topologies, Neurocomputing 173 (2016) 1322–1329.

[8] S. Zhou, W. Liu, Q. Wu, G. Yin, Leaderless consensus of linear multi-agent systems: matrix decomposition approach, in: Proc. the 7th Int. Conference on Intelligent Human-Machine Systems and Cybernetics, 2015, pp. 327–332.

[9] K. You, Z. Li, L. Xie, Consensus condition for linear multi-agent systems over randomly switching topologies, Automatica 49 (2013) 3125–3132.

[10] F. Xiao, L. Wang, Asynchronous consensus in continuous-time multi-agent systems with switching topology and time-varying delays, IEEE Trans. Autom. Control 53 (8) (2008) 1804–1816.

[11] H.J. Savino, C.R. Dos-Santos, F.O. Souza, L.C. Pimenta, M. De Oliveira, R.M. Palhares, Conditions for consensus of multi-agent systems with time-delays and uncertain switching topology, IEEE Trans. Ind. Electron. 63 (2) (2016) 1258–1267.

[12] A.V. Proskurnikov, Average consensus in switching nonlinearly coupled networks with time-varying delays, IFAC Proc. Vol. 46 (3) (2013) 457–461.

[13] N.R. Chowdhury, S. Sukumar, D. Chatterjee, A new condition for asymptotic consensus over switching graphs, Automatica 97 (2018) 18–26.

[14] Y. Zheng, L. Wang, Consensus of switched multiagent systems, IEEE Trans. Circuits Syst. II, Express Briefs 63 (3) (2016) 314–318.

[15] X. Lin, Y. Zheng, Finite-time consensus of switched multiagent systems, IEEE Trans. Syst. Man Cybern. Syst. 47 (7) (2017) 1535–1545.

[16] J. Thunberg, X. Hu, J. Goncalves, Local Lyapunov functions for consensus in switching nonlinear systems, IEEE Trans. Autom. Control 62 (12) (2017) 6466–6472.

[17] N. Barabanov, R. Ortega, Global consensus of time-varying multiagent systems without persistent excitation assumptions, IEEE Trans. Autom. Control 63 (11) (2018) 3935–3939.

[18] H.-B. Guo, H.-Y. Li, W.-C. Zhong, S.-J. Zhang, X.-B. Cao, Adaptive synchronization of networked Euler-Lagrange systems with directed switching topology, Acta Autom. Sin. 40 (11) (2014) 2541–2548.

[19] T. Li, L. Xie, Average consensus with limited data rate and switching topologies, IFAC-PapersOnLine 43 (19) (2010) 185–190.

[20] S. Arun Kumar, N.R. Chowdhury, S. Srikant, J. Raisch, Consensus analysis of systems with time-varying interactions: an event-triggered approach, IFAC-PapersOnLine 50 (1) (2017) 9349–9354.

[21] Z. Tu, H. Yu, X. Xia, Decentralized finite-time adaptive consensus of multiagent systems with fixed and switching network topologies, Neurocomputing 219 (2017) 59–67.

[22] Z. Li, M. Sun, P. Lin, Y. Jia, High-order multi-agent consensus with dynamically changing topologies and time-delays, IET Control Theory Appl. 5 (8) (2011) 976–981.

[23] S.R. Etesami, T. Başar, Convergence time for unbiased quantized consensus over static and dynamic networks, IEEE Trans. Autom. Control 61 (2) (2016) 443–455.

[24] A. Nedic, J. Liu, On convergence rate of weighted-averaging dynamics for consensus problems, IEEE Trans. Autom. Control 62 (2) (2017) 766–781.

[25] G. Duan, F. Xiao, L. Wang, Asynchronous periodic edge-event triggered control for double-integrator networks with communication time delays, IEEE Trans. Cybern. 48 (2) (2018) 675–688.

[26] D. Wang, N. Zhang, J. Wang, W. Wang, A PD-like protocol with a time delay to average consensus control for multi-agent systems under an arbitrarily fast switching topology, IEEE Trans. Cybern. 47 (4) (2017) 898–907.

[27] D. Zhao, T. Dong, Reduced-order observer-based consensus for multi-agent systems with time delay and event trigger strategy, IEEE Access 5 (2017) 1263–1271.

[28] F. Xiao, T. Chen, H. Gao, Synchronous hybrid event-and time-driven consensus in multiagent networks with time delays, IEEE Trans. Cybern. 46 (5) (2016) 1165–1174.

[29] X. Xu, L. Liu, G. Feng, Consensus of discrete-time linear multiagent systems with communication, input and output delays, IEEE Trans. Autom. Control 63 (2) (2018) 492–497.

[30] W. Liu, F. Deng, J. Liang, H. Liu, Distributed average consensus in multi-agent networks with limited bandwidth and time-delays, IEEE/CAA J. Autom. Sin. 1 (2) (2014) 193–203.

[31] L. Li, D.W. Ho, J. Lu, A unified approach to practical consensus with quantized data and time delay, IEEE Trans. Circuits Syst. I, Regul. Pap. 60 (10) (2013) 2668–2678.

[32] Z. Rui, J. Li, pth moment consensus of multi-agent systems with relative state-dependent measurement noises and time delays, IET Control Theory Appl. 12 (16) (2018) 2245–2252.

[33] R. Cepeda-Gomez, N. Olgac, An exact method for the stability analysis of linear consensus protocols with time delay, IEEE Trans. Autom. Control 56 (7) (2011) 1734–1740.

[34] Y. Zhang, Y. Sun, X. Wu, D. Sidorov, D. Panasetsky, Economic dispatch in smart grid based on fully distributed consensus algorithm with time delay, in: Chinese Control Conference, vol. 2018-July, 2018, pp. 2442–2446.

[35] H. Wu, B. An, Y. Song, On the general consensus protocol for second-order multi-agent systems with diverse time delays, in: Chinese Control Conference, CCC, vol. 2018-July, 2018, pp. 6394–6397.

[36] J. Hu, Y. Lin, Consensus control for multi-agent systems with double-integrator dynamics and time delays, IET Control Theory Appl. 2018-July (2018) 6394–6397.

[37] L. Peng, Y. Jia, Consensus control for multi-agent systems with double-integrator dynamics and time delays, IEEE Trans. Autom. Control (2010) 6394–6397.

[38] U. Münz, A. Papachristodoulou, F. Allgöwer, Nonlinear multi-agent system consensus with time-varying delays, IFAC Proc. Vol. 41 (2) (2008) 1522–1527.

[39] I.C. Morǎrescu, S.I. Niculescu, Multi-agent systems with decaying confidence and commensurate time-delays, IFAC-PapersOnLine 28 (12) (2015) 165–170.

[40] W. Lu, F.M. Atay, J. Jost, Consensus and synchronization in delayed networks of mobile multi-agents, IFAC Proc. Vol. 18 (1) (2011) 2362–2367.

[41] A.V. Proskurnikov, Consensus in networks of integrators with fixed topology and delayed nonlinear couplings, IFAC Proc. Vol. 18 (1) (2011) 8945–8950.

[42] H. Xiang Hu, W. Yu, Q. Xuan, L. Yu, G. Xie, Consensus of multi-agent systems in the cooperation-competition network with inherent nonlinear dynamics: a time-delayed control approach, Neurocomputing 158 (2015) 134–143.

[43] C. Tan, G.P. Liu, P. Shi, Consensus of networked multi-agent systems with diverse time-varying communication delays, J. Franklin Inst. 352 (7) (2015) 2934–2950.

[44] R. Sakthivel, S. Kanakalakshmi, B. Kaviarasan, Y.K. Ma, A. Leelamani, Finite-time consensus of input delayed multi-agent systems via non-fragile controller subject to switching topology, Neurocomputing 325 (2019) 225–233.

[45] W.S. Zhong, G.P. Liu, C. Thomas, Global bounded consensus of multiagent systems with nonidentical nodes and time delays, IEEE Trans. Syst. Man Cybern., Part B, Cybern. 42 (5) (2012) 1480–1488.

[46] W. Yu, G. Chen, M. Cao, W. Ren, Delay-induced consensus and quasi-consensus in multi-agent dynamical systems, IEEE Trans. Circuits Syst. I, Regul. Pap. 60 (10) (2013) 2679–2687.

[47] C.L. Chen, G.X. Wen, Y.J. Liu, F.Y. Wang, Adaptive consensus control for a class of nonlinear multiagent time-delay systems using neural networks, IEEE Trans. Neural Netw. Learn. Syst. 25 (6) (2014) 1217–1226.

[48] N. Mu, X. Liao, T. Huang, Event-based consensus control for a linear directed multi-agent system with time delay, IEEE Trans. Circuits Syst. II, Express Briefs 62 (3) (2015) 281–285.

[49] S. Giannini, A. Petitti, D. Di Paola, A. Rizzo, Asynchronous max-consensus protocol with time delays: convergence results and applications, IEEE Trans. Circuits Syst. I, Regul. Pap. 63 (2) (2016) 256–264.

[50] W. Qiao, R. Sipahi, Consensus control under communication delay in a three-robot system: design and experiments, IEEE Trans. Control Syst. Technol. 24 (2) (2016) 687–694.

[51] D.H. Nguyen, J. Khazaei, Multiagent time-delayed fast consensus design for distributed battery energy storage systems, IEEE Trans. Sustain. Energy 9 (3) (2018) 1397–1406.

[52] D. Ao, G. Yang, X. Wang, $\mathcal{H}_\infty$ consensus control of multi-agent systems: a dynamic output feedback controller with time-delays, in: Proc. the Chinese Control and Decision Conference (CCDC), 2018, pp. 4591–4596.

[53] D. Ao, X. Wang, G. Yang, C. Hu, Consensus for time-delay multi-agent systems with finite frequency-domain $\mathcal{H}_\infty$ performance, in: Proc. the 37th Chinese Control Conference (CCC), 2018, pp. 1057–1064.

[54] P. Ming, J. Liu, S. Tan, S. Li, L. Shang, X. Yu, Consensus stabilization in stochastic multi-agent systems with Markovian switching topology, noises and delay, Neurocomputing 200 (2016) 1–10.

[55] H. Qu, F. Yang, Q.-L. Han, Distributed $\mathcal{H}_\infty$-consensus filtering for a networked time-delay system with switching network topology and packet dropouts, in: Proc. the Australian & New Zealand Control Conference (ANZCC), 2018, pp. 334–339.

[56] T.S. Alireza, A. Jadbabaie, A necessary and sufficient condition for consensus over random networks, IEEE Trans. Autom. Control 53 (3) (2008) 791–796.

[57] H. Rezaee, F. Abdollahi, Consensus problem in general linear multiagent systems under stochastic topologies, IFAC-PapersOnLine 49 (13) (2016) 13–18.

[58] Y. Shang, Consensus seeking over Markovian switching networks over with time-varying delays and uncertain topologies, Appl. Math. Comput. 273 (2016) 1234–1245.

[59] W. Ni, D. Cheng, Leader-following consensus of multi-agent systems under fixed and switching topologies, Syst. Control Lett. 59 (2010) 209–217.

[60] H. Liu, H.R. Karimi, S. Du, W. Xia, C. Zhong, Leader-following consensus of discrete-time multiagent systems with time-varying delay based on large delay theory, Inf. Sci. 417 (2017) 236–246.

[61] W. Guo, H. Xiao, S. Chen, Consensus of the second-order multi-agent systems with an active leader and coupling time delay, Acta Math. Sci. 34 (2) (2014) 453–465.

[62] C.I. Aldana, E. Nuno, L. Basanez, Leader-follower pose consensus for heterogeneous robot networks with variable time-delays, IFAC Proc. Vol. 19 (2014) 6674–6679.

[63] B. Qi, K. Lou, S. Miao, B. Cui, Second-order consensus of leader-following multi-agent systems with jointly connected topologies and time-varying delays, Arab. J. Sci. Eng. 39 (2014) 1431–1440.

[64] D. Zhang, Z. Xu, Q.G. Wang, Y.B. Zhao, Leader–follower $\mathcal{H}_\infty$ consensus of linear multi-agent systems with aperiodic sampling and switching connected topologies, ISA Trans. 68 (2017) 150–159.

[65] Y.J. Pan, H. Werner, Z. Huang, M. Bartels, Distributed cooperative control of leader–follower multi-agent systems under packet dropouts for quadcopters, Syst. Control Lett. 106 (2017) 47–57.

[66] J. Ni, L. Liu, C. Liu, J. Liu, Fixed-time leader-following consensus for second-order multiagent systems with input delay, IEEE Trans. Ind. Electron. 64 (11) (2017) 8635–8646.

[67] H. Ren, F. Deng, Mean square consensus of leader-following multi-agent systems with measurement noises and time delays, ISA Trans. 71 (2017) 76–83.

[68] F. Wang, H. Yang, Z. Liu, Z. Chen, Containment control of leader-following multi-agent systems with jointly-connected topologies and time-varying delays, Neurocomputing 260 (2017) 341–348.

[69] B. Cui, C. Zhao, T. Ma, C. Feng, Leaderless and leader-following consensus of multi-agent chaotic systems with unknown time delays and switching topologies, Nonlinear Anal. Hybrid Syst. 24 (2017) 115–131.

[70] W. Zhu, Consensus of multiagent systems with switching jointly reachable interconnection and time delays, IEEE Trans. Syst. Man Cybern., Part A, Syst. Hum. 42 (2) (2012) 348–358.

[71] W. Zhu, Z.-p. Jiang, Event-based leader-following consensus of multi-agent systems with input time delay, IEEE Trans. Autom. Control 60 (5) (2015) 1362–1367.

[72] T. Xie, X. Liao, H. Li, Leader-following consensus in second-order multi-agent systems with input time delay: an event-triggered sampling approach, Neurocomputing 177 (2016) 130–135.

[73] X. Wang, H. Su, Self-triggered leader-following consensus of multi-agent systems with input time delay, Neurocomputing 330 (2018) 70–77.

[74] J. Wang, X. Li, Y. Li, X. Li, X. Luo, Event-based consensus control of multi-agent systems with input saturation constraint, in: Proc. the 30th Chinese Control and Decision Conference, CCDC 2018, 2018, pp. 4694–4699.

[75] D. Yue, J. Cao, Q. Li, X. Shi, Neuro-adaptive consensus strategy for a class of nonlinear time-delay multi-agent systems with an unmeasurable high-dimensional leader, IET Control Theory Appl. 13 (2) (2019) 230–238.

[76] Z. Yan, Y. Wu, X. Du, J. Li, Limited communication consensus control of leader-following multi-UUVs in a swarm system under multi-independent switching topologies and time delay, IEEE Access 6 (2018) 33183–33200.

[77] J. Shao, W.X. Zheng, T.Z. Huang, A.N. Bishop, On leader-follower consensus with switching topologies: an analysis inspired by pigeon hierarchies, IEEE Trans. Autom. Control 63 (10) (2018) 3588–3593.

[78] Y. Cai, H. Zhang, K. Zhang, Y. Liang, Distributed leader-following consensus of heterogeneous second-order time-varying nonlinear multi-agent systems under directed switching topology, Neurocomputing 325 (2018) 31–47.

[79] A. Petrillo, A. Salvi, S. Santini, A.S. Valente, Adaptive synchronization of linear multi-agent systems with time-varying multiple delays, J. Franklin Inst. 354 (2017) 8586–8605.

[80] J. Yu, L. Wang, Group consensus in multi-agent systems with switching topologies and communication delays, Syst. Control Lett. 59 (6) (2010) 340–348.

[81] H. Xia, T.-Z. Huang, J.-L. Shao, J.-Y. Yu, Group consensus of multi-agent systems with communication delays, Neurocomputing 171 (Supplement C) (2016) 1666–1673.

[82] G. Wen, Y. Yu, Z. Peng, H. Wang, Dynamical group consensus of heterogenous multi-agent systems with input time delays, Neurocomputing 175 (2015) 278–286.

[83] Y. Gao, J. Yu, J. Shao, M. Yu, Group consensus for second-order discrete-time multi-agent systems with time-varying delays under switching topologies, Neurocomputing 207 (2016) 805–812.

[84] J. Yu, J. Liu, L. Xiang, J. Zhou, Group consensus in networked mechanical systems with communication delays, Proc. IUTAM 22 (2017) 107–114.

[85] H. Jiang, X.K. Liu, H. He, C. Yuan, D. Prokhorov, Neural network based distributed consensus control for heterogeneous multi-agent systems, in: Proc. the American Control Conference, vol. 2018-June, 2018, pp. 5175–5180.

[86] L. Ji, X. Yu, C. Li, Group consensus for heterogeneous multiagent systems in the competition networks with input time delays, IEEE Trans. Syst. Man Cybern. Syst. (2018) 1–9.

[87] Z. Huanyu, H.P. Ju, Y. Zhang, Couple-group consensus for second-order multi-agent systems with fixed and stochastic switching topologies, Appl. Math. Comput. 232 (2014) 595–605.

[88] Y. Shang, Couple-group consensus of continuous-time multi-agent systems under Markovian switching topologies, J. Franklin Inst. 352 (11) (2015) 4826–4844.

[89] Y. Shang, L_1 group consensus of multi-agent systems with switching topologies and stochastic inputs, Phys. Lett. A 377 (25–27) (2013) 1582–1586.

[90] S. Zhai, Q. Li, Pinning bipartite synchronization for coupled nonlinear systems with antagonistic interactions and switching topologies, Syst. Control Lett. 94 (2016) 127–132.

[91] J. Hu, H. Zhu, Adaptive bipartite consensus on coopetition networks, Phys. D: Nonlinear Phenom. 307 (2015) 14–21.

[92] J. Shao, L. Shi, Y. Zhang, Y. Cheng, On the asynchronous bipartite consensus for discrete-time second-order multi-agent systems with switching topologies, Neurocomputing 316 (2018) 105–111.

[93] C. Zong, Z. Ji, L. Tian, Y. Zhang, Distributed multi-robot formation control based on bipartite consensus with time-varying delays, IEEE Access 7 (2019) 144790–144798.

[94] D. Meng, Y. Jia, Scaled consensus problems on switching networks, IEEE Trans. Autom. Control 61 (6) (2016) 1664–1669.

[95] M. Li, F. Deng, H. Ren, Scaled consensus of multi-agent systems with switching topologies and communication noises, Nonlinear Anal. Hybrid Syst. 36 (2020) 1–11.

CHAPTER 4

Distributed consensus of multiagent systems

4.1 Exponential consensus of stochastic delayed systems

The dynamic behaviors of multiagent systems have recently gained increasing research interest [1–3]. Among them, the study on synchronization or consensus problem of multiagent systems has been an active research topic in the past few years [5–8], since coordination phenomena have been found in both natural and man-made systems, such as fireflies in the forest, applause, description of hearts, distributed computing systems, and so on.

On the one hand, time-delay phenomena in spreading information through large-scale networked systems are ubiquitous in natural and technical societies because of the finite speed of signal transmission over links, as well as networked traffic congestion. Moreover, in the process of signal transmission among networks, the nodes are often subject to stochastic perturbations which arise from external random fluctuations in the process of transmission and other probabilistic factors.

A wide class of multiagent systems often have a switching topology due to link failures or a new creation, and such networks appear in many practical situations, such as communication networks [6], power grids [9], and many other fields. In addition, since the parameters of each node in networks could be time-varying, switching behaviors can take place in the dynamics of nodes, as well as in the topology. When the network topology and nodes are modeled in a switched way, multiagent systems or complex networks under consideration can be described and analyzed by the switching theory [6].

4.1.1 Introduction

Along another research direction, in switched systems the mode-dependent control is usually designed such that the resulting system is stable and follows certain performance indices, which is less conservative than the mode-independent control [11,12]. It is worth mentioning that there always exists a lag between the system switching and the switching of a mode-dependent controller [11,13]. It turns out that a majority of the research efforts on syn-

Advanced Distributed Consensus for Multiagent Systems
https://doi.org/10.1016/B978-0-12-821186-1.00012-X

chronization and/or consensus of switched networks have been devoted to networks without feedback control [6].

Hereafter, we focus on the problems of consensus of multiagent systems with asynchronously switching control. We investigate the distributed exponential consensus of stochastic switched multiagent systems with asynchronous switching. To proceed further, we initially develop a comparison principle of stochastic delayed systems with switching parameters. Based on this extended comparison principle, the distributed exponential consensus problem is examined for stochastic delayed multiagent systems with asynchronous switching by using the average dwell time approach and stochastic analysis techniques.

4.1.2 Model formulation

Consider the following stochastic delayed multiagent system with nonlinear dynamics consisting of N linearly coupled nodes:

$$dx_i(t) = \left[A_{\sigma(t)} x_i(t) + f_{\sigma(t)}(t, x_i(t), x_i(t - \tau(t))) \right.$$

$$\left. + \sum_{j=1}^{N} b_i^{\sigma(t)} j x_j(t - \tau(t)) \right] dt + g_{\sigma(t)}(t, x_i(t), x_i(t - \tau(t))) d\omega(t), \quad (4.1)$$

where $i = 1, 2, \ldots, N$, $x_i = [x_{i1}, x_{i2}, \ldots, x_{in}]^t \in \mathbb{R}^n$ is the state vector of the ith node; $\sigma(t) : [t_0, \infty) \to \Gamma = \{1, 2, \ldots, m\}$ is a piecewise constant function depending on t, continuous from the right, specifying the index of the active subsystem, that is, $\sigma(t) = k_n \in \Gamma$ for $t \in [t_n, t_{n+1})$, where t_n is the nth switching time instant; A_r is a constant matrix, $r \in \Gamma$; the time-delay $\tau(t)$ may be unknown but bounded, that is, $0 \leq \tau(t) \leq \tau$; $\omega(t)$ is a Weiner process defined on $(\Omega, \mathbf{F}, \{\mathbf{F}_t\}_t \geq 0, \mathbf{P})$, $\omega_i(t)$ is independent of $\omega_j(t)$ for $i \neq j$; $g : \mathbb{R}^+ \times \mathbb{R}^n \times \mathbb{R}^n \to \mathbb{R}^n$ is the noise intensity function matrix; $f_r = [f_{r1}, f_{r1}, \ldots, f_{rm}] : \mathbb{R}^+ \times \mathbb{R}^n \times \mathbb{R}^n \to \mathbb{R}^n$ is a continuous map; $\mathbf{B}_r = (b_{ij}^r) \in \mathbb{R}^{N \times N}$ is the coupling matrix, where b_{ij}^r is defined as follows: if there is a connection from node j to node i ($j \neq i$), then the coupling strength $b_{ij}^r \geq 0$; otherwise $b_{ij}^r = 0$. For $i = j$, b_{ij}^r is defined by

$$b_{ii}^r = - \sum_{j=1, j \neq i}^{N} b_{ij}^r$$

The initial conditions of the stochastic switched delayed multiagent system in (4.1) are assumed to be

$$x_i(t) = \xi_i(t), \quad t_0 - \tau \le t \le t_0, \ i = 1, 2, \ldots, N,$$

where $\xi_i(t) \in \mathbf{L}^2_{\mathbf{F}_{t_0}} ([-\tau, 0], \mathbb{R}^n)$.

In what follows, we consider the distributed controller adopting the following form:

$$u_i^{\sigma(t)}(t) = K_{\sigma(t-h(t))} \sum_{j=1}^N b_{ij}^{\sigma(t)} x_j(t - \tau(t)),$$

where $K_r, r \in \Gamma$ is the feedback gain to be determined and $h(t)$ is the time-delay to describe the lag between the switching of feedback control and the switching of system modes, and $0 \le h(t) \le \tau$. When $K_{\sigma(t-h(t))}$ coincides with the system mode, $u_i^{\sigma(t)}(t)$ is called a matched distributed controller; otherwise, it is called an unmatched distributed controller. Hence, the resulting closed-loop stochastic delayed multiagent system with asynchronously switching control is given by

$$dx_i(t) = \left[A_{\sigma(t)} x_i(t) + f_{\sigma(t)}(t, x_i(t), x_i(t - \tau(t))) \right.$$
$$\left. + K_{\sigma(t-h(t))} \sum_{j=1}^N b_{ij}^{\sigma(t)} x_j(t - \tau(t)) \right] dt$$
$$+ g_{\sigma(t)}(t, x_i(t), x_i(t - \tau(t))) d\omega(t). \tag{4.2}$$

Remark 4.1. In the sequel, we investigate the consensus of stochastic delayed multiagent systems with asynchronous switching control. A distributed controller is introduced to achieve coordination among agents. It is natural to extend our results to force the states of networks to a desired state, such as an equilibrium, a periodic state, or a chaotic state, by employing the techniques in leader–follower problems [15].

Remark 4.2. The following points are relevant with respect to the research results in [11,12,16]:
1. The model considered here is more general than in [11,12,16]. In this section, both stochastic disturbances and time-delays are considered for switching networks. It is worth mentioning that stochastic disturbances were overlooked in [11,12,16], and time-delays were not considered there.

2. The research problem is also different. We investigate the consensus of stochastic delayed multiagent systems by an asynchronous feedback controller. The exponential stability [11,12] and input-to-state stability [16] were provided for systems with asynchronous switching, respectively.

3. Since the model and the research problem are distinct from [11,12,16], the methods are also different. In this paper, a comparison principle for stochastic switched systems is first presented to deal with the consensus of stochastic delayed multiagent systems under asynchronous control, which differentiates from the techniques adopted in [11,12,16].

Remark 4.3. The feedback-gain switching is important from two aspects. First, the feedback-gain switching can reduce the quadratic continuous-time cost functional $J = \int_0^T (u^T(s)Pu(s) + x^T(s)Qx(s))ds$. Second, as illustrated in this chapter and in [11,16], the unmatched feedback controller may deteriorate the system's performance and even destroy stability. For example, if the coupling matrix switches between repulsive and attractive coupling, which can be observed in communication networks [8], the feedback-gain switching is useful to achieve consensus by appropriately designing K. The example of Section 4.2.2 in simulations will be provided to illustrate this point. Based on these two aspects, it is important to investigate the feedback-gain switching. For more results regarding the importance of the feedback-gain switching, please refer to [11] and [16] and the references therein.

Remark 4.4. Stochastic switched systems constitute an important generalization of nonstochastic hybrid dynamical systems, for which significant breakthroughs in the stability theory have been carved out over the last decade [17]. In order to deal with the stability of stochastic switched delayed systems, an extended comparison principle for stochastic switched delayed systems is presented in Lemma 4.2, which shows its importance in our proof and can be used in general stochastic switched delayed systems. For more details regarding the challenge of stochastic systems, please refer to the recent review [17] and the references therein.

The following definitions, assumptions, and lemmas are needed for deriving of the main results.

Definition 4.1. The stochastic delayed multiagent system with asynchronous switching and nonlinear dynamics in (4.2) is said to achieve exponential consensus in mean-square if there exist $\lambda > 0$ and $M_0 > 0$ such that for any initial values $\xi_i(s)$,

$$\mathbb{E}\|x_i(t) - x_j(t)\| \leq M_0 e^{-\lambda(t-t_0)}$$

hold for all $t \geq t_0$, and for any $i, j = 1, 2, \ldots, N$.

Definition 4.2 ([18]). For a switching signal $\sigma(t)$ and any $t > t_0$, let $N(t, t_0)$ be the switching numbers of $\sigma(t)$ over the interval $[t_0, t)$. If

$$N(t, t_0) \leq \frac{t - t_0}{T_a} + N_0$$

for $N_0 \geq 0$, $T_a > 0$, then T_a and N_0 are called the average dwell time and the chatter bound, respectively.

Assumption 4.1.1. There exist positive constants ϱ_{1r}, ϱ_{2r}, ρ_{1r}, and ρ_{2r} such that

$$\|f_r(t, x_1, y_1) - f_r(t, x_2, y_2)\| \leq \varrho_{1r}\|x_1 - x_2\| + \varrho_{2r}\|y_1 - y_2\| \tag{4.3}$$

and

$$\mathrm{trace}\left\{[g_r(t, x_1, y_1) - g_r(t, x_2, y_2)]^T[g(t, x_1, y_1) - g(t, x_2, y_2)]\right\}$$
$$\leq \rho_{1r}\|x_1 - x_2\|^2 + \rho_{2r}\|y_1 - y_2\|^2 \tag{4.4}$$

for any $x_1, x_2, y_1, y_2 \in \mathbb{R}^n$ and $t \in [t_0, +\infty)$.

Lemma 4.1 ([19]). *Let* $\mathcal{U} = (a_{ij})_{N \times N}$, $\mathcal{P} \in \mathbb{R}^{n \times n}$, $x = (x_1^T, x_2^T, \ldots, x_N^T)$ *and* $y = (y_1^T, y_2^T, \ldots, y_N^T)$ *with* $x_k, y_k \in \mathbb{R}^n$ $(k = 1, 2, \ldots, N)$. *If* $\mathcal{U} = \mathcal{U}^T$ *and each row sum of* $\mathcal{U}$ *is zero, then*

$$x^T(\mathcal{U} \otimes \mathcal{P})y = -\sum_{1 \leq i < j \leq N} a_{ij}(x_i - x_j)^T(y_i - y_j).$$

4.2 Asynchronous switching control

In this section, asynchronous switching control is considered to synchronize the stochastic delayed multiagent system with asynchronous switching in (4.2). In order to derive the results in a systematic way, we let

$$x(t) = [x_1^T(t), x_2^T(t), \ldots, x_N^T(t)]^T,$$
$$f_r(t, x(t), x(t - \tau(t))) = [f_r^T(t, x_1(t), x_1(t - \tau(t))),$$
$$\ldots, f_r^T(t, xN(t), x_N(t - \tau(t)))]^T,$$

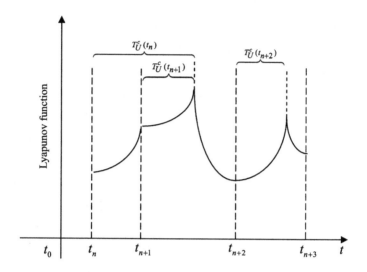

Figure 4.1 Typical case of $\mathbf{T}_U^c(t_n)$.

$$\mathbf{g}_r(t, x(t), x(t - \tau(t))) = [g_r^T(t, x_1(t), x_1(t - \tau(t))),$$
$$\ldots, g_r^T(t, x_N(t), x_N(t - \tau(t)))]^T,$$
$$\xi(t) = [\xi_1^T(t), \xi_2^T(t), \ldots, \xi_N^T(t)]^T,$$
$$\mathbf{\Omega}(t) = [\omega^T(t), \omega^T(t), \ldots, \omega^T(t)]^T, \quad r \in \Gamma.$$

Then the stochastic delayed multiagent system with asynchronous switching in (4.2) can be rewritten as

$$dx(t) = [(I_N \otimes A_{\sigma(t)})x(t) + \mathbf{f}_{\sigma(t)}(t, x(t), x(t - \tau(t)))$$
$$+ (I_N \otimes K_{\sigma(t-h(t))})(B_{\sigma(t)} \otimes I_n)x(t)]dt$$
$$+ \mathbf{g}_{\sigma(t)}(t, x(t), x(t - \tau(t)))d\mathbf{\Omega}(t),$$
$$x(s) = \xi(s), t_0 - \tau \leq s \leq t_0. \tag{4.5}$$

Without loss of generality, let $\mathbf{T}_M(t, s)$ (resp., $\mathbf{T}_U(t, s)$) denote the total activation time on the matched control (resp., the unmatched control) during $[s, t)$. Let $\mathbf{T}_U^c(t_n) = \sup_{t \geq t_n}\{t - t_n : \mathbf{T}_M(t, t_n) = 0, \forall n \in \mathbb{N}\}$, and $\tau_b = \sup\{\mathbf{T}_U^c(t_n), n \in \mathbb{N}\}$ denote the maximal time for continuously running with unmatched control. Fig. 4.1 shows a typical case of $\mathbf{T}_U^c(t_n)$.

To proceed further, we assume that $\mathbf{C}^{1,2}$ denotes the family of all nonnegative functions $\mathbf{V}(t, x, \sigma(t))$ on $[t_0 - \tau, \infty) \times \mathbb{R}^{nN} \times \Gamma$ that are continuously once differentiable in t and twice differentiable in x. For simplicity, we denote $\mathbf{V}(t, x(t), \sigma(t))$ and $\mathbf{V}(t, x(t - \tau(t)), \sigma(t))$ by $\mathbf{V}(t, \sigma(t))$ and $\mathbf{V}(t - \tau(t), \sigma(t))$, respectively. Define a diffusion operator $\mathbb{L}$ as [20]:

$$
\begin{aligned}
\mathbb{L}\mathbf{V}(t, r) = {} & \mathbf{V}_t(t, r) + \mathbf{V}_x(t, r)[(I_N \otimes A_r)x(t) + \mathbf{f}_r(t, x(t), x(t - \tau(t))) \\
& + (I_N \otimes K_{\sigma(t-h(t))})(B\sigma(t) \otimes I_n)x(t)] \\
& + \frac{1}{2}\text{trace}[\mathbf{g}_r^T(t, x(t), x(t - \tau(t)))\mathbf{V}_{xx}(t, r)\mathbf{g}_r(t, x(t), x(t - \tau(t)))],
\end{aligned}
$$

where

$$
\mathbf{V}_x(t, r) = \left(\frac{\partial \mathbf{V}(t, r)}{\partial x_{11}}, \dots, \frac{\partial \mathbf{V}(t, r)}{\partial x_{Nn}} \right),
$$

$$
\mathbf{V}_t(t, r) = \frac{\partial \mathbf{V}(t, r)}{\partial t}, \quad \mathbf{V}_{xx}(t, r) = \left(\frac{\partial^2 \mathbf{V}(t, r)}{\partial x_i \partial x_j} \right)_{nN \times nN}.
$$

In order to analyze the consensus of stochastic delayed multiagent systems with asynchronous switching, we first present a comparison principle for stochastic switched systems, which is important to present the main results.

Lemma 4.2. *Assume that a piecewise continuous function $v(t)$ is continuous on $[t_k, t_{k+1})$ for $k \in \mathbb{N}$ and $\lambda_1(t) : [t_0 - \tau, \infty) \to \mathbb{R}$ is an arbitrary function. If there exist a function $\mathbf{V}(t, \sigma(t)) \in \mathbf{C}^{1,2}$ ($\sigma(t) \in \Gamma$) and positive constants λ_2 and $\mu > 1$ such that*

$$
\begin{aligned}
D^+\mathbb{E}\mathbf{V}(t, \sigma(t)) \leq {} & \lambda_1(t)\mathbb{E}\mathbf{V}(t, \sigma(t)) \\
& + \lambda_2\mathbb{E}\mathbf{V}(t - \tau(t), \sigma(t - \tau(t))), \quad t \neq t_n, t \geq t_0, \\
\mathbb{E}\mathbf{V}(t_k, \sigma(t_k)) \leq {} & \mu\mathbb{E}\mathbf{V}(t_k^-, \sigma(t_k^-)), \quad k \in \mathbb{N},
\end{aligned} \tag{4.6}
$$

and

$$
\begin{aligned}
D^+v(t) > {} & \lambda_1(t)v(t) + \lambda_2 v(t - \tau(t)), \quad t \neq t_n, t \geq t_0, \\
v(t_k) = {} & \mu v(t_k^-), \quad k \in \mathbb{N},
\end{aligned} \tag{4.7}
$$

then $\mathbb{E}V(t, \sigma(t)) \leq v(t)$, $t \in [t_0 - \tau, t_0]$ implies that $\mathbb{E}V(t, \sigma(t)) \leq v(t)$, for $t \geq t_0$.

Proof. For $\mathbf{V}(t, \sigma(t)) \in \mathbf{C}^{1,2}$, we get that $\mathbb{E}\mathbf{V}(t, \sigma(t))$ is continuous on each interval $t \in [t_k, t_{k+1})$ for $k \in \mathbb{N}$. First, we will establish that

$$
\mathbb{E}\mathbf{V}(t, \sigma(t)) \leq v(t), \quad t \in [t_0, t_1). \tag{4.8}
$$

If (4.8) is not true, in view of $\mathbb{E}\mathbf{V}(t, \sigma(t)) \leq v(t)$, $t \in [t_0 - \tau, t_0]$, there should exist some $t \in (t_0, t_1)$ such that $\mathbb{E}\mathbf{V}(t, \sigma(t)) > v(t)$. Set $t^* = \inf\{t \in (t_0, t_1) : \mathbb{E}\mathbf{V}(t, \sigma(t)) > v(t)\}$. Since $\mathbb{E}\mathbf{V}(t, \sigma(t))$ and $v(t)$ are continuous on $t \in [t_0, t_1)$, we proceed to derive that $\mathbb{E}\mathbf{V}(t^*, \sigma(t^*)) = v(t^*)$ and $\mathbb{E}\mathbf{V}(t, \sigma(t)) > v(t)$ for $t \in (t^*, t^* + \varepsilon)$, where $\varepsilon > 0$ is sufficient small. Hence, for all $t \in (t^*, t^* + \varepsilon)$,

$$\frac{\mathbb{E}\mathbf{V}(t, \sigma(t)) - \mathbb{E}\mathbf{V}(t^*, \sigma(t^*))}{t - t^*} > \frac{v(t) - v(t^*)}{t - t^*},$$

which yields that

$$D^+\mathbb{E}\mathbf{V}(t^*, \sigma(t^*)) \geq D^+v(t^*). \tag{4.9}$$

On the other hand, according to (4.6) and (4.7), we have

$$\begin{aligned} D^+\mathbb{E}\mathbf{V}(t^*, \sigma(t^*)) &\leq \lambda_1(t^*)\mathbb{E}\mathbf{V}(t^*, \sigma(t^*)) + \lambda_2\mathbb{E}\mathbf{V}(t^* - \tau(t^*), \sigma(t^* - \tau(t^*))) \\ &< \lambda_1(t^*)v(t^*) + \lambda_2 v(t^* - \tau(t^*)) \\ &< D^+v(t^*), \end{aligned}$$

which contradicts (4.9). Thus, (4.8) holds. Assume that $\mathbb{E}\mathbf{V}(t, \sigma(t)) \leq v(t)$, $t \in [t_{k-1}, t_k)$ for $k = 1, 2, \ldots, l$. Then, $\mathbb{E}\mathbf{V}(t, \sigma(t)) \leq v(t)$, $t \in [t_l - \tau, t_l)$ and hence $\mathbb{E}\mathbf{V}(t_l^-, \sigma(t_l^-)) \leq v(t_l^-)$. According to (4.6) and (4.7),

$$\mathbb{E}\mathbf{V}(t_l, \sigma(t_l)) \leq \mu\mathbb{E}\mathbf{V}(t_l^-, \sigma(t_l^-)) \leq \mu v(t_l^-) = v(t_l).$$

Employing a similar argument as in the proof of (4.8), we can get $\mathbb{E}\mathbf{V}(t, \sigma(t)) \leq v(t)$ for $t \in [t_l, t_{l+1})$. By mathematical induction, $\mathbb{E}\mathbf{V}(t, \sigma(t)) \leq v(t)$ is true for all $t \geq t_0$. This concludes the proof. $\qquad\square$

Remark 4.5. The extended comparison principle of stochastic switched systems will be employed to investigate the distributed exponential consensus of stochastic delayed multiagent systems with asynchronous switching. Usually, the comparison principle [21] is used to study the stability of deterministic dynamical systems with impulsive effects [19]. Different from [19] and [21], we extend the comparison principle into the case of stochastic switched systems, and thus the comparison principle for deterministic systems is generalized to stochastic switched systems [10].

Before giving the main results of consensus of multiagent systems, we need a technical lemma which plays a key role in proving the distributed exponential consensus for the multiagent system in (4.2). For the stochastic delayed multiagent system with asynchronous switching in (4.2), if the effect brought by unmatched control is finally compensated by matched control, then the multiagent system in (4.2) can achieved consensus. Thus, it is crucial to guarantee that there exists enough active time of stochastic delayed multiagent systems with matched control. In the following lemma, a restriction is put on the activation time of unmatched and matched control, which is important to present our main results.

Lemma 4.3. *Suppose there exist a function* $\mathbf{V}(t, \sigma(t)) \in \mathbf{C}^{1,2}$ *and positive constants* $\alpha, \beta, \gamma, c, \mu > 1, \lambda_{2k_n}, k_n \in \Gamma$ *such that*

$$\mathbb{L}\mathbf{V}(t, k_n) \leq \begin{cases} \beta\mathbf{V}(t, k_n) + \lambda_{2k_n}\mathbf{V}(t - \tau(t), k_n), & t \in \mathbf{T}_U(tn, t_{n+1}), \\ -\alpha\mathbf{V}(t, k_n) + \lambda_{2k_n}\mathbf{V}(t - \tau(t), k_n), & t \in \mathbf{T}_M(t_n, t_{n+1}), n \in \mathbb{N}, \end{cases}$$
(4.10)

$$\mathbf{V}(t, r) \leq \mu\mathbf{V}(t, l), \quad \forall r, l \in \Gamma, r \neq l \tag{4.11}$$

$$\mathbf{T}_a > \frac{\ln \mu}{\gamma - \eta\lambda_2} > 0, \tag{4.12}$$

$$\mathbb{E}\mathbf{V}(t, r) \leq c\mathbb{E}\|\xi\|^2, \quad t \in [t_0 - \tau, t_0], \ \gamma \in \Gamma, \tag{4.13}$$

and

$$-\alpha\mathbf{T}_M(t, s) + \beta\mathbf{T}_U(t, s) \leq -\gamma(t - s) + (\gamma + \beta)\tau_b, \quad t > s > t_0. \tag{4.14}$$

Then

$$\mathbb{E}\mathbf{V}(t, \sigma(t)) \leq \mathcal{M}e^{-\lambda(t-t_0)}, \quad t \geq t_0,$$

where λ *is the unique positive solution of equation* $\lambda + \eta\lambda_2 e^{\lambda\tau} - \lambda_3 = 0$, $\lambda_2 = \max_{k_n}\{\mu\lambda_{2k_n}\}$, $\lambda_3 = \gamma - (\ln \mu)/T_a$, $\eta = e^{(\gamma+\beta)\tau_b}$ *and* $\mathcal{M} = \eta c\mathbb{E}\|\xi\|^2$.

Proof. From the Itô's differential formula [20], we have

$$d\mathbf{V}(t, k_n) = \mathbb{L}\mathbf{V}(t, k_n)dt + \mathbf{V}\mathbf{V}_x(t, k_n)\mathbf{g}k_n(t, x(t), x(t - \tau(t)))d\omega(t)$$

for $t \in [t_n, t_{n+1})$, $n \in \mathbb{N}$. It follows that

$$D^+\mathbb{E}\mathbf{V}(t, k_n) = \mathbb{E}\mathbb{L}\mathbf{V}(t, k_n), \quad t \in [t_n, t_{n+1}), n \in \mathbb{N}, \tag{4.15}$$

where $D^+\mathbb{E}\mathbf{V}(t, k_n) = \limsup_{\Delta t \to 0^+}(\mathbb{E}\mathbf{V}(t + \Delta t, k_n) - \mathbb{E}\mathbf{V}(t, k_n))/(\Delta t)$. According to (4.11), one see that for $t \in [t_n, t_{n+1})$, $n \in \mathbb{N}$,

$$\mathbf{V}(t - \tau(t), k_n) \leq \mu\mathbf{V}(t - \tau(t), \sigma(t - \tau(t))). \tag{4.16}$$

Combining (4.16) with (4.10) yields

$$D^+\mathbb{E}\mathbf{V}(t,\sigma(t)) \leq \lambda_1(t)\mathbb{E}\mathbf{V}(t,\sigma(t))$$
$$+ \lambda_2\mathbb{E}\mathbf{V}(t-\tau(t),\sigma(t-\tau(t))), \quad t \neq t_n, \ t \geq t_0,$$
$$\mathbb{E}\mathbf{V}(t_n,k_n) \leq \mu\mathbb{E}\mathbf{V}(t_n^-,k_{n-1}), \quad n \in \mathbb{N},$$
$$\mathbb{E}\mathbf{V}(t,\sigma(t)) \leq c\mathbb{E}\|\xi\|^2, \quad -\tau \leq t \leq t_0, \tag{4.17}$$

where

$$\lambda_1(t) = \begin{cases} \beta, & t \in \mathbf{T}_U(t_n, t_{n+1}), \\ -\alpha, & t \in \mathbf{T}_M(t_n, t_{n+1}), n \in \mathbb{N}. \end{cases}$$

For $\forall \varepsilon > 0$, let $v(t)$ be the unique solution of the following delayed system:

$$\dot{v}(t) = \lambda_1(t)v(t) + \lambda_2 v(t-\tau(t)) + \varepsilon, \quad t \neq t_n, \ t \geq t_0,$$
$$v(t_n) = \mu v(t_n^-), \quad n \in \mathbb{N},$$
$$v(t) = c\mathbb{E}\|\xi\|^2, \quad -\tau \leq t \leq t_0. \tag{4.18}$$

Then, by Lemma 4.3, we have

$$\mathbb{E}\mathbf{V}(t,\sigma(t)) \leq v(t), \quad t \geq t_0.$$

By the formula for the variation of parameters [28], we see that

$$v(t) = \mathbf{P}(t,t_0)v(t_0) + \int_{t_0}^t \mathbf{P}(t,s)(\lambda_2 v(s-\tau(s)) + \varepsilon)ds \tag{4.19}$$

where $\mathbf{P}(t,s)$, $t \geq s \geq t_0$ is the solution of the system

$$\dot{y}(t) = \lambda_1(t)y(t), \quad t \neq t_n,$$
$$y(t_n) = \mu y(t_n^-), \quad n \in \mathbb{N}.$$

It follows that

$$\mathbf{P}(t,s) = e^{\int_s^t \lambda_1(u)du}\mu^{N(t,s)}.$$

In view of Definition 4.1 and (4.14), we can derive that for $\forall t, s \geq t_0$,

$$P(t,s) \leq \eta e^{-\lambda_3(t-s)}, \tag{4.20}$$

where $\eta = e^{(\gamma+\beta)\tau_b}$ and $\lambda_3 = \gamma - (\ln\mu)/T_a$. Let $\mathcal{M} = \eta c \mathbb{E}\|\xi\|^2$. Then, by (4.18)–(4.20), we have

$$v(t) \le \mathcal{M}e^{-\lambda_3(t-t_0)} + \int_{t_0}^{t} \eta e^{-\lambda_3(t-s)}(\lambda_2 v(s-\tau(s)) + \varepsilon)ds. \tag{4.21}$$

Let $\varphi(\lambda) = \eta\lambda_2 e^{\lambda\tau} + \lambda - \lambda_3$. Obviously, $\varphi(+\infty) = +\infty$ and

$$\varphi'(\lambda) = 1 + \eta\lambda_2\tau e^{\lambda\tau} > 0.$$

Notice from (4.12) that

$$\eta\lambda_2 < \gamma - \frac{\ln\mu}{T_a} = \lambda_3. \tag{4.22}$$

Thus, $\varphi(0) = \eta\lambda_2 - \lambda_3 < 0$. Therefore, there exists a λ such that

$$\eta\lambda_2 e^{\lambda\tau} = \lambda_3 - \lambda.$$

The inequality $\eta = e^{(\gamma+\beta)\tau_b} > 1$ yields that

$$v(t) = c\mathbb{E}\|\xi\|^2 < \mathcal{M} = \eta c\mathbb{E}\|\xi\|^2.$$

Hence,

$$v(t) < \mathcal{M}e^{-\lambda(t-t_0)} + \frac{\eta\varepsilon}{\lambda_3 - \eta\lambda_2}, \quad t \in [t_0 - \tau, t_0].$$

In the following, we will establish that

$$v(t) < \mathcal{M}e^{-\lambda(t-t_0)} + \frac{\eta\varepsilon}{\lambda_3 - \eta\lambda_2}, \quad t > t_0. \tag{4.23}$$

Now, if (4.23) is not true, there exists a $t^* > t_0$ such that

$$v(t^*) \ge \mathcal{M}e^{-\lambda(t^*-t_0)} + \frac{\eta\varepsilon}{\lambda_3 - \eta\lambda_2} \tag{4.24}$$

and

$$v(t) < \mathcal{M}e^{-\lambda(t-t_0)} + \frac{\eta\varepsilon}{\lambda_3 - \eta\lambda_2}, \quad t < t^*. \tag{4.25}$$

Combining (4.21) with (4.25) yields

$$v(t^*) \le \mathcal{M}e^{-\lambda_3(t^*-t_0)} + \int_{t_0}^{t^*} \eta e^{-\lambda_3(t^*-s)}(\lambda_2 v(s-\tau(s)) + \varepsilon)ds$$

$$< e^{-\lambda_3(t^* - t_0)} \left\{ \mathcal{M} + \frac{\eta \varepsilon}{\lambda_3 - \eta \lambda_2} + \int_{t_0}^{t^*} \eta e^{\lambda_3(s - t_0)} \right.$$

$$\left. \times \left(\lambda_2 \mathcal{M} e^{-\lambda(s - \tau(s) - t_0)} + \frac{\eta \lambda_2 \varepsilon}{\lambda_3 - \eta \lambda_2} + \varepsilon \right) ds \right\}. \tag{4.26}$$

Taking into consideration that $\eta \lambda_2 e^{\lambda \tau} = \lambda_3 - \lambda$, we have

$$\int_{t_0}^{t^*} \eta e^{\lambda_3(s - t_0)} \left(\lambda_2 \mathcal{M} e^{-\lambda(s - \tau(s) - t_0)} + \frac{\eta \lambda_2 \varepsilon}{\lambda_3 - \eta \lambda_2} + \varepsilon \right) ds$$

$$\leq \int_{t_0}^{t^*} \eta e^{\lambda_3(s - t_0)} \left(\lambda_2 \mathcal{M} e^{\lambda \tau} e^{-\lambda(s - t_0)} + \frac{\lambda_3 \varepsilon}{\lambda_3 - \eta \lambda_2} \right) ds$$

$$\leq \int_0^{t^* - t_0} \left[\eta \lambda_2 e^{\lambda \tau} \mathcal{M} e^{(\lambda_3 - \lambda)s} + \frac{\eta \lambda_3 \varepsilon}{\lambda_3 - \eta \lambda_2} e^{\lambda_3 s} \right] ds$$

$$= \mathcal{M} e^{(\lambda_3 - \lambda)(t^* - t_0)} - \mathcal{M} + \frac{\eta \varepsilon}{\lambda_3 - \eta \lambda_2} e^{\lambda_3(t^* - t_0)} - \frac{\eta \varepsilon}{\lambda_3 - \eta \lambda_2}. \tag{4.27}$$

Therefore, it follows from (4.26) and (4.27) that

$$v(t^*) < \mathcal{M} e^{-\lambda(t^* - t_0)} + \frac{\eta \varepsilon}{\lambda_3 - \eta \lambda_2},$$

which contradicts (4.24), therefore (4.23) holds. Letting $\varepsilon \to 0$, we get from (4.23) that

$$v(t) \leq \mathcal{M} e^{-\lambda(t - t_0)}.$$

It follows that

$$\mathbb{E}\mathbf{V}(t, \sigma(t)) \leq v(t) \leq \mathcal{M} e^{-\lambda(t - t_0)}, \quad t \geq t_0.$$

This completes the proof. $\qquad\qquad\qquad\qquad\qquad\qquad\qquad\qquad\square$

Remark 4.6. Condition (4.14) is inspired by the assumption

$$(T_U(t, t_0))/(T_M(t, t_0)) \leq (\alpha - \gamma)/(\beta + \gamma)$$

in [22]. However, for $t > s \geq t_0$, $(T_U(t, s))/(_T M(t, s)) \leq (\alpha - \gamma)/(\beta + \gamma)$ is not true when $T_M(t, s) = 0$. In (4.14), $(\gamma + \beta)\tau_b$ is added to guarantee that (4.14) is still true when $T_M(t, s) = 0$. It is worth mentioning that (4.14) is easy to satisfy. For example, let the ratio of all active time of matched control $T_M(t, s)$ and all active time of unmatched control $T_U(t, s)$ be less than $(\alpha - \gamma)/(\beta + \gamma)$, and the maximum active time of multiagent systems with unmatched control be less than τ_b. Then (4.14) holds. A similar assumption can be found in [23].

In the following, Theorem 4.1 is presented to ensure the consensus of the stochastic delayed multiagent system with asynchronous switching in (4.2).

Theorem 4.1. *Under Assumption 4.1.1, if for any $r, l \in \Gamma, l \neq r$, there exist matrices $P_r > 0$, K_r and positive constants $\alpha, \beta, \upsilon_{1r}, \upsilon_{2r}, \gamma, \mu > 1$ such that*

$$2\mathbf{P}_r A_r + 2\mathbf{P}_r + 2\varrho_{1r}I - 2NB(i,j)U_l + \upsilon_{2r}\rho_{1r} - \beta\mathbf{P}_r < 0,$$
$$t \in T_U(t_{n+1}, t_n), \tag{4.28}$$
$$2\mathbf{P}_r A_r + 2\mathbf{P}_r + 2\varrho_{1r}I - 2NB(i,j)U_r + \upsilon_{2r}\rho_{1r} + \alpha\mathbf{P}_r < 0,$$
$$t \in T_M(t_{n+1}, t_n), \tag{4.29}$$
$$\mathbf{P}_r \leq \mu\mathbf{P}_l, \tag{4.30}$$
$$\upsilon_{1r}I \leq \mathbf{P}_r \leq \upsilon_{2r}I, \tag{4.31}$$
$$\mathbf{T}_a > \frac{\ln\mu}{\gamma - \eta\lambda_2} > 0 - \alpha\mathbf{T}_M(t,s) + \beta\mathbf{T}_{U(t,s)}$$
$$\leq -\gamma(t-s) + (\gamma + \beta)\tau_b, \quad t > s > t_0, \tag{4.32}$$

where $\lambda_2 = \max_{r\in\Gamma}\{\mu((2\varrho_{2r})/\upsilon_{1r} + \upsilon_{2r}/\upsilon_{1r}\rho_{2r})\}$, then the stochastic delayed multiagent system in (4.1) achieves consensus by feedback control gain $K_r = \mathbf{P}_r^{-1}U_r$.

Proof. Construct a Lyapunov function in the form

$$\mathbf{V}(t, \sigma(t)) = x^T(t)(U \otimes P_{\sigma(t)})x(t), \quad t \geq t_0 - \tau,$$

where $\sigma(t) = k_n \in \Gamma$ for $t \in [t_n, t_{n+1})$ and $\sigma(t) = \sigma(t_0)$ for $t \in [t_0 - \tau, t_0)$. Then, we get $\mathbb{E}\mathbf{V}(t, \sigma(t)) \leq c\mathbb{E}\|\xi\|^2$, $t_0 - \tau \leq t \leq t_0$, with $c = \max_r\{\lambda_{\max}(U \otimes \mathbf{P}_r), r \in \Gamma\}$. For simplicity, denote $\mathbf{f}_{k_n}(t, x(t), x(t - \tau(t)))$ and $\mathbf{g}_{k_n}(t, x(t), x_t)$ by $\mathbf{f}_{k_n}(t)$ and $\mathbf{g}_{k_n}(t)$, respectively. For $t \in [t_n, t_{n+1})$, $n \in \mathbb{N}$, we have

$$\mathbb{L}\mathbf{V}(t, k_n) = 2x^T(t)(U \otimes \mathbf{P}_{k_n})\big[(I_N \otimes A_{k_n})x(t) + \mathbf{f}_{k_n}(t)$$
$$+ (I_N \otimes K_{\sigma(t-h(t))})(B_{k_n} \otimes I_n)x(t)\big]$$
$$+ \text{trace}\{\mathbf{g}_{k_n}(t)^T(U \otimes \mathbf{g}_{k_n})(t)\}. \tag{4.33}$$

It is readily evident that

$$2x^T(t)(U \otimes \mathbf{P}_{k_n})[(I_N \otimes A_{k_n})x(t) + \mathbf{f}_{k_n})(t) + (I_N \otimes K_{k_n})(B_{k_n} \otimes I_n)x(t)]$$
$$= 2x^T(t)[U \otimes (\mathbf{P}_{k_n}A_{k_n})x(t) + (U \otimes \mathbf{P}_{k_n})\mathbf{f}_{k_n}(t) + (U \otimes \mathbf{P}_{k_n})(B_{k_n} \otimes K_{k_n})x(t)]$$

$$= 2x^T(t)[U \otimes (\mathbf{P}_{k_n} A_{k_n}) x(t) + (U \otimes \mathbf{P}_{k_n}) \mathbf{f}_{k_n}(t) + N B_{k_n} \otimes (\mathbf{P}_{k_n} K_{k_n}) x(t)].$$

(4.34)

Now, let $x_{ij}(t) = x_i(t) - x_j(t)$, $f_{k_n}^{ij}(t) = f_{k_n i}(t) - f_{k_n j}(t)$, and $g_{k_n}^{ij}(t) = g_{k_n i}(t) - g_{k_n j}(t)$. It follows that

$$
\begin{aligned}
\mathbb{L}\mathbf{V}(t, k_n) \leq 2 \sum_{1 \leq i < j \leq N} x_{ij}^T(t) \Big[& P_{k_n} A_{k_n} x_{ij}(t) + P_{k_n} f_{k_n}^{ij}(t) \\
& - N B(i,j) (P_{k_n} K_{\sigma(t-h(t))}) x_{ij} \Big] \\
& + \text{trace} \Big\{ \sum_{1 \leq i < j \leq N} (g_{k_n}^{ij}(t))^T P_{k_n} g_{k_n}^{ij}(t) \Big\}.
\end{aligned}
$$

(4.35)

It follows from Assumption 4.1.1, Young's inequality, and the results of [29] that

$$
\begin{aligned}
\sum_{1 \leq i < j \leq N} & x_{ij}^T(t) \mathbf{P}_{k_n} f_{k_n}^{ij}(t) \\
\leq & \sum_{1 \leq i < j \leq N} x_{ij}^T(t) \mathbf{P}_{k_n} x_{ij}(t) + \sum_{1 \leq i < j \leq N} (f_{k_n}^{ij}(t))^T f_{k_n}^{ij}(t) \\
\leq & \sum_{1 \leq i < j \leq N} x_{ij}^T(t) \mathbf{P}_{k_n} x_{ij}(t) + \sum_{1 \leq i < j \leq N} x_{ij}^T(t) \varrho_{1 k_n} x_{ij}(t) \\
& + \frac{\varrho_{2 k_n}}{\upsilon_{1 k_n}} \sum_{1 \leq i < j \leq N} x_{ij}^T(t - \tau(t)) \mathbf{P}_{k_n} x_{ij}(t - \tau(t))
\end{aligned}
$$

(4.36)

and

$$
\begin{aligned}
\text{trace} & \Big\{ \sum_{1 \leq i < j \leq N} (g_{k_n}^{ij}(t))^T \mathbf{P}_{k_n} g_{k_n}^{ij}(t) \Big\} \\
\leq & \sum_{1 \leq i < j \leq N} \upsilon_{2 k_n} \text{trace} \big\{ (g_{k_n}^{ij}(t))^T g_{k_n}^{ij}(t) \big\} \\
\leq & \sum_{1 \leq i < j \leq N} \Big[\upsilon_{2 k_n} \rho_{1 k_n} x_{ij}^T(t) x_{ij}(t) \\
& + \frac{\upsilon_{2 k_n}}{\upsilon_{1 k_n}} \rho_{2 k_n} x_{ij}^T(t - \tau(t)) \mathbf{P}_{k_n} x_{ij}(t - \tau(t)) \Big].
\end{aligned}
$$

(4.37)

Substituting (4.36) and (4.37) into (4.35) yields

$$
\mathbb{L}\mathbf{V}(t_{k_n}) \leq \sum_{1 \leq i < j \leq N} \Big\{ x_{ij}^T(t) (2 \mathbf{P}_{k_n} A_{k_n} + 2 P_{k_n}
$$

$$+ 2\varrho_{1k_n} - 2NB(i,j)\mathbf{P}_{k_n}K_{k_n} + \upsilon_{2k_n}\rho_{1k_n})x_{ij}(t)$$

$$+ \lambda_2 x_{ij}^T(t - \tau(t))\mathbf{P}_{\sigma(t-\tau(t))}x_{ij}(t - \tau(t))\Big\},$$

where $\lambda_2 = \max_{r\in\Gamma}\{\mu((2\varrho_{2r})/\upsilon_{1r} + \upsilon_{2r}/\upsilon1r\rho_{2r})\}$. Letting $U_r \triangleq P_r^{-1}K_r$, (4.28) and (4.29) yield

$$2\mathbf{P}_rA_r + 2\mathbf{P}_r + 2\varrho_{1r}I - 2NB(i,j)\mathbf{P}_rK_l$$
$$+ \upsilon_{2r}\rho_{1r} - \beta\mathbf{P}_r < 0, \quad t \in \mathbf{T}_U[t_{n+1}, t_n], \tag{4.38}$$
$$2P_rA_r + 2P_r + 2\varrho_{1r}I - 2NB(i,j)\mathbf{P}_rK_r$$
$$+ \upsilon_{2r}\rho_{1r} + \alpha P_r < 0, \quad t \in \mathbf{T}_M[t_{n+1}, t_n]. \tag{4.39}$$

For $t \in \mathbf{T}_U[t_{n+1}, t_n]$, it can be checked that

$$\mathbb{L}\mathbf{V}(t_{k_n}) \leq \sum_{1\leq i<j\leq N} \big[\beta x_{ij}^T(t)\mathbf{P}_rx_{ij}(t)$$
$$+ \lambda_2 x_{ij}^T(t - \tau(t))\mathbf{P}_rx_{ij}(t - \tau(t))\big]$$
$$\leq \beta\mathbf{V}(t, \sigma(t)) + \lambda_2\mathbb{E}\mathbf{V}(t - \tau(t), \sigma(t - \tau(t))).$$

Similarly, it can be derived that for $t \in T_M[t_{n+1}, t_n]$,

$$\mathbb{L}V(t_{k_n}) \leq -\alpha V(t, \sigma(t)) + \lambda_2 V(t - \tau(t), \sigma(t - \tau(t))).$$

Thus, the conditions in Lemma 4.3 hold, and so we have

$$\mathbb{E}\mathbf{V}(t, \sigma(t)) \leq \mathcal{M}e^{-\lambda(t-t_0)}, \quad t \geq t_0.$$

This follows that

$$\mathbb{E}\|x_i - x_j\|^2 \leq \sum_{1\leq i<j\leq N} \mathbb{E}\|x_i - x_j\|^2 = \mathbb{E}\mathbf{V}(t, \sigma(t)),$$

from which follows that $\mathbb{E}\|x_i - x_j\| \leq \sqrt{\mathcal{M}e^{-(\lambda/2)(t-t_0)}}$, $t \geq t_0$, $i,j = 1,2, \ldots, N$.

In view of Definition 4.1, the stochastic switched network with asynchronous switching in (4.2) is exponentially achieved consensus in mean-square. This completes the proof. $\qquad\square$

Remark 4.7. When the feedback control is synchronous with system switching, Theorem 4.1 becomes the consensus criterion of multiagent systems with synchronously switching control. In model (4.2), the switchings are assumed to exist not only in nodes but also in topology, and stochastic disturbances are also considered. Thus, the model (or the controller) here is more general than the models considered in [6]. On the other hand, the stability of switched stochastic systems without time-delay was investigated in [14], and the method used in [14] can hardly be used to examine switched stochastic systems with time-delays. Obviously, the presented comparison principle for stochastic switched systems presented in this paper can be utilized to study the stability of switched stochastic systems with time-delays. Here we investigate the consensus of large-scale multiagent systems with state feedback control, in which the control implementation is asynchronous.

Remark 4.8. Note that in the case of $b_{ij}^{\sigma(t)} = 0$ for $\sigma(t) \in \Gamma$, the stochastic delayed multiagent system with asynchronous switching in (4.2) are uncoupled, and the dynamics of each single node is independent of the other nodes. Hence, by means of Theorem 4.1, sufficient conditions can be obtained to guarantee the global exponential stability for each single node with asynchronous switching. The stability of switched systems with asynchronous switching was studied in [11]. However, in these works, time-delays and stochastic disturbances were not considered.

We can design a procedure for the results obtained in Theorem 4.1 as follows:

Step 1.   Input matrices A_k and B_k as well as constants ρ_{1k}, ρ_{2k}, ϱ_{1k}, μ, ϱ_{2k}, $k \in \Gamma$.

Step 2.   Solve the linear matrix inequality (LMI) conditions in (4.20)–(4.23) to obtain the constants α, β, υ_{11}, υ_{12}, υ_{21}, and υ_{22} and matrices U_r and P_r^{-1}. Get the feedback control gain $K_r = P_r^{-1} U_r$.

Step 3.   Check whether the switching rule satisfies conditions (4.24) and (4.25) or not.

Then the stochastic delayed multiagent system in (4.1) achieves consensus by the feedback control gain $K_r = P_r^{-1} U_r$.

In the following, two numerical examples are given to illustrate the results in the previous section.

4.2.1 Numerical example 4.1

Consider the following stochastic delayed multiagent system with four nodes:

$$dx_i(t) = \left[A_{\sigma(t)} x_i(t) + f_{\sigma(t)}(t, x_i(t), x_i(t - \tau(t))) \right.$$

$$+ K_{\sigma(t-h(t))} \sum_{j=1}^{N} b_{ij}^{\sigma(t)} x_j(t - \tau(t)) \Bigg] dt$$

$$+ g_{\sigma(t)}(t, x_i(t), x_i(t - \tau(t))) d\omega(t), \tag{4.40}$$

where $i = 1, 2, 3, 4, \sigma(t) \in \Gamma = \{1, 2\}$. Let

$$f_r(t, x_i(t), x_i(t - \tau(t))) = D_{1r}h(x_i(t)) + D_{2r}h(x_i(t - \tau(t))),$$

$$g_r(t, x_i(t), x_i(t - \tau(t))) = 0.3 \begin{bmatrix} x_i(t) & 0 \\ 0 & x_i(t - \tau(t)) \end{bmatrix},$$

and

$$A_1 = \begin{bmatrix} 1 & 0 \\ 0 & 1 \end{bmatrix}, \quad A_2 = \begin{bmatrix} 1.1 & 0 \\ 0 & 1.1 \end{bmatrix},$$

$$D_{11} = D_{12} = \begin{bmatrix} 2 & -0.1 \\ -5 & 3.2 \end{bmatrix},$$

$$D_{21} = D_{22} = \begin{bmatrix} -1.6 & -0.1 \\ -0.26 & -2.5 \end{bmatrix},$$

$$B_1 = B_2 = \begin{bmatrix} -0.36 & 0.12 & 0.12 & 0.12 \\ 0.24 & -0.72 & 0.24 & 0.24 \\ 0.12 & -.12 & -0.36 & 0.12 \\ 0.06 & 0.06 & 0.06 & -0.18 \end{bmatrix},$$

where $h(x_i(t)) = (\tanh(x_{i1}(t)), \tanh(x_{i2}(t)))$ and $\tau(t) = e^t/((1 + e^t))$. By using this set of parameters, the subsystem exhibits chaotic behaviors [24].

Step 1. Input constant $\mu = 1.002$ and matrices A_1, A_2, B_1, and B_2. It can be obtained that $\varrho_{11} = \varrho_{12} = 0.1419$, $\varrho_{21} = \varrho_{22} = 0.0618$, and $\rho_{1r} = \rho_{2r} = 0.09$.

Step 2. By solving LMIs (4.28)–(4.31), we get constants $\alpha = 1.5$, $\beta = 0.25$, $\upsilon_{21} = 0.6945$, $\upsilon_{22} = 0.7532$, $\upsilon_{11} = 0.2602$, and $\upsilon_{12} = 0.3134$, as well as matrices

$$K_1 = \begin{bmatrix} 49.4143 & 0 \\ 0 & 49.4143 \end{bmatrix} \quad \text{and} \quad K_2 = \begin{bmatrix} 97.5356 & 0 \\ 0 & 97.5356 \end{bmatrix}.$$

It can be derived that $\lambda_2 = 0.7152$.

Step 3. Suppose that the ratio of stochastic delayed multiagent systems with matched control to unmatched control is $3 : 1$. Set $\gamma = 1.1$ and

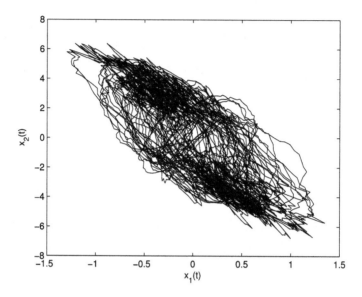

Figure 4.2 State trajectories of subsystem 1 of Example 4.1.

$\tau_b = 0.15$, which yields that $\eta = e^{1.15} = 1.3002$ and $T_a > (\ln \mu)/(\gamma - \eta\lambda_2) = 0.0154$. Conditions (4.24) and (4.25) hold. Thus, the stochastic delayed multiagent system in (4.40) achieves consensus.

Figs. 4.2 and 4.3 depict the state trajectories of respective subsystems. We set that the stochastic delayed multiagent system in (4.40) runs on each subsystem with period 0.6 s, and each subsystem runs with unmatched control in the first 0.15 s, then runs with matched control in the following 0.45 s. Fig. 4.4 shows state trajectories of x_{ij}, and Fig. 4.5 depicts the time response of consensus error $E(t) = (1/N) \sum_{i=1}^{n} \sqrt{\sum_{j=1}^{N} (x_{1i} - x_{ji})^2}$.

4.2.2 Numerical example 4.2

Consider the following stochastic delayed multiagent system with 20 nodes:

$$dx_i(t) = \left[A_{\sigma(t)} x_i(t) + f_{\sigma(t)}(t, x_i(t), x_i(t - \tau(t))) \right.$$

$$\left. + k_{\sigma(t)} K_{\sigma(t-h(t))} \sum_{j=1}^{N} b_{ij}^{\sigma(t)} x_j(t - \tau(t)) \right] dt$$

$$+ g_{\sigma(t)}(t, x_i(t), x_i(t - \tau(t))) d\omega(t) \tag{4.41}$$

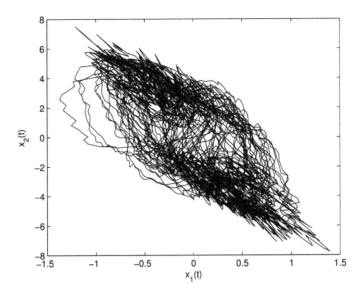

Figure 4.3 State trajectories of subsystem 2 of Example 4.1.

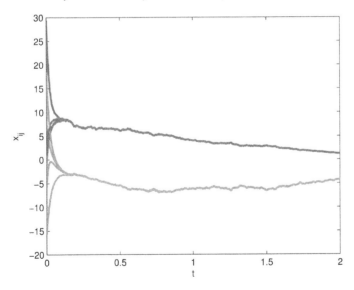

Figure 4.4 State trajectories of x_{ij} of Example 4.1.

where $i = 1, 2, \ldots, 20$, $\sigma(t) \in \Gamma = \{1, 2\}$, $k_1 = 1$, $k_2 = -1$, and

$$A_1 = -\begin{bmatrix} 3 & 0 \\ 0 & 2 \end{bmatrix}, \quad A_2 = -\begin{bmatrix} 0.5 & 0 \\ 0 & 1 \end{bmatrix},$$

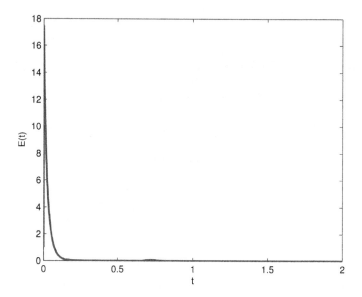

Figure 4.5 $E(t)$ of Example 4.1.

$$B_1 = B_2 = 0.12 \cdot \begin{bmatrix} -20 & 1 & 1 & \cdots & 1 \\ 1 & -20 & 1 & \cdots & 1 \\ \vdots & \vdots & \ddots & \ddots & \vdots \\ 1 & 1 & 1 & \ddots & 1 \\ 1 & 1 & 1 & \cdots & -20 \end{bmatrix}_{20 \times 20},$$

where $k_1 = 1$ and $k_2 = -1$ are coefficients, which can be absorbed into the coupling matrix to represent attractive and repulsive coupling, respectively. This model is slightly different from (4.2), and it is easy to get the corresponding conditions from Theorem 4.1. Other matrices are defined the same as in the example of Section 4.2.1.

Step 1. Input constant $\mu = 1.002$ and matrices A_1, A_2, B_1, and B_2. It can be obtained that $\varrho_{11} = \varrho_{12} = 0.1419$, $\varrho_{21} = \varrho_{22} = 0.0618$, and $\rho_{1r} = \rho_{2r} = 0.09$.

Step 2. By solving LMIs (4.28)–(4.31), we get constants $\alpha = 1.5$, $\beta = 0.25$, $\upsilon_{21} = 1.8448$, $\upsilon_{22} = 1.8450$, $\upsilon_{11} = 1.5339$, and $\upsilon_{12} = 1.5342$, and matrices

$$K_1 = \begin{bmatrix} 2.0835 & 0 \\ 0 & 2.7089 \end{bmatrix}, \quad \text{and} \quad K_2 = \begin{bmatrix} -4.1663 & 0 \\ 0 & -3.5409 \end{bmatrix}.$$

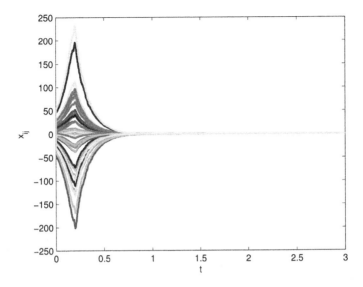

Figure 4.6 State trajectories of x_{ij} of Example 4.2.

It can be derived that $\lambda_2 = 0.1892$.

Step 3. Suppose that the ratio of stochastic delayed multiagent systems with matched control to unmatched control is $4:1$. Set $\gamma = 1.1$ and $\tau_b = 1$, which yields that $\eta = 5.7546$ and $T_a > (\ln \mu)/(\gamma - \eta\lambda_2) = 0.1784$. Conditions (4.24) and (4.25) hold. Thus, the stochastic delayed multiagent system in (4.41) achieves consensus.

Assume that the stochastic delayed multiagent system in (4.41) runs on each subsystem with period 1 s, and each subsystem runs with unmatched control in the first 0.2 s, then runs with matched control in the following 0.8 s. Fig. 4.6 depicts state trajectories of x_{ij}, and Fig. 4.7 shows the time response of consensus error $E(t) = (1/N) \sum_{i=1}^{n} \sqrt{\sum_{j=1}^{N} (x_{1i} - x_{ji})^2}$. The simulations indicate that the matched controller is beneficial to reduce consensus errors, while the unmatched controller enlarges the consensus errors.

4.3 H_∞ and H_2 consensus in directed networks

In recent years, distributed coordination control of multiagent systems has attracted increasing attention from various communities because of its broad applications in sensor networks [30], automated highway systems [31], air traffic control [32], and so on. As a fundamental and interesting issue, distributed consensus control has received particular attention when designing

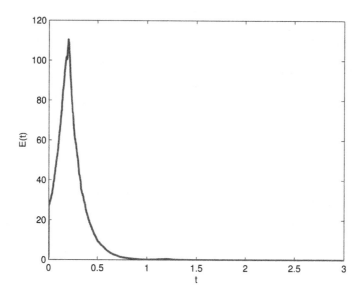

Figure 4.7 $E(t)$ of Example 4.2.

a distributed protocol based only on the local information exchange to enable the states of all agents to converge to the same values [33].

Theoretical results on distributed consensus with linear multiagent systems are presented including first-order linear dynamics [34,35], second-order linear dynamics [36,37], high-order linear dynamics [38,39], and general linear dynamics [40]. Note that intensive research on consensus problems with more practical nonlinear dynamics is still ongoing today. Consensus for higher-order multiagent systems with nonlinear dynamics and directed topologies was addressed in [41]. In [42], consensus tracking problems with group dispersion and cohesion behaviors were discussed for a group of Lagrange systems. Distributed attitude containment control for multiple rigid bodies was addressed in [43]. Li et al. [44] studied the global consensus problem of multiagent systems with Lur'e type dynamics.

In reality, the dynamics of agents are often subject to various external disturbances. The existence of external disturbances might deteriorate the system's performance or even destabilize the closed-loop multiagent systems. Thus, many researchers have focused on consensus seeking of multiagent systems with performance requirements. The synthesizing problem of a linear quadratic regulator was investigated in [45].

4.3.1 Introduction

The consensus problem for multiagent systems of first-order integrators with a directed topology and external disturbances was studied in [46] using H_∞ control theory. The robust H_∞ consensus problem for a class of multiagent systems with second-order dynamics was investigated in [47] with the consideration of parameter uncertainty and external disturbances. Furthermore, the H_∞ consensus performance problem for linear multiagent systems with undirected topologies was studied in [48–52]. To solve the distributed H_∞ and H_2 control of identical coupled linear systems, a decomposition approach was proposed by Massioni and Verhaegen [48]. In [49], a distributed state feedback protocol was proposed to ensure consensus of the multiagent systems with a prescribed H_∞ disturbance attenuation index and time delays. The H_∞ and H_2 performance regions, as an extension of the consensus region, were derived and analyzed in [50]. In [51], the H_∞ consensus control and H_2 robust control synthesized with transient performance problems were investigated for linear multiagent systems. Furthermore, a novel observer-type consensus protocol was provided using dynamic output feedback approach in [52] dealing with the H_∞ consensus performance problem. Wen et al. [53] studied the consensus problem with prescribed L_2-gain performance in networked multiagent systems with nonlinear dynamics subject to intermittent communications. The H_∞ global consensus problem for nonlinear multiagent systems with strongly connected balanced communication graphs was investigated in [54]. Until now, it is still unclear how to achieve H_∞ consensus in linear multiagent systems with strongly connected directed topologies without the balanced condition.

Motivated by the aforementioned results, we find that the disturbance rejection problems of linear multiagent systems with strongly connected directed communication graphs have not been solved by now. We also note that the robustness of a system to the worst case of external disturbances is indicated by the H_∞ norm, while the disturbance rejection performance of a system to the bounded external disturbances is shown by the H_2 norm. Thus, to sufficiently consider the disturbance rejection performance, both the H_∞ and H_2 consensus problems of linear multiagent systems under directed networks are studied in this chapter. To do this, a distributed consensus protocol is proposed, based only on the relative states of the neighboring agents. A two-step algorithm is further presented to construct a protocol to make the closed-loop multiagent systems reach H_∞

consensus for a strongly connected communication graph. For the H_2 consensus problem, another theorem and algorithm are then given to design a protocol which achieves consensus with a guaranteed H_2 performance index. Considering that neither the Laplacian matrix $\mathbb{L}$ nor matrix $\mathbb{L} + \mathbb{L}^T$ of a strongly connected directed graph is positive semidefinite, the similarity transformation employed in [50] is not applicable here. Therefore by constructing an appropriate Lyapunov function, some sufficient conditions for achieving consensus are provided and analyzed. Specifically, the H_∞ and H_2 consensus problems of a group of identical agents under controllers is converted to the feasibility of a low-dimensional linear matrix inequality (LMI), thereby reducing the computational complexity significantly. It is also worth noting that the H_∞ and H_2 consensus problems under consideration include that addressed in [50] as a special case.

4.3.2 Notations

Let $\mathbb{R}^{n \times n}$ be the set of $n \times n$ real matrices, $\mathbb{R}_+$ be the set of positive real numbers and I_N be the identity matrix of order N, respectively. Matrices, if not explicitly stated, are assumed to have compatible dimensions. Let $\mathbb{L}_2^r[0, \infty)$ represent the space of r-dimensional square-integrable vector functions over $[0, \infty)$. Notation $\| \cdot \|$ denotes L_2-norm of the corresponding function. Denote by $\mathbf{1}$ the column vector with all entries being one. Also, $\mathrm{diag}(A_1, A_2, \ldots, A_n)$ denotes a block-diagonal matrix with matrices A_i, $i = 1, 2, \ldots, n$, being its diagonal blocks. Moreover, for real symmetric matrices X and Y, $X > Y(X \geq Y)$ means that matrix $X - Y$ is positive (semi-) definite; $A \otimes B$ denotes the Kronecker product of matrices A and B. Finally, a matrix is Hurwitz if and only if all of its eigenvalues have negative real parts.

4.3.3 Graph theory

A directed graph $\mathbb{G}$ is a pair $(\mathbb{V}, \mathbb{E})$, where $\mathbb{V} = \{1, 2, \ldots, N\}$ is a set of nodes and $\mathbb{E} \subset \mathbb{V} \times \mathbb{V}$ is a set of edges. The associated adjacency matrix of $\mathbb{G} = (\mathbb{V}, \mathbb{E})$ is denoted by $\mathbb{A} = [a_{ij}] \in \mathbb{R}^{N \times N}$, where $_{ij} > 0$ if $(i, j) \in \mathbb{E}$; otherwise $a_{ij} = 0$. The Laplacian matrix $\mathbb{L} = [\ell_{ij}] \in \mathbb{R}^{N \times N}$, associated with adjacency matrix $\mathbb{A}$, is defined as $\ell_{ii} = \sum_{j \neq i}^n a_{ij}$, $\ell_{ij} = -a_{ij}$ for $i \neq j$. For a directed graph, the Laplacian matrix $\mathbb{L}$ has the following properties.

Lemma 4.4 ([35]). *Zero is an eigenvalue of $\mathbb{L}$ with $\mathbb{I}$ as the corresponding right eigenvector and all nonzero eigenvalues have positive real parts. Furthermore, zero is a simple eigenvalue of $\mathbb{L}$ if and only if the graph contains a directed spanning tree.*

Lemma 4.5 ([56]). *Suppose that* $\mathbb{G}$ *is strongly connected. Let* $\xi = [\xi_1, \xi_2, \ldots, \xi_N]^T$ *be the positive left eigenvector of* $\mathbb{L}$ *associated with zero eigenvalue. Then,* $\Xi\mathbb{L} + \mathbb{L}^T\Xi \geq 0$, *where* $\Xi = \mathrm{diag}(\xi_1, \xi_2, \ldots, \xi_N)$.

Lemma 4.6 ([55]). *For a strongly connected graph* $\mathbb{G}$ *with Laplacian matrix* $\mathbb{L}$, *define its generalized algebraic connectivity as*

$$a(\mathbb{L}) = \min_{\xi^T x = 0, x \neq 0} \frac{x^T(\Xi\mathbb{L} + \mathbb{L}^T\Xi)}{x^T \Xi x},$$

where ξ *and* Ξ *are defined as in Lemma 4.5. Then,* $a(\mathbb{L}) > 0$.

4.3.4 H_∞ consensus

Consider a network of N identical linear systems

$$\dot{x}_i = Ax_i + Bu_i + D\omega_i, \quad i = 1, 2, \ldots, N, \tag{4.42}$$

where $x_i \in \mathbb{R}^n$ and $u_i \in \mathbb{R}^p$ are the state and the control input of the ith agent, respectively; A, B, and D are constant matrices with compatible dimensions; $\omega_i \in \mathbf{L}_2^n[0, \infty)$ is the external disturbance. The directed communication topology graph $\mathbb{G}$ is assumed to be strongly connected.

In order to achieve consensus, the following protocol is adopted:

$$u_i = -cK \sum_{j=1}^{N} a_{ij}(x_i - x_j), \quad i = 1, 2, \ldots, N, \tag{4.43}$$

where $c > 0$ denotes the coupling strength, $K \in \mathbb{R}^{p \times n}$ is the feedback gain matrix to be designed, and $\mathbb{A} = [a_{ij}]_{N \times N}$ is the adjacency matrix associated with $\mathbb{G}$.

The objective is to find an appropriate protocol (4.43) for system (4.42) to reach consensus, meanwhile maintaining a desirable disturbance rejection performance. To this end, define the performance variables as $y_i = Cz_i$, where $y_i \in \mathbb{R}^m$; $z_i = x_i - \sum_{j=1}^{N} \xi_j x_j$, $i = 1, 2, \ldots, N$; $C \in \mathbb{R}^{m \times n}$ is a constant matrix; $\xi = [\xi_1, \xi_2, \ldots, \xi_N]^T > 0$, satisfying $\xi^T \mathbb{L} = 0$, $\sum_{j=1}^{N} \xi_j = 1$. Let $z = [z_1^T, z_2^T, \ldots, z_N^T]^T$ and $x = [x_1^T, x_2^T, \ldots, x_N^T]^T$. Then, one obtains

$$z = [(I_N - \mathbf{1}_N \xi^T) \otimes I_N]x. \tag{4.44}$$

By definition it is easy to check that 0 is a simple eigenvalue of $I_N - \mathbf{1}\xi^T$ with $\mathbb{I}$ as its right eigenvector, and 1 is the other eigenvalue with algebraic multiplicity $N - 1$. Thus, it follows from (4.44) that $z = 0$ if and only if $x_1 = x_2 = \cdots = x_N$.

The H_∞ consensus for (4.42) under the protocol (4.43) is now defined.

Definition 4.3. System (4.42) is said to reach H_∞ consensus, if the following conditions are satisfied:

1. System (4.42) with $\omega \equiv 0$ can achieve consensus in the sense that $\lim_{t \to \infty} \|x_i - x_j\| = 0$, $\forall i, j = 1, 2, \ldots, N$, for any given initial conditions.
2. Under the zero-initial conditions, the performance index satisfies $\|\mathbb{T}_{\omega y}\|_\infty < \gamma$ for $\omega \neq 0$, where $\|\mathbb{T}_{\omega y}\|_\infty$ is the H_∞ norm of $\mathbb{T}_{\omega y}$, defined as $\|\mathbb{T}_{\omega y}\|_\infty = \sup_{\omega \in \mathbb{R}} \sigma(\mathbb{T}_{\omega y}(j w))$.

According to the definition of z_i, $i = 1, 2, \ldots, N$, and the fact that $\xi^T \mathbb{L} = 0$, one obtains the following error dynamical systems:

$$\dot{z}_i = A z_i - cBK \sum_{j=1}^{N} \ell_{ij} x_j + D\left(\omega_i - \sum_{j=1}^{N} \xi_j \omega_j\right),$$

$$y_i = C z_i, \quad i = 1, 2, \ldots, N, \tag{4.45}$$

where $\mathbb{L} = [\ell_{ij}]_{N \times N}$ is the Laplacian matrix associated with $\mathbb{G}$.

Let $y = [y_1^T, y_2^T, \ldots, y_N^T]^T$. Then, the error dynamical systems can be rewritten as

$$\dot{z} = [I_N \otimes A - cL \otimes (BK)]z + [(I_N - \mathbf{1}_N \xi^T) \otimes D]\omega,$$

$$y = [I_N \otimes C]z. \tag{4.46}$$

Obviously, protocol (4.43) solves the H_∞ consensus problem if the system (4.46) is asymptotically stable and satisfies $\|\mathbb{T}_{\omega y}\|_\infty < \gamma$.

The following theorem presents a sufficient condition for achieving H_∞ consensus of (4.45).

Theorem 4.2. *Suppose that the communication topology is strongly connected. Then, system (4.42) with controller (4.43) achieves H_∞ consensus with disturbance attenuation $\gamma > 0$, if there exist a positive definite matrix $X > 0$ and a positive constant $\tau > 0$ such that*

$$\begin{bmatrix} XA^T + AX - \tau BB^T & XC^T & D \\ \bullet & -\frac{\xi_{\min}}{\xi_{\max}} I_N & 0 \\ \bullet & \bullet & -\gamma^2 \xi_{\min} I_N \end{bmatrix} < 0, \tag{4.47}$$

and the coupling strength $c \in S$, where $S = (c_{th}, +\infty)$, $c_{th} = [\tau \xi_{\max}]/[2a(\mathcal{L})]$, $\xi_{\min} = \min_i \xi_i$, $\xi_{\max} = \max_i \xi_i$, and $K = (1/\xi_{\max})B^T X^{-1}$.

Proof. Multiply both sides of (4.47) by $\mathrm{diag}([1/\sqrt{\xi_{max}}]X^{-1}, \sqrt{\xi_{max}}I_n,$ $[1/\sqrt{\xi_{max}}]I_n)$ and let $P = [1/\xi_{max}]X^{-1}$. Then, one obtains

$$\begin{bmatrix} A^T P + PA - \tau\xi_{max}PBB^T P & C^T & PD \\ \bullet & -\xi_{min}I_n & 0 \\ \bullet & \bullet & -\frac{\gamma^2\xi_{min}}{\xi_{max}}I_n \end{bmatrix} < 0. \qquad (4.48)$$

By Schur's complement lemma, the above inequality implies that

$$A^T P + PA - \tau\xi_{max}PBB^T P + \frac{1}{\xi_{min}}C^T C + \frac{\xi_{max}}{\gamma^2\xi_{min}}PDD^T P < 0. \qquad (4.49)$$

It then follows from (4.49) and $c > [\tau\xi_{max}]/[2a(\mathcal{L})]$ that

$$A^T P + PA - 2ca(\mathbb{L})PBB^T P + \frac{1}{\xi_{min}}C^T C + \frac{\xi_{max}}{\gamma^2\xi_{min}}PDD^T P < 0. \qquad (4.50)$$

Consider the Lyapunov function candidate

$$\mathbf{V} = \frac{1}{2}z^T[\Xi \otimes P]z$$

where $\Xi = \mathrm{diag}(\xi_1, \xi_2, \dots, \xi_N)$, $\xi = [\xi_1, \xi_2, \dots, \xi_N]^T$ is the positive left eigenvector of Laplacian matrix $\mathbb{L}$ associated with zero eigenvalue, satisfying $\xi^T\mathbf{1}_N = 1$.

Differentiating V with respect to t along the trajectories of (4.46) and substituting $K = B^T P$ gives

$$\begin{aligned}
\dot{\mathbf{V}} &= z^T\left[\Xi \otimes (PA) - (c\mathbb{L}) \otimes (PBB^T P)\right]z \\
&\quad + z^T[(\Xi(I_N - \mathbf{1}_N\xi^T)) \otimes (PD)]\omega \\
&= z^T\left[\Xi \otimes (PA) - \frac{c\mathbb{L} + \mathbb{L}^T\Xi}{2} \otimes (PBB^T P)\right]z \\
&\quad + z^T[(\Xi(I_N - \mathbf{1}_N\xi^T)) \otimes (PD)]\omega \\
&= z^T[\Xi \otimes (PA)]z \\
&\quad - cz^T\left[(I_N \otimes (PB))\left(\frac{c\Xi\mathbb{L} + \mathbb{L}^T\Xi}{2} \otimes I_N\right)(I_N \otimes (B^T P))\right]z \\
&\quad + z^T[(\Xi(I_N - \mathbf{1}_N\xi^T)) \otimes (PD)]\omega. \qquad (4.51)
\end{aligned}$$

Since $(\xi^T \otimes I_N)z = 0$, so that $(\xi^T \otimes I_n)[I_N \otimes (PB)]z = 0$, one obtains from Lemma 4.6 that

$$z^T\left[(I_N \otimes (PB))\left(\frac{c\Xi\mathbb{L} + \mathbb{L}^T\Xi}{2} \otimes I_N\right)(I_N \otimes (B^T P))\right]z$$

$$\geq a(\mathbb{L})z^T[(I_N \otimes (PB))(\Xi \otimes I_n)(I_N \otimes (B^T P))]z, \tag{4.52}$$

where $a(\mathbb{L}) > 0$. In light of (4.52), it follows from (4.51) that

$$\begin{aligned}
\dot{V} &\leq z^T[\Xi \otimes (PA)]z \\
&\quad - ca(\mathcal{L})z^T[(I_N \otimes (PB))(\Xi \otimes I_N)(I_N \otimes (B^T P))]z \\
&\quad + z^T[(\Xi(I_N - \mathbf{1}_N \xi^T)) \otimes (PD)]\omega.
\end{aligned}$$

In case of $\omega(t) \equiv 0$, it follows from the above analysis that

$$\dot{V} \leq z^T \left[\Xi \otimes \left(\frac{PA + A^T P}{2} - ca(\mathcal{L})PBB^T P \right) \right] z < 0.$$

This guarantees the achievement of the consensus in system (4.46) and therefore the first condition in Definition 4.3 is satisfied.

Next, we consider the H_∞ performance of the multiagent systems with $\omega(t) \neq 0$. Similar to the foregoing analysis, we obtain

$$\begin{aligned}
\dot{V} &\leq z^T[\Xi \otimes (PA - ca(\mathbb{L})PBB^T P)]z \\
&\quad + z^T[(\Xi(I_N - \mathbf{1}_N \xi^T)) \otimes (PD)]\omega \\
&\leq z^T[\Xi \otimes (PA - ca(\mathbb{L})PBB^T P)]z \\
&\quad + z^T[(\Xi(I_N - \mathbf{1}_N \xi^T)) \otimes (PD)]\omega \\
&\quad + \frac{1}{2\xi_{min}} z^T[\Xi \otimes (C^T C)]z \\
&\quad - \frac{1}{2}\gamma^T \gamma - \frac{\gamma^2}{2\xi_{max}} \omega^T[\Xi \otimes I_N]\omega + \frac{\gamma^2}{2}\omega^T \omega \\
&= \frac{1}{2} \begin{bmatrix} z \\ \omega \end{bmatrix}^T \Omega \begin{bmatrix} z \\ \omega \end{bmatrix} - \frac{1}{2}\gamma^T \gamma + \frac{\gamma^2}{2}\omega^T \omega, \tag{4.53}
\end{aligned}$$

where

$$\Omega = \begin{bmatrix} \Omega_1 & (\Xi(I_N - \mathbf{1}_N \xi^T)) \otimes (PD) \\ ((I_N - \mathbf{1}_N \xi^T)^T \Xi) \otimes D^T P & -\Xi \otimes (\frac{\gamma^2}{\xi_{max}} I_N) \end{bmatrix}, \tag{4.54}$$

with $\Omega_1 = \Xi \otimes (A^T P + PA - 2ca(\mathcal{L})PBB^T P + \frac{1}{\xi_{min}}C^T C)$. The case of $\Omega < 0$ is guaranteed if and only if the following inequalities hold:

$$-\Xi \otimes \left(\frac{\gamma^2}{\xi_{max}} I_N \right) < 0, \tag{4.55}$$

$$\Xi \otimes \left(A^T P + PA - 2ca(\mathbb{L})PBB^T P + \frac{1}{\xi_{min}}C^T C \right)$$
$$+ (\Xi(I_N - \mathbf{1}_N\xi^T)^2) \otimes \left(\frac{\xi_{max}}{\gamma^2}PDD^T P \right) < 0. \qquad (4.56)$$

It follows that

$$\Xi \otimes \left(A^T P + PA - 2ca(\mathbb{L})PBB^T P + \frac{1}{\xi_{min}}C^T C \right)$$
$$+ (\Xi(I_N - \mathbf{1}_N\xi^T)^2) \otimes \left(\frac{\xi_{max}}{\gamma^2}PDD^T P \right)$$
$$\leq \Xi \otimes \left(A^T P + PA - 2ca(\mathbb{L})PBB^T P + \frac{1}{\xi_{min}}C^T C \right)$$
$$+ (\max_i \lambda_i(\Xi(I_N - \mathbf{1}_N\xi^T)^2)I_N) \otimes \left(\frac{\xi_{max}}{\gamma^2}PDD^T P \right)$$
$$\leq \Xi \otimes \left(A^T P + PA - 2ca(\mathbb{L})PBB^T P + \frac{1}{\xi_{min}}C^T C \right)$$
$$+ (\max_i \lambda_i(\Xi(I_N - \mathbf{1}_N\xi^T)^2)\Xi) \otimes \left(\frac{\xi_{max}}{\gamma^2}\xi_{min}PDD^T P \right).$$

By the Gershgorin theorem, one has $\max_i \lambda_i(\Xi(I_N - \mathbf{1}_N\xi^T)^2) \leq 1$. Therefore, we have

$$\Xi \otimes \left(A^T P + PA - 2ca(\mathbb{L})PBB^T P + \frac{1}{\xi_{min}}C^T C \right)$$
$$+ (\max_i \lambda_i(\Xi(I_N - \mathbf{1}_N\xi^T)^2)\Xi) \otimes \left(\frac{\xi_{max}}{\gamma^2\xi_{min}}PDD^T P \right)$$
$$\leq \Xi \otimes \left(A^T P + PA - 2ca(\mathbb{L})PBB^T P + \frac{1}{\xi_{min}}C^T C \right)$$
$$+ \Xi \otimes \left(\frac{\xi_{max}}{\gamma^2\xi_{min}}PDD^T P \right). \qquad (4.57)$$

Invoking (4.50) it gives

$$\dot{V} + \frac{1}{2}\gamma^T\gamma - \frac{\gamma^2}{2}\omega^T\omega < 0, \qquad (4.58)$$

for all z and ω such that $|z|^2 + |\omega|^2 \neq 0$.

Integrating inequality (4.58) over infinite horizon yields

$$\|\gamma\|_2^2 - \gamma^2\|\omega\|_2^2 - z^T(0)[\Xi \otimes P]z(0) < 0.$$

Recalling that $z(0) = 0$, one has

$$\|\gamma\|_2^2 - \gamma^2\|\omega\|_2^2 < 0,$$

that is,

$$\|\mathbb{T}_{w\gamma}\|_\infty < \gamma. \qquad (4.59)$$

Thus, the second condition in Definition 4.3 is satisfied, which means that the H_∞ consensus problem is solved. The proof is completed. $\quad\square$

Remark 4.9. Note that contrary to that of an undirected communication graph, the Laplacian matrix $\mathbb{L}$ associated with a directed communication graph is not positive semidefinite. Especially for a strongly connected graph without balanced conditions which is of concern herein, matrix $\mathbb{L} + \mathbb{L}^T$ is not positive semidefinite either. Thus, this unfavorable feature does not facilitate in applying a similarity transformation employed in [49,54] here. Therefore, the key tool leading to Theorem 4.2 is relying on the properly designed Lyapunov function and the generalized algebraic connectivity associated with Laplacian matrix $\mathbb{L}$.

Remark 4.10. Theorem 4.2 converts the H_∞ consensus problem of the coupled N identical agents in (4.42) with protocol (4.43) to the feasibility problem of a low-dimensional LMI. The effects of the communication topology on the achievement of consensus are characterized by the generalized algebraic connectivity of the corresponding Laplacian matrix $\mathbb{L}$, thereby reducing the computational complexity significantly. Observe that the requirement for the communication topology in the present work is much relaxed.

Remark 4.11. By using Schur's complement lemma, it is not difficult to obtain that there exist a matrix $X > 0$ and a scalar $\tau > 0$ such that LMI (4.47) holds if and only if there exist a matrix X and a scalar τ such that

$$XA^T + AX - \tau BB^T + \frac{\xi_{\max}}{\xi_{\min}}XC^TCX + \frac{1}{\gamma^2\xi_{\min}}DD^T < 0. \qquad (4.60)$$

Since $(\xi_{\max}/\xi_{\min})XC^TCX \geq XC^TCX$ and $1/\xi_{\min} \geq 1$, it implies

$$XA^T + AX - \tau BB^T + XC^TCX + \frac{1}{\gamma^2}DD^T < 0. \qquad (4.61)$$

Note also that the LMI (4.61) is corresponding to the condition for the existence of a controller for an undirected topology in [50]. Thus, the H_∞ consensus problem of directed topology is much more complicated than that of the undirected topology as it is related to the positive left eigenvector of $\mathbb{L}$ associated with the zero eigenvalue.

Remark 4.12. It is interesting to compare Theorem 4.2 to the results in [47], where the authors also derived LMI-based conditions to the H_∞ consensus problem of a group of agents with second-order dynamics under directed networks containing a directed spanning tree. The basis of [47] and the earlier developed results relies on the Lyapunov stability analysis and algebraic graph theory. In [47], however, both the parameter uncertainty of the communication protocol and external disturbances were taken into consideration. It should be noted that the strongly connected directed topology is a special case of the directed topology containing a spanning tree. This makes the conditions derived here systematic and simpler than those of [47].

Next, an algorithm is proposed to construct protocol (4.43) for achieving H_∞ consensus.

Algorithm 4.1. For a given scalar $\gamma > 0$, protocol (4.43) can be constructed as follows:
1. Solve the LMI (4.47) to obtain a matrix $X > 0$ and a scalar $\tau > 0$. Then, choose $K = 1/\xi_{\max} B^T X^{-1}$.
2. Select the coupling strength $c > c_{th}$, where $c_{th} = [\tau \xi_{\max}]/[2a(\mathbb{L})]$.

4.4 H_2 consensus

In the last section, the H_∞ performance index indicates the robustness of a system to the worst case of external disturbances. Here, we examine the H_2 performance index as a criterion showing the disturbance rejection performance of a system to the bounded external disturbances. To appropriately consider the consensus control, the H_2 consensus problem for system (4.42) is discussed in this section.

The H_2 consensus for (4.42) under controller (4.43) is defined first.

Definition 4.4. System (4.42) is said to reach H_2 consensus, if the following conditions are satisfied:
1. Network (4.46) is stable when $\omega(t) \equiv 0$.

2. When the initial conditions are zero, that is, $z(0) = 0$, and the external disturbance is excited by a pulse signal, that is, $\omega(t) = \omega_0 \delta(t)$, one has $\mathbf{J}_2 < \beta$, where $\mathbf{J}_2$ is the H_2 performance index defined as

$$\mathbf{J}_2 = \sup_{\substack{\omega(t)=\omega_0\delta(t) \\ \|\omega_0\|\leq 1}} \|y\|^2. \tag{4.62}$$

The following theorem presents a sufficient condition for achieving H_2 consensus of (4.45).

Theorem 4.3. *Suppose that the communication topology is strongly connected. Then, system (4.42) with controller (4.43) achieves H_2 consensus with performance index satisfying $\mathbf{J}_2 < \beta$, if there exist positive matrices $Q > 0$, $Z > 0$, and a positive constant $\tilde{\tau} > 0$ such that*

$$\begin{bmatrix} QA^T + AQ - \tilde{\tau}BB^T & QC^T \\ \bullet & -\xi_{min}I_n \end{bmatrix} < 0, \tag{4.63}$$

$$\begin{bmatrix} -Z & D^T \\ \bullet & -Q \end{bmatrix} < 0, \tag{4.64}$$

$$\mathrm{trace}(Z) < \frac{\beta_2}{N}, \tag{4.65}$$

and the coupling strength $c \in \mathbf{S}$, where $\mathbf{S} = (\tilde{c}_{th}, +\infty)$, $\tilde{c}_{th} = \tilde{\tau}/[2a(\mathbb{L})]$ and $K = B^T Q^{-1}$.

Proof. Multiply both sides of (4.63) by $\mathrm{diag}(Q^{-1}, I_n)$ and let $P = Q^{-1}$. Then, one obtains

$$\begin{bmatrix} A^TP + PA - \tilde{\tau}PBB^TP & C^T \\ \bullet & -\xi_{min}I_n \end{bmatrix} < 0, \tag{4.66}$$

$$\begin{bmatrix} -Z & D^T \\ \bullet & -P^{-1} \end{bmatrix} < 0, \tag{4.67}$$

$$\mathrm{trace}(Z) < \frac{\beta_2}{N}. \tag{4.68}$$

By Schur's complement lemma, the above inequalities imply that

$$A^TP + PA - \tilde{\tau}PBB^TP + \frac{1}{\xi_{min}}C^TC < 0, \tag{4.69}$$

$$D^TPD < Z, \tag{4.70}$$

$$\text{trace}(Z) < \frac{\beta_2}{N}. \tag{4.71}$$

It follows from (4.69) and $c > [\tilde{\tau}]/[2a(\mathcal{L})]$ that

$$A^T P + PA - 2ca(\mathbb{L})PBB^T P + \frac{1}{\xi_{\min}} C^T C < 0. \tag{4.72}$$

Consider the Lyapunov function candidate

$$\mathbf{V}_2 = \frac{1}{2} z^T [\Xi \otimes P] z. \tag{4.73}$$

By following similar steps to those in the proof of Theorem 4.2, one can obtain the time derivative of V_2 along the trajectory of (4.46) as

$$\begin{aligned}
\dot{\mathbf{V}}_2 &\le z^T [\Xi \otimes (PA)] z \\
&\quad - ca(\mathbb{L}) z^T [(I_N \otimes (PB))(\Xi \otimes I_n)(I_N \otimes (B^T P))] z \\
&\quad + z^T [(\Xi (I_N - \mathbf{1}_N \xi^T)) \otimes (PD)] \omega.
\end{aligned} \tag{4.74}$$

By comparing Theorem 4.3 with Theorem 4.2, it follows from Theorem 4.2 that the first condition in Definition 4.3 is satisfied.

Next, to analyze the H_2 performance of the multiagent systems with $z(0) = 0$ and $\omega(t) = \omega_0 \delta(t)$, we obtain from (4.74) that

$$\begin{aligned}
\dot{\mathbf{V}}_2 &\le z^T [\Xi \otimes (PA)] z \\
&\quad - ca(\mathbb{L}) z^T [(I_N \otimes (PB))(\Xi \otimes I_n)(I_N \otimes (B^T P))] z \\
&\quad + z^T [(\Xi (I_N - \mathbf{1}_N \xi^T)) \otimes (PD)] \omega \\
&\quad + \frac{1}{2\xi_{\min}} z^T [\Xi \otimes (C^T C)] z - \frac{1}{2} \gamma^T \gamma.
\end{aligned}$$

Note that the equations $z(0) = 0$ and $\omega(t) = \omega_0 \delta(t)$ are equivalent to $z(0) = [(I_N - \mathbf{1}_N \xi^T) \otimes D] \omega_0$ and $\omega(t) = 0$. Invoking (4.69) gives that

$$\dot{\mathbf{V}}_2 + \frac{1}{2} \gamma^T \gamma < 0. \tag{4.75}$$

Integrating inequality (4.75) over infinite horizon yields

$$\begin{aligned}
\|\gamma\|_2^2 &< \omega_0^T [((I_N - \mathbf{1}_N \xi^T)^T \Xi (I_N - \mathbf{1}_N \xi^T)) \otimes (D^T PD)] \omega_0 \\
&\le \omega_0^T [(\max_i \lambda_i ((I_N - \mathbf{1}_N \xi^T)^T \Xi (I_N - \mathbf{1}_N \xi^T)) I_N) \otimes (D^T PD)] \omega_0. \tag{4.76}
\end{aligned}$$

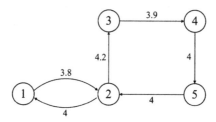

Figure 4.8 Communication topology.

Using the Gershgorin theorem, one has that $\max_i \lambda_i((I_N - \mathbf{1}_N \xi^T)^T \Xi (I_N - \mathbf{1}_N \xi^T)) \leq 1$. It follows that

$$\|\gamma\|_2^2 \leq \omega_0^T [I_N \otimes (D^T PD)] \omega_0. \tag{4.77}$$

Considering (4.70) and (4.71), we obtain from (4.77) that

$$\|\gamma\|_2^2 < \beta^2, \tag{4.78}$$

when $\|\omega_0\| \leq_1$. Therefore, the H_2 consensus problem is solved.   □

Next, an algorithm is presented to construct protocol (4.43) for achieving H_2 consensus.

Algorithm 4.2. For a given scalar $\beta > 0$, protocol (4.43) can be constructed as follows:

1. Solve the LMI (4.63)–(4.65) to obtain matrices $Q > 0$, $Z > 0$, and a scalar $\tilde{\tau} > 0$. Then, choose $K = B^T Q^{-1}$.
2. Select the coupling strength $c > \tilde{c}_{th}$, where $\tilde{c}_{th} = \tilde{\tau}/[2a(\mathbb{L})]$.

Remark 4.13. Theorem 4.3 and Algorithm 4.2 extend the results of H_2 consensus control of networked multiagents with undirected topology in [50] to the case where the communication topology is strongly connected. The properly designed Lyapunov function and the generalized algebraic connectivity associated with Laplacian matrix $\mathbb{L}$ still play an important role in the proof of Theorem 4.3 as mentioned in Remark 4.9.

In what follows, simulation examples are provided to illustrate the effectiveness of the theoretical results.

4.4.1 Numerical example 4.3

Consider a networked system consisting of five agents under a communication topology as shown in Fig. 4.8, where the weights (communication bandwidths) are indicated on the edges. The dynamics of the ith agent is described by (4.42), with

$$x_i = \begin{bmatrix} x_{i1} \\ x_{i2} \end{bmatrix}, \quad A = \begin{bmatrix} -0.4 & -19.5998 \\ 4.333 & 0.4 \end{bmatrix}, \quad B = \begin{bmatrix} 1.5 & 0.3 \\ 0 & 1 \end{bmatrix},$$

$$C = \begin{bmatrix} 1 & 0 \end{bmatrix}, \quad D = \begin{bmatrix} 1 \\ 1 \end{bmatrix}.$$

The external disturbance is $\omega = \begin{bmatrix} 2w, & 5w, & 4w, & w, & 1.5w \end{bmatrix}^T$, where $w(t)$ is a ten-period square wave starting at $t = 0$ with width 5 and height 1.

From Fig. 4.8, it is easy to see that the network $\mathbb{G}$ is strongly connected. Some simple calculations give that both the minimum and maximum values of vector ξ are equal to 0.4472. Then, one obtains that $a(\mathbb{L}) = 2.2244$. Choose the H_∞ performance index $\gamma = 1$. Solving LMI (4.47) by using the LMI toolbox of Matlab® gives a feasible solution

$$P = \begin{bmatrix} 0.21 & 0.08 \\ 0.08 & 0.03 \end{bmatrix}, \quad \tau = 7.52.$$

Thus, by Algorithm 4.1, the feedback gain matrix of (4.43) is given as

$$K = \begin{bmatrix} -23.52 & 57.73 \\ 33.78 & -83.06 \end{bmatrix}.$$

Then, according to Theorem 4.2 and Algorithm 4.1, protocol (4.43) with K chosen as above could solve H_∞ consensus with performance index $\gamma = 1$, if the coupling strength $c \geq 3.52$. For the case of $\omega = 0$, the state trajectories of the agents are respectively shown in Figs. 4.9 and 4.10, with initial conditions $x_1(0) = [1, 2]^T$, $x_2(0) = [4, -1]^T$, $x_3(0) = [5, 7]^T$, $x_4(0) = [-3, 6]^T$, and $x_5(0) = [-5, 3]^T$. It can be seen that the H_∞ consensus is indeed achieved. Furthermore, the trajectories of the performance variables y_i, $i = 1, 2, \ldots, 5$, in the presence of disturbances under the zero initial conditions are shown in Fig. 4.11.

4.4.2 Numerical example 4.4

In this example, the H_2 consensus will be taken into consideration. The agent dynamics are the same as in the numerical example in Section 4.4.1, and the communication graph is defined in Fig. 4.8. The external disturbance is Gaussian white noise with zero mean.

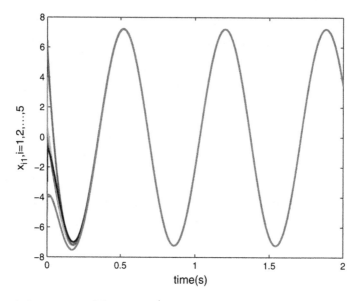

Figure 4.9 First states of five agents under controller (4.43) constructed via Algorithm 4.1.

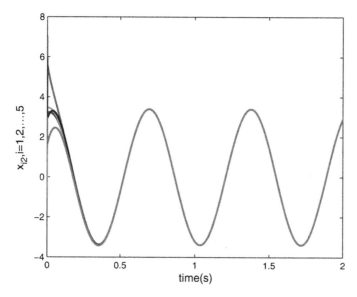

Figure 4.10 Second states of five agents under controller (4.43) constructed via Algorithm 4.1.

Choose the H_2 performance index $\beta = 1$. By Algorithm 4.2, the feedback gain matrix of (4.43) is given as

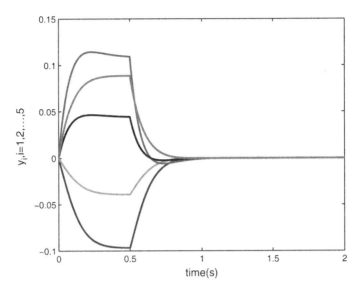

Figure 4.11 Performance variables under controller (4.43) constructed via Algorithm 4.1.

$$K_2 = \begin{bmatrix} -0.129 & -0.089 \\ -0.085 & -91.980 \end{bmatrix}.$$

According to Theorem 4.3 and Algorithm 4.2, protocol (4.43) with K_2 chosen as above could solve the H_2 consensus problem with performance index $\beta = 1$, if the coupling strength $c \geq 11.4$. For the case without disturbances, the state trajectories of the agents are respectively depicted in Figs. 4.12 and 4.13, with initial conditions $x_1(0) = [7, -2]^T$, $x_2(0) = [-4, 3]^T$, $x_3(0) = [8, 1]^T$, $x_4(0) = [3, -6]^T$, and $x_5(0) = [5, 2]^T$. It can be seen that the H_2 consensus is indeed achieved. With the zero-initial conditions and Gaussian white noise disturbance, the trajectories of the states are shown in Figs. 4.14 and 4.15.

4.5 Notes

In this chapter, we have shown that

1. The distributed exponential consensus of stochastic delayed multiagent systems with asynchronous switching can be dealt with. The model (or the controller) considered here is more general and encompasses some recently well-studied results of switched networks (multiagent systems) or networked control systems [6].

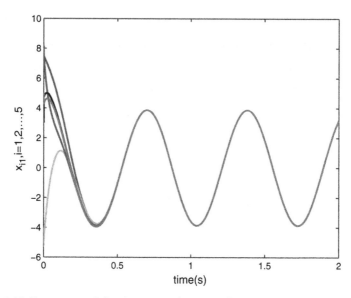

Figure 4.12 First states of five agents under controller (4.43) constructed via Algorithm 4.2.

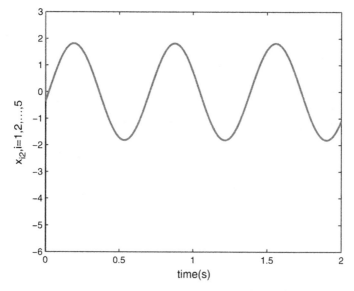

Figure 4.13 Second states of five agents under controller (4.43) constructed via Algorithm 4.2.

2. A new consensus criterion of stochastic delayed multiagent systems with asynchronous switching is obtained by assuming that the ratio

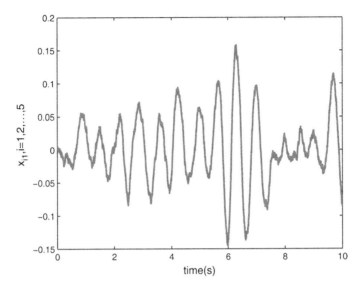

Figure 4.14 First states of five agents under controller (4.43) constructed via Algorithm 4.2 subject to Gaussian white noise disturbance.

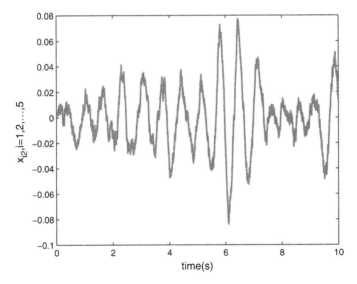

Figure 4.15 Second states of five agents under controller (4.43) constructed via Algorithm 4.2 subject to Gaussian white noise disturbance.

of the total time running on unmatched control and that of matched control is less than some upper bound.

3. The comparison principle for stochastic switched systems is first presented to deal with the consensus analysis of stochastic delayed multiagent systems with asynchronous switching, which makes the derivation easy to follow and understand.

4. The results developed here are not confined to multiagent systems and can be easily extended to networked control systems [11,13,14], which shows the applicability of our results to other kinds of control systems.

In addition, by using tools from Lyapunov stability analysis and algebraic graph theory, the H_∞ consensus problem of a group of linear multiagent systems with strongly connected directed communication graph has been studied in this chapter. To achieve consensus, a distributed consensus protocol based only on the relative states of the neighboring agents has been proposed and employed. Some sufficient conditions have been provided to achieve H_∞ consensus with a guaranteed H_∞ performance for a group of linear agents subject to disturbances. Another theorem and algorithm have been given to design a protocol which can achieve consensus with a guaranteed H_2 performance index. Future work will focus on the design of distributed output feedback control protocol for achieving H_∞ consensus problem of linear multiagent systems under communication delays with fixed or switching directed communication topologies.

References

[1] S. Seifzadeh, B. Khaleghi, F. Karray, Distributed soft-data-constrained multi-model particle filter, IEEE Trans. Cybern. 45 (3) (Mar. 2015) 384–394.

[2] T. Liu, D.J. Hill, J. Zhao, Synchronization of dynamical networks by network control, IEEE Trans. Autom. Control 57 (6) (Jun. 2012) 1574–1580.

[3] M. Li, Z. Li, A. Vasilakos, A survey on topology control in wireless sensor networks: taxonomy comparative study and open issues, Proc. IEEE 101 (12) (Dec. 2013) 2538–2557.

[4] M. Khan, H. Tembine, A. Vasilakos, Game dynamics and cost of learning in heterogeneous 4G networks, IEEE J. Sel. Areas Commun. 30 (1) (Jan. 2012) 198–213.

[5] Y. Tang, F. Qian, H. Gao, J. Kurths, Synchronization in complex networks and its application–a survey of recent advances and challenges, Annu. Rev. Control 38 (2) (2014) 184–198.

[6] J. Zhao, D.J. Hill, T. Liu, Synchronization of complex dynamical networks with switching topology: a switched system point of view, Automatica 45 (11) (2009) 2502–2511.

[7] C. Zhou, H. Chen, N. Xiong, X. Huang, A. Vasilakos, Model-driven development of reconfigurable protocol stack for networked control systems, IEEE Trans. Syst. Man Cybern., Part C, Appl. Rev. 42 (6) (Nov. 2012) 1439–1453.

[8] G. Shi, M. Johansson, K.H. Johansson, How agreement and disagreement evolve over random dynamic networks, IEEE J. Sel. Areas Commun. 31 (6) (Jun. 2013) 1061–1071.

[9] C. Maia, M. Goncalves, Application of switched adaptive system to load forecasting, Electr. Power Syst. Res. 78 (4) (2008) 721–727.

[10] W. Xing, Y. Zhao, H.R. Karimi, Convergence analysis on multi-AUV systems with leader–follower architecture, IEEE Access 5 (2017) 853–868.

[11] X. Zhao, P. Shi, L. Zhang, Asynchronously switched control of a class of slowly switched linear systems, Syst. Control Lett. 61 (12) (2012) 1151–1156.

[12] L. Zhang, N. Cui, M. Liu, Y. Zhao, Asynchronous filtering of discrete-time switched linear systems with average dwell time, IEEE Trans. Circuits Syst. I, Regul. Pap. 58 (5) (May 2011) 1109–1118.

[13] H. Lin, P.J. Antsaklis, Stability and stabilizability of switched linear systems: a survey of recent results, IEEE Trans. Autom. Control 54 (2) (Feb. 2009) 308–322.

[14] F. Wei, T. Jie, Z. Ping, Stability analysis of switched stochastic systems, Automatica 47 (1) (2011) 148–157.

[15] Y. Hong, J. Hu, L. Gao, Tracking control for multi-agent consensus with an active leader and variable topology, Automatica 42 (7) (2006) 1177–1182.

[16] Y. Wang, X. Sun, P. Shi, J. Zhao, Input-to-state stability of switched nonlinear systems with time delays under asynchronous switching, IEEE Trans. Cybern. 43 (6) (Dec. 2013) 2261–2265.

[17] A.R. Teel, A. Subbaraman, A. Sferlazza, Stability analysis for stochastic hybrid systems: a survey, Automatica 50 (10) (2014) 2435–2456.

[18] J. Hespanha, A.S. Morse, Stability of switched systems with average dwell time, in: Proc. the 38th IEEE Conf. Decision Control, vol. 3, 1999, pp. 2655–2660.

[19] Z.-H. Guan, Z. Liu, G. Feng, Y. Wang, Synchronization of complex dynamical networks with time-varying delays via impulsive distributed control, IEEE Trans. Circuits Syst. I, Regul. Pap. 57 (8) (Aug. 2010) 2182–2195.

[20] X. Mao, Stochastic Differential Equations and Applications, Elsevier, Chichester, UK, 2007.

[21] Z. Yang, D. Xu, Stability analysis and design of impulsive control systems with time delay, IEEE Trans. Autom. Control 52 (8) (Aug. 2007) 1448–1454.

[22] G. Zhai, B. Hu, K. Yasuda, A.N. Michel, Piecewise Lyapunov function for switched systems with average dwell time, Asian J. Control 2 (3) (2000) 192–197.

[23] M.A. Muller, D. Liberzon, Input/output-to-state stability and state-norm estimators for switched nonlinear systems, Automatica 48 (9) (2012) 2029–2039.

[24] H. Lu, Chaotic attractors in delayed neural networks, Phys. Lett. A 298 (Jun. 2002) 109–116.

[25] X. Ban, X. Gao, X. Huang, A. Vasilakos, Stability analysis of the simplest Takagi–Sugeno fuzzy control system using circle criterion, Inf. Sci. 177 (20) (2007) 4387–4409.

[26] S. Yin, Z. Huang, Performance monitoring for vehicle suspension system via fuzzy positivistic c-means clustering based on accelerometer measurements, IEEE/ASME Trans. Mechatron. 20 (5) (2015) 2613–2620.

[27] S. Yin, X. Zhu, O. Kaynak, Improved PLS focused on key-performance-indicator-related fault diagnosis, IEEE Trans. Ind. Electron. 62 (3) (Mar. 2015) 1651–1658.

[28] V. Lakshmikantham, D.D. Bainov, P.S. Simeonov, Theory of Impulsive Differential Equations, World Scientific, Singapore, 1989.

[29] Z.-H. Guan, D.J. Hill, X. Shen, On hybrid impulsive and switching systems and application to nonlinear control, IEEE Trans. Autom. Control 50 (7) (Jul. 2005) 1058–1062.

[30] J. Stankovic, T. Abdelzaher, C. Lu, S. Lui, J. Hou, Real-time communication and coordination in embedded sensor networks, Proc. IEEE 91 (7) (2003) 1002–1022.

[31] J. Bender, An overview of systems studies of automated highway systems, IEEE Trans. Veh. Technol. 40 (1) (1991) 82–99.

[32] C. Tomlin, G. Pappas, S. Sastry, Conflict resolution for air traffic management: a study in multi-agent hybrid systems, IEEE Trans. Veh. Technol. 43 (4) (1998) 509–521.

[33] A. Jadbabaie, A. Morse, Coordination of groups of autonomous mobile agents using nearest rules, IEEE Trans. Autom. Control 48 (6) (2003) 988–1001.

[34] R. Olfati-Saber, R. Murray, Consensus problems in networks of agents with switching topology and time-delays, IEEE Trans. Autom. Control 49 (9) (2004) 1520–1533.

[35] W. Ren, R. Beard, Consensus seeking in multiagent systems under dynamically changing interaction topologies, IEEE Trans. Autom. Control 50 (5) (2005) 655–661.

[36] W. Yu, G. Chen, M. Cao, Some necessary and sufficient conditions for second-order consensus in multi-agent dynamical systems, Automatica 46 (6) (2010) 1089–1095.

[37] W. Ren, On consensus algorithms for double-integrator dynamics, IEEE Trans. Autom. Control 53 (6) (2008) 1503–1509.

[38] W. Ren, K. Moore, Y. Chen, High-order and model reference consensus algorithms in cooperative control of multivehicle systems, J. Dyn. Syst. Meas. Control 129 (5) (2007) 678–688.

[39] W. Yu, G. Chen, W. Ren, J. Kurths, W. Zheng, Distributed higher-order consensus protocols in multi-agent dynamical systems, IEEE Trans. Circuits Syst. I, Regul. Pap. 58 (8) (2011) 1924–1932.

[40] G. Wen, Z. Duan, G. Chen, Distributed consensus of multi-agent systems with general linear node dynamics through intermittent communications, in: Proc. 24th Chinese Control and Decision Conf., 2012, pp. 1–5.

[41] G. Wen, Z. Duan, G. Chen, W. Yu, Consensus tracking of multi-agent systems with Lipschitz-type node dynamics and switching topologies, IEEE Trans. Circuits Syst. I, Regul. Pap. 61 (2) (2014) 499–511.

[42] Z. Meng, Z. Lin, W. Ren, Leader–follower swarm tracking for networked Lagrange systems, Syst. Control Lett. 61 (1) (2012) 117–126.

[43] Z. Meng, W. Ren, Z. You, Distributed finite-time attitude containment control for multiple rigid bodies, Automatica 46 (12) (2010) 2092–2099.

[44] Z. Li, Z. Duan, G. Chen, Global consensus regions of multi-agent systems with nonlinear dynamics, in: World Cong. Intelligent Control and Automation, 2010, pp. 4877–4882.

[45] V. Gupta, B. Hassibi, R. Murray, A sub-optimal algorithm to synthesize control laws for a network of dynamic agents, IEEE Trans. Autom. Control 78 (16) (2005) 1302–1313.

[46] P. Lin, Y. Jia, L. Li, Distributed robust H_∞ consensus control in directed networks of agents with time-delay, Syst. Control Lett. 57 (8) (2008) 643–653.

[47] P. Lin, Y. Jia, Robust H_∞ consensus analysis of a class of second-order multi-agent systems with uncertainty, IET Control Theory Appl. 4 (3) (2010) 487–498.

[48] P. Massioni, M. Verhaegen, Distributed control for identical dynamically coupled systems: a decomposition approach, IEEE Trans. Autom. Control 55 (1) (2009) 124–135.

[49] Y. Liu, Y. Jia, H_∞ consensus control for multi-agent systems with linear coupling dynamics and communication delays, Int. J. Syst. Sci. 43 (1) (2012) 50–62.

[50] Z. Li, Z. Duan, G. Chen, On H_∞ and H_2 performance regions of multi-agent systems, Automatica 57 (1) (2010) 213–224.

[51] J. Wang, Z. Duan, Y. Zhao, G. Qin, Y. Yan, H_∞ and H_2 control of multi-agent systems with transient performance improvement, Int. J. Control 86 (12) (2013) 2131–2145, https://doi.org/10.1080/00207179.2013.802371.

[52] Y. Zhao, Z. Duan, G. Wen, G. Chen, Distributed H_∞ consensus of multi-agent systems: a performance region-based approach, Int. J. Control 85 (3) (2012) 332–341.

[53] G. Wen, Z. Duan, Z. Li, G. Chen, Consensus and its L_2-gain performance of multi-agent systems with intermittent information transmissions, Int. J. Control 85 (4) (2012) 384–396.

[54] Z. Li, X. Liu, M. Fu, L. Xie, Global H_∞ consensus of multiagent systems with Lipschitz nonlinear dynamics, IET Control Theory Appl. 6 (13) (2012) 2041–2048.

[55] W. Yu, G. Chen, M. Cao, J. Kurths, Second-order consensus for multiagent systems with directed topologies and nonlinear dynamics, IEEE Trans. Syst. Man Cybern., Part B, Cybern. 40 (3) (2010) 881–891.

[56] Z. Qu, Cooperative Control of Dynamical Systems: Applications to Autonomous Vehicles, Springer-Verlag, London, 2009.

CHAPTER 5

Consensus over vulnerable networks

5.1 Introduction

Reaching agreement in a distributed network is a phenomenon strongly predicated on the health of the communication network on which information is exchanged among agents. In coordination control, the network is defined by a graph $\mathbb{G}(\mathbb{V}, \mathbb{E})$, which is mostly considered fixed or switching. In reality, the dynamics of the network is determined by some external factors often beyond the control of the agents participating in the coordination process. Network failures may result from faulty nodes, intruder attacks, or environment, thus making the network vulnerable. In fact, any system of systems interacting over a network can be considered vulnerable. In this regard, it is necessary to design networks which are robust to vulnerabilities such that the coordination objective is reached despite the loss of a node or link failures.

Definition 5.1 (*r*-reachable subset, [1]). In graph $\mathbb{G}(\mathbb{V}, \mathbb{E})$, a subset $\mathbb{S} \subset \mathbb{V}$ is *r*-reachable if there exists a node in $\mathbb{S}$ that has at least *r* neighbors in $\mathbb{N}_i \setminus \mathbb{S}$.

Definition 5.2 (*r*-robustness, [1]). A graph $\mathbb{G}(\mathbb{V}, \mathbb{E})$ is *r*-robust if for every pair of nonempty, disjoint subsets of $\mathbb{V}$, one of the subsets is *r*-reachable. The notion of *r*-robustness can be generalized as follows: Let $r \in \mathbb{Z}_+$, then define $\mathbb{X}_s^r \subset \mathbb{S}$ as a subset of nodes in $\mathbb{S}$, each of which has at least *r* neighbors outside of $\mathbb{S}$, that is,

$$\mathbb{X}_s^r = \{i \in \mathbb{S} : |\mathbb{N}_i \setminus \mathbb{S}| \geq r\}. \tag{5.1}$$

Definition 5.3 ((*r*, *s*)-robustness, [1]). A graph $\mathbb{G}(\mathbb{V}, \mathbb{E})$ is (*r*, *s*)-robust for $r, s \in \mathbb{Z}_+$ if for every pair of nonempty, disjoint subsets of $\mathbb{V}$, say $\mathbb{S}_1$ and $\mathbb{S}_2$, at least one of the following is true:
1. $\mathbb{X}_{s1}^r = |\mathbb{S}_1|$;
2. $\mathbb{X}_{s2}^r = |\mathbb{S}_2|$;
3. $\mathbb{X}_{s1}^r + \mathbb{X}_{s1}^r \geq |\mathbb{S}|$.

Definition 5.4 (*r*-robust graph, [12]). A digraph $\mathbb{G}$ is *r*-robust for if for every pair $\mathbb{A}$ and $\mathbb{B}$ of nonempty disjoint subsets of nodes, at least one of

Advanced Distributed Consensus for Multiagent Systems
https://doi.org/10.1016/B978-0-12-821186-1.00013-1

the subsets is r-reachable. In other words, $\forall A, B \subset V : A \cap B = \emptyset \implies A$ and/or B is r-reachable.

Lemma 5.1 ([12]). *For an (r, s)-robust graph G, the following conditions hold:*
- G *is (r', s')-robust where $0 \le r' \le r$ and $1 \le s' \le s$, and in particular, it is r-robust.*
- G *is $(r - 1, s + 1)$-robust.*
- G *is at least r-connected, but an r-connected graph is not necessarily r-robust.*
- G *has a directed spanning tree.*
- $r \le \lceil n/2 \rceil$. *Also, if G is a complete graph, then it is (r', s)-robust for all $0 < r' \le \lceil n/2 \rceil$ and $1 \le s \le n$.*

Definition 5.5 (*f-local model*, [12]). A graph G is f-local if for each normal agent i, the number of malicious nodes in its neighbor set N_i is at most f, that is, $|N_i \cap M| \le f$, $\forall i \in VM$.

Definition 5.6 (*f-total model*, [12]). A graph G is f-total malicious if the number n_m of faulty agents is at most f, that is, $n_m \le f$.

Definition 5.7 (Trusted nodes, [1]). The set of nodes $T \subset V$ which are insusceptible to removals are called trusted nodes. In practice, these nodes are reinforced to be resilient to failures or intruder attacks.

Definition 5.8 (Byzantine node, [11]). A node $i \in A$ is said to be Byzantine if it does not send the same value to all of its out neighbors at some time-step or if it applies some other update function $f_i'(t)$.

Definition 5.9 (Malicious node, [11]). A node $i \in A$ is said to be malicious if it sends the same value to all of its out neighbors at some time-step and it applies another update function $f_i'(t)$.

All malicious nodes are Byzantine, but not all Byzantine nodes are malicious [11].

Definition 5.10 (Vulnerable networks). A network is considered vulnerable if it can be attacked or can fail due to other external factors either permanently or for a short period of time.

Definition 5.11 (Crash adversary). An agent $k \in A$ is a crash adversary (or node) if there exists a $t_k \in \mathbb{R}_{\ge 0}$ selected by the adversary such that
- agent k behaves normally before $t = t_k$ according to its prescribed update rule, that is, $\dot{x}_k = f_{k, \sigma(t)}(t, x_N, x_{(A,k)})$ $\forall t < t_k$,
- agent k stops changing its state for all $t \ge t_k$, that is, $x_k(t) = x_k(t_k)$,

- agent k conveys the same state to each out neighbor, that is,

$$x_{(k,i)} = x_{(k,j)}, \quad i, j \in \mathbb{N}_k^{\text{out}}.$$

The notion of f-fraction local model was introduced in [3] to define the scope of threats for each node $i \in \mathbb{V}$. Contrary to the f-local model defined earlier, the f-fraction local model specifies an upper bound on the fraction of adversaries in each node's neighbors.

Definition 5.12 (f-fraction local set). A set $\mathbb{S} \subset \mathbb{V}$ is f-fraction local if it contains at most a fraction f of agents in the neighborhood of the other agents for all t, that is, $|\mathbb{N}_i^{\text{in}} \backslash \mathbb{S}| \leq \lfloor f |\mathbb{N}_i^{\text{in}}(t)| \rfloor \ \forall i \in \mathbb{S}, f \in [0, 1]$. The f-fraction local model defines the scenario where the set of adversaries is an f-fraction local set.

Definition 5.13 (p-fraction edge reachable set). Consider a nonempty digraph $\mathbb{D}$ and a nonempty subset $\mathbb{S}$ of nodes of $\mathbb{D}$. We say that $\mathbb{S}$ is a p-fraction edge reachable set if there exists $i \in \mathbb{S}$, such that $|\mathbb{N}_i^{\text{in}}| > 0$ and $|\mathbb{N}_i^{\text{in}} \backslash \mathbb{S}| \geq \lceil p|\mathbb{N}_i^{\text{in}}\rceil$ where $0 < p \leq 1$. If $|\mathbb{N}_i^{\text{in}} \backslash \mathbb{S}| > 0$ for all $i \in \mathbb{S}$, then $\mathbb{S}$ is 0-fraction reachable.

Definition 5.14 (p-fraction robustness). A nonempty, nontrivial digraph $\mathbb{D} = (\mathbb{V}, \mathbb{E})$ is p-fraction robust with $0 \leq p \leq 1$ if for every pair of nonempty, disjoint subsets of $\mathbb{V}$, at least one of the subsets is p-fraction edge reachable. If $\mathbb{D}$ is empty or trivial, then $\mathbb{D}$ is 0-fraction robust.

Definition 5.15 (Resilient consensus). A network of MASs under intruder or faulty conditions is said to be resilient provided the following conditions are satisfied:

- there exist $C \in \mathbb{R}$ such that $\lim_{t \to \infty} x_i[t] = C$;
- $x_i[t] \in \{x_i^{\min}, x_i^{\max}\} \ \forall i \in \mathbb{N}, t \in \mathbb{Z}_{\geq 0}$.

5.2 Discrete-time resilient consensus algorithms

Different resilient consensus algorithms have been proposed for an MAS to reach consensus in the presence of adversaries. Some of the techniques which have been proposed require the knowledge of some nonlocal information or completeness of the network [8]. Recently some resilient distributed algorithms were proposed based on local information, including W-MSSR [8], ARC-P [9], SW-MSSR [10], DP-MSSR [13], and MCA [11] algorithms.

5.2.1 Mean subsequence reduced (MSSR) algorithm

The mean subsequence reduced algorithm was originally introduced by [15] and is mathematically given as

$$F(N_k) = \text{mean}(Sel_\sigma(Red^t(N_k))). \tag{5.2}$$

It can be described sequentially as follows:

- The Red^t function denotes reduction which removes the t largest and t smallest elements from the multiset N_k at instant k, where t denotes the maximum number of faulty processes to be tolerated.
- The function Sel_σ selects a submultiset of σ elements from the reduced multiset $Red^t(N_k)$.
- If the algorithm is restricted such that Sel_σ produces a subsequence of $Red^t(N_k)$, then $F(N_k)$ is the mean of a subsequence of reduced multiset.

5.2.2 Weighted mean subsequence reduced (W-MSSR) algorithm

- At each time k, each normal node obtains the values of its neighbors and forms a sorted list.
- The F values strictly larger or smaller than its own value, $x_i[t]$, are removed.
- If $R[k]$ denotes the set of values that are removed, then the following update rule is applied to the system:

$$x_i[k+1] = \sum_{j \in \mathbb{N}_i[t] \setminus \mathbb{R}_i[k]} \omega_i[k] x_j^i[k], \tag{5.3}$$

where the weights form a convex combination at each time instant.

5.2.3 Median-based consensus algorithm (MCA)

This algorithm was proposed in [11] and it goes as follows:

- At each time t, each normal node obtains the values of its neighbors and forms a sorted list, $\mathbb{L}_i$.
- The node i computes the median $\tilde{x}_i[k]$ of $\mathbb{L}_i$, that is,

$$\tilde{x}_i[k] = \frac{\mathbb{L}_i\left(\left\lceil \frac{L+1}{2} \right\rceil\right) + \mathbb{L}_i\left(\left\lfloor \frac{L+1}{2} \right\rfloor\right)}{2}, \tag{5.4}$$

where $L = |\mathbb{L}_i|$ and $\mathbb{L}_i(j)$ denotes the jth element of $\mathbb{L}_i$.

- Each normal node i applies the update

$$x_i[k+1] = \omega_i[k]x_i[k] + \tilde{\omega}_i[k]\tilde{x}_i, \tag{5.5}$$

where $\omega_i + \tilde{\omega}_i = 1$ and there exists a constant such that $\omega_i, \tilde{\omega}_i > \alpha$.

5.2.4 Synchronous DP-MSSR algorithm

The double-integrator position-based mean squared subsequence reduced (DP-MSSR) algorithm was proposed in [12] based on the W-MSSR algorithm for a double-integrator MAS system. Consider an MAS with the following double-integrator dynamics:

$$\dot{r}_i(t) = v_i(t),$$
$$\dot{v}_i(t) = v_i(t), \tag{5.6}$$

where $r_i(t) \in \mathbb{R}$ and $v_i(t) \in \mathbb{R}$ are position and velocity values, respectively, and $u_i(t)$ is the control input of the ith agent. A discretized version of (5.6) over sampling period T is given by [12]:

$$r_i[k+1] = r_i[k] + Tv_i[k] + \frac{T^2}{2}u_i[k],$$
$$v_i[k+1] = v_i[k] + Tu_i[k]. \tag{5.7}$$

The following distributed consensus protocol (5.8) was proposed in [12]:

$$u_i[k] = -\sum_{j=1}^{n} a_{ij}[k][(r_i[k] - \delta_i) - (r_j[k] - \delta_j)] - \alpha v_i[k], \tag{5.8}$$

where $\delta_i \in \mathbb{R}$ is a constant scalar representing desired position of agent i in a formation, $\alpha > 0$ and $a_{ij}[k] \in \{0, 1\}$ is an entry of the adjacency matrix $\mathbb{A}$ defining the interaction topology. In [12], a formation consensus problem was considered with the objective that the agents reach a defined formation and then stop asymptotically, that is,

$$r_i[k] - r_j[k] \to \delta_{ij}, \quad v_i[k] \to 0 \quad \text{as } k \to \infty. \tag{5.9}$$

Definition 5.16 (Second-order resilient consensus, [12]). Considering the second-order MAS (5.7), under the f-local model, normal nodes achieve resilient consensus provided the following conditions are satisfied for any initial position values and zero initial velocities:

- (Agreement) There exists $c \in \mathbb{R}$ such that

$$\lim_{k \to \infty} \hat{r}_i[k] = c; \quad \lim_{k \to \infty} \hat{v}_i[k] = 0 \quad \forall i \in \mathbb{V}\backslash\mathbb{M}; \tag{5.10}$$

- (Safety)

$$\hat{r}_i[k] \in [r_{\min}, r_{\max}] \quad \forall i \in \mathbb{V} \ \mathbb{M}, \ \forall k \in \mathbb{Z}_+. \tag{5.11}$$

Applying the following state transformation $\hat{x}_i[k] = x_i[k] - \delta_i$, (5.7) can be rewritten as

$$\hat{x}[k+1] = \hat{x}[k] + Tv[k] + \frac{T^2}{2}u[k],$$

$$v[k+1] = v[k] + Tu[k], \tag{5.12}$$

and the control law $u[k]$ is given by

$$u[k+1] = -\mathbb{L}[k]\hat{x}[k] - \alpha v[k], \tag{5.13}$$

where $\hat{x}[k] = [\hat{x}_1[k], \ldots, \hat{x}_n[k]]^T$, $\hat{v}[k] = [\hat{v}_1[k], \ldots, \hat{v}_n[k]]^T$ and $\mathbb{L}[k]$ is the Laplacian of graph $\mathbb{G}[k]$.

For the MAS defined by (5.12) under protocol (5.13), the DP-MSSR algorithm is thereby formulated as follows:

- Each normal node i receives the position values of its neighbors at each time step k, and sorts them from largest to smallest.
- If there are fewer than f agents which have position values strictly larger or smaller than $r_i[k]$, then the normal node i ignores the incoming edges from those nodes, otherwise, it ignores the incoming edges from f agents which have position values larger or smaller than $r_i[k]$.
- Control input (5.8) is then applied by substituting $a_{ij}[k] = 0$ for ignored edges (j, i) in step 2.

Theorem 5.1 ([12]). *For an f-local model, suppose the network is $(2f + 1)$-robust, then the normal agents using the DP-MSSR algorithm with parameter $2f$ achieve resilient consensus.*

Theorem 5.2 ([12]). *For an f-local model, if the normal agents based on the DP-MSSR algorithm with parameter $2f$ achieve resilient consensus, then the network is $(f + 1)$ robust.*

5.2.5 Asynchronous DP-MSSR algorithm

The following distributed consensus algorithm (5.14) was proposed in [13] as an extension of the work in [12] considering partial asynchrony and delays:

$$u_i[k] = \sum_{j \in N_i[k]} a_{ij}[k](\hat{x}_j[k - \tau_{ij}] - \hat{x}_i[k]) - \alpha v_i[k], \tag{5.14}$$

where $\tau_{ij} \in \mathbb{Z}_+$ denotes delay in the edge (j, i) at time k satisfying

$$0 \le \tau_{ij}[k] \le \tau. \tag{5.15}$$

The following asynchronous DP-MSSR algorithm was proposed to achieve resilient consensus:

- At each time step k, each normal agent i decides whether to make an update or not.
- If it decides to do so, then it uses the relative position values of its neighbors $j \in N_i$ based on the most recent values in the form $\hat{x}_j[k - \tau_{ij}[k]] - \hat{x}_i[k]$ and then follows step 2 of the synchronous DP-MSSR algorithm based on these values. Afterward, it applies the control input (5.14) by substituting $a_{ij} = 0$ for edges (j, i) which are ignored in step 2 of DP-MSSR.
- Otherwise, it applies the control (5.14) where the first term of position values of its neighbors remains the same as the previous time step, and for the second term, its own current velocity is used.

Theorem 5.3. *Under the f-total malicious model, the network of agents with second-order dynamics using the control in (5.14) and the asynchronous DP-MSSR algorithm reaches resilient consensus only if the underlying graph is $(f + 1, f + 1)$-robust. Moreover, if the underlying graph is $(2f + 1)$-robust, then the resilient consensus is attained with a safety interval given by*

$$S_\tau = \left[\min z^N[0] + \min \left\{ 0, \left(T - \frac{\alpha T^2}{2} \right) v^N[0] \right\}, \right.$$
$$\left. \max z^N[0] + \max \left\{ 0, \left(T - \frac{\alpha T^2}{2} \right) v^N[0] \right\} \right]. \tag{5.16}$$

Proposition 5.1. *There exists a $(2f)$-robust network with the minimum degree $2f + 1$ under which normal agents may not achieve resilient consensus by asynchronous DP-MSSR algorithm.*

Corollary 5.1. *Under the f-local malicious model, the network of agents with the second-order dynamics using the control in (5.14) and the asynchronous DP-MSSR algorithm if the underlying graph is (2f + 1)-robust with safety interval (5.16).*

Definition 5.17 (Robust time varying graph, [13]). The time-varying graph $\mathbb{G}_0[k] = (\mathbb{V}, \mathbb{E}[k])$ is said to be jointly $(2f + 1)$-robust if there exists a fixed h such that the union of $\mathbb{G}_0[k]$ over each consecutive h steps is $(2f + 1)$-robust.

Corollary 5.2 (Robust time varying graph, [13]). *Under the f-total/f-local malicious model, the time-varying network $\mathbb{G}_0[k]$ of agents with second-order dynamics using control in (5.14) and the asynchronous DP-MSSR algorithm reaches resilient consensus if $\mathbb{G}_0[k]$ is jointly $(2f + 1)$-robust with the condition $h \leq \tau$ and safety interval given by (5.16).*

5.2.6 Heterogeneous HP-MSSR algorithm

The heterogeneous position-based mean–subsequence–reduced (HP-MSSR) algorithm was proposed in [14] based on the DP-MSSR algorithm. In this case, the MAS is modeled using heterogeneous agent dynamics, that is, one considers a network of MASs with heterogeneous double (5.17) and single (5.18) integrator dynamics:

$$\dot{x}_i(t) = v_i(t),$$
$$\dot{v}_i(t) = u_i(t), \quad i = 1, \ldots m, \tag{5.17}$$
$$\dot{x}_i(t) = u_i(t), \quad i = m + 1, \ldots, n, \tag{5.18}$$

where $x_i(t) \in \mathbb{R}$, $v_i(t) \in \mathbb{R}$, and $u_i(t) \in \mathbb{R}$ represent position, velocity, and control inputs of agent i, respectively.

Upon discretization over sampling period $T > 0$, the double integrator and single integrator dynamics becomes:

$$x_i[k + 1] = x_i[k] + Tv_i[k] + \frac{T^2}{2}u_i[k],$$
$$v_i[k + 1] = x_i[k] + Tu_i[k], \quad i = 1, \ldots, m, \tag{5.19}$$
$$x_i[k + 1] = x_i[k] + Tu_i[k], \quad i = m + 1, \ldots, n, \tag{5.20}$$

where $x_i[k] \in \mathbb{R}$, $v_i[k] \in \mathbb{R}$, and $u_i[k] \in \mathbb{R}$ represent position, velocity, and control inputs of agent i, respectively, at time $t = kT$. In [14], the following distributed control protocols were proposed:

$$u_i[k] = -\alpha v_i[k] + \sum_{j \in \mathbb{N}_i[k]} a_{ij}[k](x_j[k] - x_i[k]), \quad i = 1 \ldots, m, \tag{5.21}$$

$$u_i[k] = \sum_{j \in \mathbb{N}_i[k]} a_{ij}[k](x_j[k] - x_i[k]), \quad i = m+1, \ldots, n. \tag{5.22}$$

Definition 5.18 ([14]). For a fixed period $P, P \geq 0$, the switching graph $\mathbb{G}[k] = (\mathbb{V}, \mathbb{E}[k])$, is (P, r, s)-robust with trusted nodes if the union of graphs is (r, s)-robust with trusted nodes in the bounded time $[k - P, k]$, $k \geq P$, the union of graphs is denoted as $\mathbb{G}^P[k]$.

Hence, the HP-MSSR algorithm goes as follows [14]:

- At each time step k, each normal agent i receives full information from its neighboring agents during a bounded time interval $[k - P, k]$. Neighboring agents are the second- or first-order agents which contain different information. During the interval $[k - P, k]$, normal agents i may receive several values transmitted from the neighboring agent, e.g., agent j, at different time steps. The maximum position value is represented by $x_i[k_{ij}]$ from the received data, while the time stamp is denoted as $k_{ij} \in [k - P, k]$; $x_i[k_{ij}]$ is the standard value used for sorting, the neighboring set of agent i is denoted as $\mathbb{N}_i[k]$ at the time-step k.

- Neighboring agents are sorted by relocating their largest to smallest values based on the position values, $x_i[k_{ij}]$, of neighbors $j \in \mathbb{N}_i[k]$. If there are fewer than F agents with position values strictly larger than $x_i[k]$, agent i disregards all information from these agents, otherwise agent i removes the largest F value in the sorted list. Similarly, if there are fewer than F agents with position values strictly smaller than $x_i[k]$, agent i disregards all information from these agents, otherwise it removes F smallest value. The set of neighbors which are filtered by the remove operation is represented by $\mathbb{N}_i'[k]$.

- Let $\mathbb{T}_i[k]$ be the set of the trusted agents in the neighbors of agent i at the time-step k, that is, $\mathbb{T}_i[k] = \mathbb{N}_i[k] \cap \mathbb{T}$. Combine the trusted agents set $\mathbb{T}_i[k]$ and the neighbor set $\mathbb{N}_i'[k]$ which is given in step 2 to form a new set $\mathbb{R}_i[k] = \mathbb{T}_i[k] \cup \mathbb{N}_i'[k]$. Then the next time-step value of agent i is updated by its neighbors in $\mathbb{R}_i[k]$, including second- and first-order control laws defined in (5.21) and (5.22) with $a_{ij} = 0$ if the neighboring agent j is precluded in the removed set $\mathbb{R}_i[k]$.

Theorem 5.4 ([14]). *Consider a switching network modeled by $\mathbb{G}[k] = (\mathbb{V}, \mathbb{E}[k])$, each normal agent of which updates its value by using control laws (5.21) and (5.22) and HP-MSSR. Based upon the F-local attack model, the heterogeneous MAS achieves consensus if the switching network is $(P, 2F + 1)$-robust with trusted agents.*

Theorem 5.5 ([14]). *Consider a switching network modeled by* $\mathbb{G}(k) = (\mathbb{V}, \mathbb{E}[k])$, *each normal agent of which updates its value by using control laws (5.21) and (5.22) and HP-MSSR. Based upon the F-total attack model, the heterogeneous MAS achieves consensus if the switching network is* $(P, F+1, F+1)$-*robust with trusted agents.*

5.2.7 Resilient group consensus algorithm

The authors of [5] discussed extensively the resilient group consensus algorithm for an MAS over graphs with primary and secondary layers [6,7]. Consider a directed network $\mathbb{D} = (\mathbb{V}, \mathbb{E})$ consisting on n agents. Let $\mathbb{N}_i^+ = \{j \in \mathbb{V} : (j, i) \in \mathbb{E}\}$, $\mathbb{N}_i^- = \{j \in \mathbb{V} : (i, j) \in \mathbb{E}\}$, and $\mathbb{N}_i = \mathbb{N}_i^+ \cup \{i\}$ denote the set of neighbors, out-neighbors, and inclusive neighbors of node i. An MAS with the following update rule was studied in [5]:

$$x_i[k+1] = \sum_{j \in \mathbb{N}_i} w_{ij} x_j[k], \qquad (5.23)$$

where the weighting coefficient satisfy the following:

- $w_{ij} = 0$, wherever $j \notin \mathbb{N}_i$, $i \in \mathbb{V}$;
- $w_{ij} > 0$, $\forall j \in \mathbb{N}_i$, $i \in \mathbb{V}$;
- $\sum_{j=1}^{n} w_{ij} = 1$, $i \in \mathbb{V}$.

Definition 5.19 (Group consensus, [5]). Consider a digraph $\mathbb{D} = (\mathbb{V}, \mathbb{E})$ where $\mathbb{V}$ and $\mathbb{E}$ represent the set of nodes and edges, respectively. The MAS network (5.23) achieves group consensus if there exist K distinct constants c_l, $l = 1, \ldots, K$ and K distinct, nonoverlapping, nonempty sets $\mathbb{S}_l$ such that $\bigcup_{l=1}^{K} \mathbb{S}_l = \mathbb{V}$, $\mathbb{S}_l \cap \mathbb{S}_m = \emptyset$ for $l \neq m$ and $l, m = 1, \ldots, K$, and $\lim_{k \to \infty} x_i[k] = c_l$ holds for all $i \in \mathbb{S}_l$, $l = 1, \ldots, K$ and $x[0] \in \mathbb{R}^n$.

Definition 5.20 (Structured Byzantine (StrByz), [5]). A node $i \in \mathbb{F}$, where $\mathbb{F}$ is the set of faulty (or Byzantine) agents, is said to be StrByz (structured Byzantine) if it sends different information to its out neighbors while employing $|\mathbb{N}_i^-|$ different update rules starting with $|\mathbb{N}_i^-|$ different initial conditions.

Definition 5.21 (Unstructured Byzantine (uStrBYZ), [5]). A node $i \in \mathbb{F}$ is said to be uStrBYZ if it does not send the same information to its out neighbors at time step k and it applies an arbitrary function to update its state values.

Definition 5.22 (Primary layer subgraphs, [6,7]). Let $\mathbb{G} = (\mathbb{V}, \mathbb{E})$ be a digraph such that there exist l_1 disjoint subsets $\mathbb{V}_{l,s}$, $s = 1, \ldots, l_1$ in the vertex set $\mathbb{V}$ where each $\mathbb{V}_{l,s}$ is the largest possible subset that has a spanning tree for its subgraph $\mathbb{G}_{l,s}$ and for all $i \in \mathbb{V}_{l,s}$ and $j \in \mathbb{V}_{l,s}$, we have $(j,i) \notin \mathbb{E}$. Then, we can say that $\mathbb{G}_{1,s}$, $s = 1, \ldots, l_1$ are the primary layer subgraphs.

Definition 5.23 (Secondary layer subgraphs, [6,7]). Let $\bar{\mathbb{V}}$ be the set which consists of the vertices that are not in the primary layer disjoint subgraphs, that is, $\bar{\mathbb{V}} = \mathbb{V} \backslash \bigcup_{s=1}^{l_1} \mathbb{V}_{l,s}$. Then there exists l_2 disjoint subsets $\mathbb{V}_{2,s}$, $s = 1, \ldots, l_2$ in $\mathbb{V}$, called secondary layer subgraphs, such that $\mathbb{V}_{2,s}$ has a spanning tree for its subgraph and there exists exactly one node $i \in \mathbb{V}_{2,s}$ which satisfies the following:

- i is the root of a spanning tree in $\mathbb{V}_{2,s}$;
- For all $j \in \mathbb{V}_{2,s} \backslash i$ and $m \in \mathbb{V} \backslash \mathbb{V}_{2,s}$, we have $(m, j) \notin \mathbb{E}$;
- There exists at least two vertices p and v in two different subgraphs (either primary or secondary layer) such that $(p, i) \in \mathbb{E}$ and $(v, i) \in \mathbb{E}$.

Definition 5.24 (Approximate Byzantine group consensus, [5]). Consider a network represented by $\mathbb{G} = (\mathbb{V}, \mathbb{E})$, where $\mathbb{V} = \mathbb{V}_n \cup \mathbb{V}_f$, with $\mathbb{V}_n$ and $\mathbb{V}_f$ denoting the set of normal and faulty (or Byzantine) nodes, respectively. The network achieves approximate Byzantine group consensus if the following conditions are satisfied simultaneously:

- (k-group agreement) There exist k distinct constants α_l and k distinct nonoverlapping, nonempty sets $\mathbb{S}_l$ such that $\bigcup_{l=}^{k} \mathbb{S}_l = \mathbb{V}_n$, $\mathbb{S}_l \cap \mathbb{S}_m = \emptyset$ for $l \neq m$ and $l, m = 1, \ldots, k$ and for the set $\mathbb{S}_l$, $\lim_{k \in \infty} x_i[k] = \alpha_l$ holds for all $i \in \mathbb{S}_l \cap \mathbb{V}_n$ and $x[0] \subset \mathbb{R}^n$.
- (validity) For all $k > 0$, $m[0] \le m[k] \le M[k] \le M[0]$ where $M[k]$ and $m[k]$ denote the maximum and minimum state values of the set of nonfaulty nodes at time step k.

5.2.7.1 Layered/rooted mean-select-reduced (MSR) algorithm

The layered/rooted MSR algorithm was proposed for resilient group consensus under unstructured Byzantine faults in [5]. Consider the MAS (5.23) over a digraph $\mathbb{D} = (\mathbb{V}, \mathbb{E})$. The set of nodes $\mathbb{V}$ is defined as $\mathbb{V} = (\mathbb{V}_n, \mathbb{V}_f)$, where $\mathbb{V}_n$ and $\mathbb{V}_f$ denote the set of normal and faulty nodes, respectively. Suppose there are l_1 and l_2 primary and secondary layers in the digraph $\mathbb{D}$, and the set of unstructured Byzantine nodes is denoted by $\mathbb{F}_{m,s}$. Let $f_{1,s}$, $s = 1, 2, \ldots, l_1$ be the maximum number of unstructured Byzantine nodes in the sth primary layer, and likewise let $f_{2,s}$, $s = 1, 2, \ldots, l_1$ be the maximum number of unstructured Byzantine nodes in the sth secondary layer. The exact number of unstructured Byzantine nodes is assumed to be unknown. However, the there exists an upper bound $f_{m,s}$ of unstructured

Byzantine nodes in each layer, which is assumed to be known. The L-MSR algorithm for the set of nonfaulty agents is presented as follows [5]:

- Each nonfaulty agent $i \in \mathbb{V}_n$ sends its state value $x_i[k]$ to its neighbors and saves the values that it has received from its in-neighbors at time step k and creates a list $X_{\mathbb{N}_i^+}[k]$ of received values in ascending order.
- Each nonfaulty node i then computes the following parameter

$$\tilde{f}_i = \begin{cases} |\mathbb{N}_i^+|, & \text{if } |\mathbb{N}_i^+| \geq f_{m,s} + 3. \\ f_{m,s}, & \text{otherwise.} \end{cases} \tag{5.24}$$

- Each node $i \in \mathbb{N}$ eliminates the largest $\tilde{f}_i$ values in the sorted list that are greater than its current state value. If the number of greater values is less than f_i, it eliminates all of them. Similarly, it deletes the smallest $\tilde{f}_i$ values in the list that are less than its state value. If the number of smaller values is less than f_i, it eliminates all of them.
- The normal nodes then apply the following update rule for their states:

$$x_i[k+1] = \sum_{j \in (\mathbb{N}_i \cap \mathbb{V}_n) \backslash \mathbb{R}_i[k]} w_{ij}[k]x_j[k] + \sum_{j \in (\mathbb{N}_i \cap \mathbb{V}_f) \backslash \mathbb{R}_i[k]} w_{ij}[k]x_j[k], \tag{5.25}$$

where $\mathbb{R}_i[k]$ denotes the set of removed values at time instant k. The averaging coefficients are chosen such that the following conditions are satisfied:

- $w_{ij}[k] = 0$ whenever $j \notin \mathbb{N}_i \backslash \mathbb{R}_i[k]$, $k \in \mathbb{Z}_{\geq 0}$;
- $0 < \alpha \leq w_{ij}[k] \leq 1$ whenever $j \in \mathbb{N}_i \backslash \mathbb{R}_i[k]$, $k \in \mathbb{Z}_{\geq 0}$;
- $\sum_{j=1}^{n} w_{ij}[k] = 1$ whenever $i \in \mathbb{V}, j \in \mathbb{N}_i \backslash \mathbb{R}_i[k]$, $k \in \mathbb{Z}_{\geq 0}$.

By applying the following parameter definition in step 2 for $\tilde{f}_i$,

$$\tilde{f}_i = \begin{cases} |N_i^+|, & \text{if } i \in \mathbb{G}_{2,s} \text{ is the root node,} \\ f_{m,s}, & \text{otherwise,} \end{cases} \tag{5.26}$$

another algorithm, tagged rooted-MSR or R-MSR, is defined [5].

5.3 Continuous-time resilient consensus algorithms

In this section, we examine some algorithms for resilient consensus with agents exhibiting continuous-time dynamics.

Definition 5.25. Consider an MAS consisting of normal, $\mathbb{N}$, and adversarial, $\mathbb{H}$, agents. The normal agents are said to achieve continuous-time resilient asymptotic consensus (CTRAC) in the presence of adversarial agents if the following agreement and safety conditions are satisfied:

- $\exists x_\alpha \in \mathbb{R}$ such that $\lim_{t \to \infty} x_i(t) = x_\alpha \; \forall i \in \mathbb{N}$;
- $x_i(t) \in \mathbb{I}_0 = [x_{min}^N(0), x_{max}^N(0)] \; \forall t \in \mathbb{R}_{\geq 0}, \; i \in \mathbb{N}$, for any choice of initial values $x_N(0) \in \mathbb{R}^N$.

5.3.1 Adversarial robust consensus protocol (ARC-P)

An adversarial robust consensus protocol was proposed for an MAS with continuous-time single integrator dynamics [16]. Consider a multiagent system with single-integrator dynamics

$$\dot{x}_i(t) = u_i(t). \tag{5.27}$$

It is assumed that the MAS network features both cooperative and adversarial agents, that is, given a graph $\mathbb{G} = \{\mathbb{V}, \mathbb{E}\}$, the vertex set is partitioned as $\mathbb{V} = \mathbb{N} \cup \mathbb{W}$, where $\mathbb{N}$ is the set of normal agents and $\mathbb{W}$ is the set of adversarial agents. For the set of cooperative agents, the following agreement and safety condition needs to be satisfied:

$$\lim_{t \to \infty} |x_i(t) - x_j(t)| = 0 \quad \forall \, x_i \in \mathbb{X}_c, \tag{5.28}$$

where $\mathbb{X}_c$ is the set of the states of all cooperative nodes in the network. Equivalently, we can see that (5.28) implies that the states of all cooperative agents span the agreement subspace $\mathbb{A} = \{x \in \mathbb{R}^p | x_i = x_j \; \forall i, j \in \mathbb{N}\}$. The safety (or validity) condition ensures that the states of the agents remain bounded over a specified time interval, that is, letting $\mathbb{I} = [\min_{i \in \mathbb{N}} x_i(0) \; \max_{j \in \mathbb{N} x_j(0)}]$, then the safety condition is written as

$$\lim_{t \to \infty} x_i(t) \in \mathbb{I} \quad \forall \, i \in \mathbb{N}, \; x_i \in \mathbb{X}_c. \tag{5.29}$$

Now, consider a simple linear consensus protocol

$$u_i(t) = \sum_{j \in \mathbb{N}_i} (x_j(t) - x_i(t)). \tag{5.30}$$

A necessary condition for the robustness of the ARC-P algorithm is that there are more cooperative agents than adversaries. If the number of adversaries in the network is denoted by F, then the requirement is that the number of cooperative agents is $n > 2F$. Under this assumption, the algorithmic steps for the ARC-P algorithm are:

- Each cooperative agent sorts the state values and filters the F largest and smallest values so that the remaining values lie within the range of cooperative states.
- The state of each agent i is then subtracted from each of the remaining values to form $m = n - 2F$ relative state values. A relative state value is negative if the state of each agent i is greater than the filtered state value and nonnegative otherwise.
- The rate of change of the state of agent i is the sum of these m relative state values. As a result, the stage of each cooperative agent i increases (or decreases) whenever it is smaller or larger than the average of the n filtered values and remains constant if it is equal. Intuitively, this forces the cooperative agents to converge to the average of the filtered values.
- In the extreme instance where we have $n = 2F + 1$, only the median of the state values remains and the cooperative agents are driven towards the median of the state values.

The ARC-P algorithm is more precisely described in mathematical terms as follows:

- The concatenation function $\chi_{p,q} : \mathbb{R}^p \times \mathbb{R}^q \rightarrow \mathbb{R}^{p+q}$ is defined as

$$\chi_{p,q}(\gamma, z) = \begin{bmatrix} \gamma \\ z \end{bmatrix}. \tag{5.31}$$

- The (ascending) sorting function on n elements $\rho_n : \mathbb{R}^n \rightarrow \mathbb{R}^n$ is defined by $\eta = \rho_n(x)$ such that η is a permutation of x satisfying

$$\eta_1 \leq \eta_2 \leq \cdots \leq \eta_n. \tag{5.32}$$

- The reducing function with respect to $F \in \mathbb{Z}^+$ is defined by $r_F : \mathbb{R}^n \rightarrow \mathbb{R}^{n-2F}$, $n > 2F$, satisfying

$$r_F(\zeta) = [\zeta_{F+1}, \zeta_{F+2}, \ldots, \zeta_{n-F}]^T. \tag{5.33}$$

- The summation function $s : \mathbb{R}^m \rightarrow \mathbb{R}$ is defined by

$$s(x) = \sum_{j=1}^{m} x_j. \tag{5.34}$$

- The composition of the concatenation, sorting, reducing and summation functions is defined by $\phi : \mathbb{R}^p \times \mathbb{R}^q \rightarrow \mathbb{R}$, satisfying for all

$(y, z) \in \mathbb{R}^p \times \mathbb{R}^q$:

$$\phi(y, z) = (s \circ r_F \circ \rho \circ \chi_{p,q})(y, z). \tag{5.35}$$

Finally, the control protocol for the ith agent is written as

$$u_i(t) = -mx_i(t) + \phi(x_c(t), x_m(t)), \tag{5.36}$$

where $x_c(t)$ and $x_m(t)$ are state vectors of cooperative and malicious agents, respectively.

The authors of [3] show that f-fraction robustness is a necessity in the presence of crash adversaries, and $2f$-fraction robustness is sufficient in the presence of Byzantine adversaries, thus providing the following lemmas and theorems.

Lemma 5.2 ([3]). *Given a time-varying network where each normal node updates its value according to ARC-P with parameter $f \in [0, 1/2]$ under the f'-fraction local (Byzantine) model with $f' \leq f$, the safety condition of the CTRAC problem is ensured.*

Theorem 5.6 ([3]). *For a time-invariant network modeled by a digraph $\mathbb{D} = (\mathbb{V}, \mathbb{E})$ where each normal node updates its value according to ARC-P with parameter $f \in [0, 1/2]$ under the f-fraction local crash model, $\mathbb{D}$ is f-fraction robust if CTRAC is achieved.*

Theorem 5.7 ([3]). *For a time-invariant network modeled by a digraph $\mathbb{D} = (\mathbb{V}, \mathbb{E})$ where each normal node updates its value according to ARC-P with parameter $f \in [0, 1/2]$ under the f-fraction local Byzantine model, CTRAC is achieved if the network topology is p-fraction robust where $2f < p \leq 1$.*

Theorem 5.8 ([3]). *Consider a time-varying network modeled by $\mathbb{D}(t) = (\mathbb{V}, \mathbb{E}(t))$ under the f-fraction local Byzantine model. Let $\{t_k\}$ denote the switching times of $\sigma(t)$ and assume there exists $\tau \in \mathbb{R}_{\geq 0}$ such that $t_{k+1} - t_k \geq \tau$ for all k. Suppose each normal node updates itself based on ARC-P with parameter $f \in [0, 1/2]$. Then CTRAC is achieved if there exists $t' > 0$ such that $\mathbb{D}(t)$ is p-fraction robust where $2f < p < 1$, for all $t \geq t'$.*

5.3.2 k-connectivity-based robust algorithms

Some robust consensus algorithms have been proposed in the literature based on the notion of k-connectivity. Here we define the notion of k-connectivity and proceed to examine some robust algorithms based on this notion in consensus literature.

Definition 5.26 (k-connected graph, [2]). A graph $\mathbb{G}$ is said to be k-connected if its minimum cuts partition the nodes of the graph in two sets joined by at least k edges.

Definition 5.27 (k-safe network, [2]). A connected network $\mathbb{G}$ whose nodes are partitioned into the sets $\mathbb{V}_{safe}$ and $\mathbb{V}_{unsafe}$, where $\mathbb{V}_{safe}$ and $\mathbb{V}_{unsafe}$ represent cooperative and uncooperative nodes, respectively, is said to be k-safe if the corresponding subgraph $\mathbb{G}_{safe} \subset \mathbb{G}$ is k-connected and all nodes in the set $\mathbb{V}_{unsafe}$ are connected only to nodes in the set V_{safe}.

5.3.2.1 Problem statement: median value consensus

Consider a network of n MASs defined by the following single-integrator dynamics:

$$\dot{x}_i(t) = u_i(t), \quad x_i(0) = z_i(0), \quad i = 1, 2, \ldots, n, \tag{5.37}$$

where $x_i(t) \in \mathbb{R}$ is the state of the agent and $u_i(t)$ is a local control input to be designed. Let the set of initial conditions be

$$z = [z_1, z_2, \ldots, z_n] \tag{5.38}$$

such that $z_i \leq z_{i+1}$, $i = 1, 2, \ldots, n-1$. The median value between the agents' initial values, stored in the vector z, is

$$m(z) = \arg\min_{l \in \mathbb{R}} \sum_{i=1}^{n} |z_i - l|. \tag{5.39}$$

The median value $m(z)$ of vector z is such that

$$m(z) \begin{cases} = z_{\frac{n+1}{2}} & \text{if } n \text{ is odd,} \\ \in \left[z_{\frac{n}{2}}, z_{\frac{n}{2}+1} \right] & \text{if } n \text{ is even.} \end{cases} \tag{5.40}$$

The state variables $x_i(t) \in \mathbb{R}$, $i \in V$, of a networked MAS are said to reach a finite-time consensus if there exist $T > 0$ and $c(t) \in \mathbb{R}$ such that

$$x_i(t) = c(t) \quad \forall i \in V, \; t \geq T, \tag{5.41}$$

where $c(t)$ is referred to as a *consensus function*.

In [2], the following consensus protocol is defined for the MAS:

$$u_i(t) = -\alpha_i \text{sgn}(x_i(t) - z_i) - \sum_{j \in N_i} \lambda_{ij} \text{sgn}(x_i(t) - x_j(t)), \tag{5.42}$$

where $\alpha_i \in \mathbb{R}^+$, $i \in \mathbb{V}$ and $\lambda_{ij} \in \mathbb{R}^+$ are some tuning parameters, thereby resulting in:

$$x_i(t) = -\alpha_i \text{sgn}(x_i(t) - z_i) - \sum_{j \in \mathbb{N}_i} \lambda_{ij} \text{sgn}(x_i(t) - x_j(t)), \tag{5.43}$$

$$x_i(0) = z_i. \tag{5.44}$$

According to [2], the following constants dictate the speed of convergence and robustness

$$\lambda_{\max} = \max_{(i,j) \in \mathbb{E}} \lambda_{ij}, \quad \lambda_{\min} = \min_{(i,j) \in \mathbb{E}} \lambda_{ij}, \tag{5.45}$$

$$\alpha_{\max} = \max_{(i,j) \in \mathbb{E}} \alpha_i, \quad \alpha_{\min} = \min_{(i,j) \in \mathbb{E}} \alpha_i. \tag{5.46}$$

For vulnerable networks with cooperative and noncooperative nodes, the following equivalent constants are defined:

$$\lambda_{\max}^{\text{safe}} = \max_{(i,j) \in \mathbb{E}_{\text{safe}}} \lambda_{ij}, \quad \lambda_{\min}^{\text{safe}} = \min_{(i,j) \in \mathbb{E}_{\text{safe}}} \lambda_{ij}, \tag{5.47}$$

$$\lambda_{\max}^{\text{unsafe}} = \max_{(i,j) \in \mathbb{E} \setminus \mathbb{E}_{\text{safe}}} \lambda_{ij}, \quad \lambda_{\min}^{\text{unsafe}} = \min_{(i,j) \in \mathbb{E} \setminus \mathbb{E}_{\text{safe}}} \lambda_{ij}. \tag{5.48}$$

Theorem 5.9. *Consider an undirected k-safe network $\mathbb{G}$ of single-integrator agents with $\bar{\delta}$ maximal number of uncooperative agents whose arbitrary trajectories cross trajectories with the cooperative agents at isolated instants of time. The consensus protocol (5.42) with tuning parameters satisfying:*

$$\lambda_{\min}^{\text{safe}} > \lambda_{\max}^{\text{unsafe}}, \tag{5.49}$$

$$\lambda_{ij} = \lambda_{ji} > 0 \quad \forall (i,j) \in \mathbb{E}, \tag{5.50}$$

$$\alpha_{\max} < 2 \frac{k \lambda_{\min}^{\text{safe}} - \Delta_{\max} \bar{\delta} \lambda_{\max}^{\text{unsafe}}}{n - \bar{\delta}}, \tag{5.51}$$

$$\frac{\lambda_{\min}^{\text{safe}}}{\lambda_{\min}^{\text{unsafe}}} > \frac{\Delta_{\max} \bar{\delta}}{k}, \tag{5.52}$$

where $\Delta_{\max}$ is the maximum degree of nodes in the set $\mathbb{V}_{\text{unsafe}}$, solves the median finite consensus problem in finite time $T > 0$ such that

$$T \leq T_1 - \frac{\max_{i \in \mathbb{V}_c} x_i(0) - \min_{i \in \mathbb{V}_c} x_i(0)}{\beta^2}, \tag{5.53}$$

$$\beta^2 = 2 \left(\frac{k \lambda_{\min}^{\text{safe}} - \Delta_{\max} \bar{\delta} \lambda_{\max}^{\text{unsafe}}}{n - \bar{\delta}} - \alpha_{\max} \right). \tag{5.54}$$

Theorem 5.10. *Consider an undirected k-safe network* $\mathbb{G}$ *of single integrator agents with consensus protocol* (5.42) *with tuning parameters such that*

$$\lambda_{ij} = \lambda_{ji} > 0 \quad \forall (i, j) \in \mathbb{E}, \tag{5.55}$$

$$\frac{2(\Delta_{max}\bar{\delta}\lambda_{max}^{unsafe})}{n - \bar{\delta}} < \alpha_{max}, \tag{5.56}$$

$$\frac{\alpha_{max} - \alpha_{min}}{\alpha_{min}} < \frac{1}{n}, \tag{5.57}$$

where $\bar{\delta}$ *is the maximal number of cooperative agents. The MAS solves the finite-time median consensus problem provided that the above conditions are satisfied and there exists a finite time* $\bar{T} > 0$ *such that the consensus function is bounded:*

$$c(t) \in [z_{min}^{c}, z_{max}^{c}] \quad \forall t \le T + \bar{T}, \tag{5.58}$$

where

$$z_{max}^{c} = \max_{i \in \mathbb{V}_{c}} z_{i}, \quad z_{min}^{c} = \min_{i \in \mathbb{V}_{c}} z_{i}, \tag{5.59}$$

and

$$\bar{T} \le \frac{\left| c(T) - \dfrac{z_{max}^{c} + z_{min}^{c}}{2} \right| - \dfrac{z_{max}^{c} - z_{min}^{c}}{2}}{\beta^2}, \tag{5.60}$$

$$\beta^2 = \frac{\alpha_{max}}{2} - \lambda_{max}^{unsafe} \frac{\Delta_{max}\bar{\delta}}{n - \bar{\delta}}. \tag{5.61}$$

5.3.3 Absolute weighted MSR (AW-MSR) algorithm

The absolute weighted MSR algorithm was proposed in [4] for bipartite consensus in MAS. Consider the following single-integrator MAS:

$$\dot{x}_i(t) = u_i(t). \tag{5.62}$$

The normal nodes $i \in \mathbb{N}$ are updated here using a distributed consensus algorithm which is designed based on antagonistic interactions:

$$u_i(t) = \sum_{k \in \mathbb{N}_i} |a_{ik}(t)|((\text{sgn}(a_{ik})(t))x_k(t) - x_i(t)), \quad i \in \mathbb{I}. \tag{5.63}$$

Definition 5.28 (Resilient bipartite consensus). An MAS defined by (2.9) reaches resilient bipartite consensus under F-total or F-local model if all the normal nodes reaches bipartite consensus, that is,

$$\lim_{t \to +\infty} |x_i(t)| = x^* > 0, \quad i \in I, x^* \in \mathbb{R}, \tag{5.64}$$

for any choice of initial values.

The algorithmic steps of the AW-MSR are as follows:
- Each node obtains values from neighbors according to (5.63) and forms a sorted list.
- Each node removes nodes whose values are strictly larger or smaller than its own value $x_i(t)$. If there are fewer than F values strictly larger (or smaller) than its own value, then all values strictly larger or smaller are removed. Otherwise, it removes precisely the largest(or smallest) F values.
- Each normal node updates according to the new decision rule

$$\dot{x}_i = \sum_{k \in N_i \setminus \mathbb{R}_i} |a_{ik}(t)|(\mathrm{sgn}(a_{ik}(t))x_k(t) - x_i(t)), \quad i \in \mathbb{I}, \tag{5.65}$$

where $\mathbb{R}_i$ stands for the set of removed nodes.
The AW-MSR algorithm inherently induces switching, although the network topology is fixed. However, for a signed graph, the switching topology phenomena may change the connectivity properties of the network and thus bipartite consensus may not be achieved unless some sufficient robustness conditions are satisfied.

5.3.4 Secure linear consensus protocols

Consider an MAS with the following general linear dynamics:

$$\dot{x}_i(t) = Ax_i(t) + Bu_i(t), \tag{5.66}$$

where $x_i(t)$ and $u_i(t)$ are state and control inputs; A and B are matrices of appropriate dimensions. Some results were presented for reaching consensus over networks with denial-of-service attacks in [17].

Definition 5.29 (Denial-of-service attacks). Denial-of-service attacks are network attacks that render some components of the network inaccessible. In the context of MAS, DoS attacks render communication channels between agents unavailable for short periods of time. In some cases, the nature of the attack may be distributed, thus rendering more than one channel inaccessible at any given time, a phenomenon defined as *distributed denial-of-service attack, DDoS.*

Consider the following distributed consensus protocol:

$$u_i(t) = K \sum_{j \in \mathbb{N}_i \setminus \Gamma(t)}^{N} a_{ij}(x_j(t) - x_i(t)), \tag{5.67}$$

where K represents a feedback gain to be designed, $\Gamma(t)$ represents the attack modes of the communication graph.

Assumption 5.3.1 (DoS duration, [17]). There exist positive scalars ζ_{ij} and $\mu_{ij} < 1$ such that

$$|\mathbb{D}_{(i,j)}| < \eta_{ij} + \mu_{ij}(t - s) \tag{5.68}$$

where $\mathbb{D}_{(i,j)}$ is the union of DoS intervals over channel $(i,j) \in \mathbb{E}$ over $[s, t)$, $i < j$, and μ_{ij} is the attack intensity for edge (i, j).

Theorem 5.11 ([17]). *For a connected undirected graph* $\mathbb{G}$, *with agents defined by* (5.66), *protocol* (5.67) *solves the secure consensus problem given positive scalars* α_Γ *if there exist matrices* X *and* R *such that*

$$I_{N-1} \otimes [(XA + A^T X) - \alpha_\Gamma X] - 2(\Lambda - \Lambda_\Gamma) \otimes BRB^T < 0 \tag{5.69}$$

where $\Gamma = \text{diag}\{\lambda_2, \ldots, \lambda_N\}$, $\Lambda_\Gamma = v^T \mathbb{L}_\Gamma v$, $\Gamma \subseteq \mathbb{E}$, *and the Lyapunov function satisfies* $\dot{\mathbf{V}} \le \alpha_\Gamma \mathbf{V}$ *and the feedback matrix is* $K = RB^T P$, *with* $P = X^{-1}$.

5.4 Asymptotic consensus value

Distributed consensus algorithms have attracted a significant amount of attention in the past few years. Besides their wide range of applications in distributed and parallel computation [23], distributed control [24,25], and robotics [26], they have also been used as models of opinion dynamics and belief formation in social networks [27,28]. The central focus in this vast body of literature is to study whether a group of agents in a network, with *local* communication capabilities can reach a *global* agreement, using simple, deterministic information exchange protocols.

5.4.1 Introduction

In recent years, there has also been some interest in understanding the behavior of consensus algorithms in random settings [29]. The randomness can be either due to the choice of a randomized network communication protocol or simply caused by the potential unpredictability of the environment in which the distributed consensus algorithm can be implemented. It is shown that consensus algorithms over i.i.d. random networks lead to a global agreement on a possibly random value, as long as the network is connected in expectation. Different aspects of consensus algorithms over random switching networks, such as conditions for convergence, the speed of convergence have been widely studied and a characterization of the distribution of the asymptotic consensus value has attracted little attention. In [30], the asymptotic behavior of the random consensus value is studied in the special case of symmetric networks and in [31], the mean and variance of the consensus value for general independent and identically distributed (i.i.d.) graph processes are computed.

In the sequel, we build on the results of [31] and examine asymptotic properties of consensus algorithms over a general class of switching, directed random graphs. We aim to derive closed-form expressions for the mean and an upper-bound for the variance of the asymptotic consensus value, when the underlying network evolves according to an i.i.d. *directed* random graph process. At each time period, a directed communication link is established between two agents with some exogenously specified probability. Due to the potential asymmetry in pairwise communications between different agents, the asymptotic value of consensus is not guaranteed to be the average of the initial conditions. Instead, agents will asymptotically agree on some random value in the convex hull of the initial conditions.

5.4.2 Consensus over switching directed graphs

Consider the discrete-time linear dynamical system

$$\mathbf{x}(k+1) = W_k \mathbf{x}(k), \tag{5.70}$$

where $k \in \{1, 2, \ldots\}$ is the discrete time index, $\mathbf{x}(k) \in \mathbb{R}^n$ is the state vector at time k, and $\{W_k\}_{k=0}^{\infty}$ is a sequence of stochastic matrices. We interpret (5.70) as a distributed scheme where a collection of agents, labeled 1 through n, update their state values as a convex combination of the state values of their neighbors at the previous time step. Given this interpretation, $\mathbf{x}_i(k)$ corresponds to the state value of agent i at time k, and W_k captures the neighborhood relation between different agents at time k: the (i, j) element of W_k is positive only if agent i has access to the state of agent j. For the remainder of the chapter, we assume that the weight matrices W_k are randomly generated by an independent and identically distributed (i.i.d.) matrix process.

Remark 5.1. The dynamical system (5.70) reaches **consensus** asymptotically along some path $\{W_k\}_{k=0}^{\infty}$, if on that path there exists $x^* \in \mathbb{R}$ such that $x_i(k) \to x^*$ for all i as $k \to \infty$. The quantity x^* is termed the **asymptotic consensus value**.

Remark 5.2. It is well-known that for i.i.d. random networks, the dynamical system (5.70) reaches consensus on almost all paths if and only if the graph corresponding to the communications between agents is connected in expectation.

In [31], it is established that the conditional mean and the conditional variance of the random consensus value are given by the random consensus value are given by

$$\mathbb{E}x^* = \mathbf{x}(0)^T \mathbf{v}_1(\mathbb{E}W_k), \tag{5.71}$$

$$\text{var}(x^*) = [\mathbf{x}(0) \otimes \mathbf{x}(0)]^T \text{vec}(\text{cov}(d)) \tag{5.72}$$

$$= [\mathbf{x}(0) \otimes \mathbf{x}(0)]^T \mathbf{v}_1(\mathbb{E}[W_k \otimes W_k]) - [\mathbf{x}(0)^T \mathbf{v}_1(\mathbb{E}W_k)]^2$$

where $\mathbf{v}_1(\cdot)$ denotes the normalized left eigenvector corresponding to the unit eigenvalue.

5.4.3 Mean analysis

Consider a connected undirected graph $\mathbb{G}_c = (\mathbb{V}, \mathbb{E}_c)$ with a fixed set of vertices $\mathbb{V} = [n]$, and unweighted edges (no self-loops allowed). Each undirected edge in $\mathbb{E}_c$ represents a potential communication channel between nodes i and j, where this channel can be used to send information in both directions.

In the sequel, we focus on directed communications, that is, the event of node i sending information towards node j is independent of the event of node j sending information towards i. In this regard, it is convenient to interpret an undirected edge $\{i,j\} \in \mathbb{E}_c$ as the union of two independent directed edges, $\{(i,j), (j,i)\}$, where the ordered pair (i,j) represents a directed link from node i to node j.

In particular, in each discrete time slot $k \geq 1$, we construct a random *directed* graph $\mathbb{G}_k = (\mathbb{V}, \mathbb{E}_k)$, with $\mathbb{E}_k \subseteq \mathbb{E}_c$, such that the existence of a directed edge $(u,v) \in \mathbb{E}_k$ is determined randomly and independently of all other directed edges (including the reciprocal edge (v,u)) with a probability $p_{uv} \in (0,1)$ for $(u,v) \in \mathbb{E}_c$, and $p_{uv} = 0$ for $(u,v) \notin \mathbb{E}_p$.

To this end, in each time slot, we randomly select a subset $\mathbb{E}_k$ of directed links chosen from a set of candidate (directed) links in $\mathbb{E}_c$. We are especially interested in the case in which the probability p_{uv} of existence of a directed link (u, v) depends exclusively on the node that receives information via that link, i.e., $\text{Pr}\,((u, v) \in \mathbb{E}_k) = p_v$, where $p_v \in (0, 1)$.

Let us denote by $\mathbb{A}_c$ the symmetric adjacency matrix of the graph $\mathbb{G}_c$, where entries are $a_{ij} = 1$ if $\{i, j\} \in \mathbb{E}_c$, and $a_{ij} = 0$ otherwise. We define the degree of node i as $d_i = \sum_{j=1}^{n} a_{ji}$, and the associated degree matrix as $\mathbb{D}_c = \text{diag}(d_i)$. We also denote the random (asymmetric) adjacency matrix associated with $\mathbb{G}_k$ as $\tilde{A}_k = \left[\tilde{a}_{uv}^{(k)}\right]$, which can be described as

$$\tilde{a}_{uv}^{(k)} = \begin{cases} a_{uv}, & \text{with probability } p_v, \\ 0, & \text{with probability } 1 - p_v. \end{cases} \tag{5.73}$$

We denote the in-degrees of $\mathbb{G}_k$ as $\tilde{d}_v^{(k)} = \sum_u \tilde{a}_{uv}^{(k)}$, and the in-degree matrix as $\tilde{\mathbb{D}}_k = \text{diag}(\tilde{d}_v^{(k)})$. In terms of $\mathbb{G}_k$, the in-degrees are independent Bernoulli random variables $\tilde{d}_v^{(k)} \sim \text{Ber}\left(d_v, p_v\right)$, that is,

$$\text{Pr}\left(\tilde{d}_v^{(k)} = d\right) = \binom{d_v}{d} p_v^d \left(1 - p_v\right)^{d_v - d}.$$

We describe the consensus dynamics in (5.70) via a sequence of stochastic matrices $\{W_k\}_{k=0}^{\infty}$ associated to the sequence of random directed graphs $\{\mathbb{G}_k = (\mathbb{V}, \mathbb{E}_k)\}_{k=0}^{\infty}$ as follows:

$$W_k = \left(\tilde{\mathbb{D}}_k + I\right)^{-1} \left(\tilde{\mathbb{A}}_k + I\right)^{T}, \quad \text{for } k \geq 1. \tag{5.74}$$

Remark 5.3. In order to avoid singularities associated with the existence of isolated nodes in $\mathbb{G}_k$ (for which $\tilde{d}_i = 0$, and $\tilde{\mathbb{D}}_k$ would not be invertible), we added the identity matrix to the adjacency matrix in (5.74), which creates self-loops. Since $\mathbb{G}_c$ is connected and $p_v > 0$ for all $v \in \mathbb{V}$, the expected communications graph is connected and the stochastic dynamical system in (5.70) reaches consensus on almost all paths, although the asymptotic consensus value x^* is a random variable (not the initial average).

5.4.4 Mean of consensus value

We use (5.71) to study the mean of the consensus value. We first derive an expression for $\mathbb{E}W_k$, and then study its dominant left eigenvector $\mathbf{v}_1\,(\mathbb{E}W_k)$. For notational convenience, we define the random variable $z_i \triangleq 1/\left(\tilde{d}_i + 1\right)$ where $\tilde{d}_i \sim \text{Ber}\left(d_i, p_i\right)$, and denote its first and second moments as $M_i^{(1)} \triangleq$

$\mathbb{E}(z_i)$ and $M_i^{(2)} \triangleq \mathbb{E}(z_i^2)$. The diagonal entries of $\mathbb{E}W_k$ are then given by $\mathbb{E}w_{ii} = \mathbb{E}\left[1/\left(\tilde{d}_i + 1\right)\right]$, which present the following explicit expression:

$$\mathbb{E}w_{ii} = M_i^{(1)} = \frac{1 - q_i^{d_i+1}}{p_i\left(d_i + 1\right)}, \tag{5.75}$$

where $q_i = 1 - p_i$. Furthermore, the off-diagonal entries of $\mathbb{E}W_k$ are given by

$$\mathbb{E}w_{ij} = a_{ji}\frac{q_i^{d_i+1} + p_i\left(d_i + 1\right) - 1}{p_i\left(d_i + 1\right)d_i} = a_{ji}\frac{1 - M_i^{(1)}}{d_i}. \tag{5.76}$$

Algebraic manipulations of $\mathbb{E}W_k$ using (5.75) and (5.76) lead to

$$\mathbb{E}W_k = \Sigma + (I - \Sigma)D_c^{-1}A_c^T, \tag{5.77}$$

where $\Sigma \triangleq \text{diag}\left[M_i^{(1)}\right]$. It is readily seen that $\mathbb{E}W_k$ is a stochastic matrix, i.e., $(\mathbb{E}W_k)\mathbb{I}_n = \Sigma\mathbb{I}_n + (I - \Sigma)D_c^{-1}A_c^T\mathbb{I}_n = \mathbb{I}_n$. The following result is established:

Theorem 5.12. *Consider the random adjacency matrix $\tilde{A}_k$ in (5.73) and the associated (random) stochastic matrix W_k in (5.74). The expectation of the asymptotic consensus value of (5.70) is given by*

$$\mathbb{E}\left(x^*\right) = \sum_{i=1}^{n}\rho w_i\, x_i\left(0\right), \tag{5.78}$$

where

$$w_i\left(p_i, d_i\right) \triangleq \frac{p_i\left(d_i + 1\right)d_i}{p_i\left(d_i + 1\right) - 1 - q_i^{d_i+1}} \quad and \quad \rho\left(p_i, d_i\right) \triangleq \left(\sum_i w_i\left(p_i, d_i\right)\right)^{-1}.$$

Proof. We compute $v_1\left(\mathbb{E}W_k\right)$ and apply (5.71). Define $\mathbf{v} \triangleq v_1\left(\mathbb{E}W_k\right)$ and $\mathbf{w} \triangleq (I - \Sigma)\mathbf{v}$. From (5.77), we have that the eigenvalue equation corresponding to the dominant left eigenvector of $\mathbb{E}W_k$ is $\mathbf{v}^T(\Sigma + (I - \Sigma)D_c^{-1}A_c^T) = \mathbf{v}^T$, which can be rewritten as $\mathbf{v}^T(I - \Sigma)D_c^{-1}A_c^T = \mathbf{v}^T(I - \Sigma)$. The latter equation can be written as $\mathbf{w}^T D_c^{-1}A_c^T = \mathbf{w}^T$. The solution to this equation is the stationary distribution of the Markov chain with transition matrix $D_c^{-1}A_c^T$, which is equal to $\pi = \mathbf{d}/\sum_i d_i$, where $\mathbf{d} = (d_1, \ldots, d_n)^T$. Hence, the solution to the eigenvalue equation is

$$v_1\left(\mathbb{E}W_k\right) = \sigma\left(I - \Sigma\right)^{-1}\mathbf{d},$$

where we include the normalizing parameter

$$\sigma = \left(\sum_i \frac{d_i}{1 - M_i^{(1)}} \right)^{-1}$$

such that $\|v_1 (\mathbb{E} W_k)\|_1 = 1$. Hence, from (5.71) we have

$$\mathbb{E}\left(x^*\right) = \sum_{i=1}^{n} \sigma \frac{d_i}{1 - M_i^{(1)}} x_i(0).$$

Substituting the expression for $M_i^{(1)}$ into (5.75), we reach (5.78) via simple algebraic simplifications. $\qquad\square$

Remark 5.4. In general, the asymptotic mean $\mathbb{E}(x^*)$ does not coincide with the initial average $\bar{x}_0 = \frac{1}{n}\sum_i x_i(0)$. There is a simple technique, based on Theorem 5.12, that allows us to make the expected consensus value to be equal to the initial average. This technique consists of using $y_i(0) = (\rho w_i)^{-1} x_i(0)$ as initial conditions in (5.70). Hence, one can easily check that the asymptotic consensus value $\mathbb{E}\left(y^*\right)$ equals the initial average $\bar{x}_0$.

5.5 Variance of the asymptotic consensus value

In this section, we derive an expression that explicitly relates the variance $\text{var}(x^*)$ with the three elements that influences it, namely, the set of initial conditions $\{x_u(0)\}_{u \in V}$, the nodes properties $\{(p_u, d_u)\}_{u \in V}$, and the network structure (via the eigenvalues of the expected matrix $\mathbb{E} W_k$). In the sequel, we denote

$$\mathbf{R} \stackrel{\Delta}{=} \mathbb{E}(W_k \otimes W_k), \quad \mathbf{Q} = (\mathbb{E} W_k \otimes \mathbb{E} W_k).$$

To start the analysis, we express (5.72) as

$$\text{var}\left(x^*\right) = [\mathbf{x}(0) \otimes \mathbf{x}(0)]^T \left[v_1(\mathbf{R}) - v_1(\mathbf{Q}) \right].$$

We now establish an upper-bound for the variance of the asymptotic consensus value as follows:

$$\text{var}\left(x^*\right) \leq \|\mathbf{x}(0) \otimes \mathbf{x}(0)\|_1 \left\| v_1(\mathbf{R}) - v_1(\mathbf{Q}) \right\|_\infty. \qquad (5.79)$$

From the rules of Kronecker product, we can write the first factor in terms of the initial conditions as:

$$\|\mathbf{x}(0) \otimes \mathbf{x}(0)\|_1 = \sum_{1 \leq i,j \leq n} |x_i(0) x_j(0)|. \tag{5.80}$$

In the following, we derive an upper bound for the second factor, namely $\|v_1(\mathbf{R}) - v_1(\mathbf{Q})\|_\infty$, in terms of the node properties and the network structure.

The approach to bound $\|v_1(\mathbf{R}) - v_1(\mathbf{Q})\|_\infty$ hinges on the observation that both $\mathbf{R}$ and $\mathbf{Q}$ are $n^2 \times n^2$ stochastic matrices, and the dominant left eigenvectors $\mathbf{v}_1(\mathbf{R})$ and $\mathbf{v}_1(\mathbf{Q})$ are stationary distributions of the Markov chains with transition matrices $\mathbf{R}$ and $\mathbf{Q}$. We denote these distributions by $\mathbf{v}_1(\mathbf{R}) \triangleq \tilde{\pi}$ and $\mathbf{v}_1(\mathbf{Q}) \triangleq \pi$, respectively. The following lemma from [32] is recalled:

Lemma 5.3. *Consider two Markov chains with transition matrices $\mathbf{Q}$ and $\mathbf{R}$, and stationary distributions π and $\tilde{\pi}$, respectively. We define $\mathbf{G} = I - \mathbf{Q}$, and denote its pseudoinverse by $\mathbf{G}^\dagger = [g_{ij}^\dagger]$. Then,*

$$\|\tilde{\pi} - \pi\|_\infty \leq \kappa_s \|\mathbf{R} - \mathbf{Q}\|_\infty, \tag{5.81}$$

where $\kappa_s = \max_{i,j} |g_{ij}^\dagger|$ is called the condition number of the chain described by the transition matrix $\mathbf{Q}$.

5.5.1 Infinity norm $\|\mathbf{R} - \mathbf{Q}\|_\infty$

We proceed by comparing the respective entries of the $n^2 \times n^2$ matrix

$$\mathbf{R} = \mathbb{E}[W_k \otimes W_k] \triangleq \mathbb{E}(w_{ij})\mathbb{E}(w_{rs})$$

with the corresponding entries of

$$\mathbf{Q} = \mathbb{E}W_k \otimes \mathbb{E}W_k \triangleq \mathbb{E}(w_{ij}w_{rs}),$$

where i, j, r, and s range from 1 to n.

For the sake of clarity, we illustrate this pattern for $n = 3$ in Fig. 5.1, where the numbers in parenthesis correspond to each one of different seven cases.

Since the entries of $\mathbf{R}$ follow the same pattern as the entries of $\mathbf{Q}$ (although these entries are different), the perturbation matrix $\Delta = \mathbf{R} - \mathbf{Q}$ also

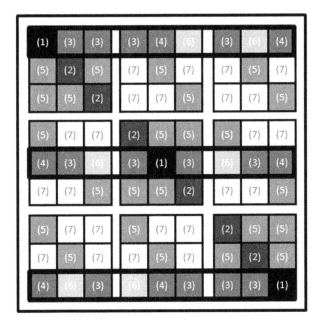

Figure 5.1 The pattern of $\mathbb{E}[W_k \otimes W_k]$ for $n = 3$. The numbers in parentheses represent the value of each entry.

follows the same pattern as $\mathbf{R}$ and $\mathbf{Q}$. Hence, comparing the respective entries, we can easily deduce the following seven cases for the entries of Δ:

$$\Delta_2 = \Delta_5 = \Delta_7 = 0,$$

$$\Delta_1 = M_2(z_i) - M_1^2(z_i),$$

$$\Delta_3 = \frac{a_{ji}}{d_i}\left(-M_2(z_i) + M_1^2(z_i)\right),$$

$$\Delta_4 = a_{ji}\left(\frac{M_1(z_i) - M_2(z_i)}{d_i} - \frac{(1 - M_1(z_i))^2}{d_i^2}\right),$$

$$\Delta_6 = \begin{cases} \left(\frac{(1-M_i)^2}{d_i^2} - \frac{1+2V_i-3M_i}{d_i(d_i-1)}\right), & \text{for } d_i > 1, \\ -a_{ji}a_{si}(1 - M_1(z_i))^2, & \text{for } d_i = 1, \end{cases}$$

where $M_i^{(2)} \triangleq \mathbb{E}\left[1/(\tilde{d}_i + 1)^2\right]$ can be written as a hypergeometric function that depends on p_i and d_i. From the above entries, we can compute the following expression for the infinity norm of the perturbation:

$$\|R - Q\|_\infty = \|\Delta\|_\infty = \max_{1 \le i \le n}\{S_i\}, \tag{5.82}$$

where

$$S_i = 2(1 - M_i^{(1)}) \left[M_i^{(1)} + \frac{1}{d_i} \left(M_i^{(1)} - 1 \right) \right].$$

Note that S_i is a function that depends exclusively on the sequence of node properties $\{(p_u, d_u)\}_{u \in V}$; hence, $\|R - Q\|_\infty$ depends exclusively on the set of node degrees and probabilities, but it is independent of how these nodes interconnect. Next, we show how the pattern of interconnection among nodes influences the upper-bound of the variance in (5.79) via the condition number κ_s in (5.81).

5.5.2 Perturbation-based bound for the variance

We now derive an explicit relationship between the condition number κ_s and the network structure via the spectral properties of $\mathbb{E}W_k = \Sigma + (I - \Sigma)\mathbb{D}_c^{-1}\mathbb{A}_c^T$. This result will then be used to bound the variance of x^* in (5.79). We base our analysis on a the following bound [33], relating κ_s with the eigenvalues of $\mathbf{Q}$, denoted as $\{\mu_i\}_{i=1}^{n^2}$:

$$\max_{i,j} \left| g_{ij}^\dagger \right| < \frac{2\left(n^2 - 1\right)}{\prod_{k=2}^{n^2}\left(1 - \mu_k\right)}. \tag{5.83}$$

Denote by $\{\lambda_i\}_{i=1}^n$ and $\{\mu_j\}_{j=1}^{n^2}$ the set of eigenvalues of $\mathbb{E}W_k$ and $\mathbb{E}W_k \otimes \mathbb{E}W_k$, respectively. The ordering of the eigenvalue sequences is determined by their distance to 1, that is, $|1 - \lambda_i| \le |1 - \lambda_j|$ for $i \le j$. The main result is stated below.

Theorem 5.13. *The variance of the asymptotic consensus value of (5.70) can be upper-bounded by*

$$\mathrm{var}\left(x^*\right) \le \underbrace{\|\mathbf{x}(0) \otimes \mathbf{x}(0)\|_1}_{(A)} \underbrace{\left(\max_{1 \le i \le n} \{S_i\} \right)}_{(B)} \underbrace{\left(\frac{2\left(n^2 - 1\right)}{\prod\left(1 - \lambda_i \lambda_j\right)} \right)}_{(C)}, \tag{5.84}$$

where $\{\lambda_i\}_{i=1}^n$ are the eigenvalues of $\mathbb{E}W_k$, and the product

$$\prod\left(1 - \lambda_i \lambda_j\right) = \prod_{(i,j) \ s.t. \ (i,j) \ne (1,1)}\left(1 - \lambda_i \lambda_j\right).$$

Proof. Considering (5.79) and using Lemma 5.3, we get

$$\mathrm{var}\left(x^*\right) \le \left\| v_1\left(\mathbf{R}\right) - v_1\left(\mathbf{Q}\right) \right\|_\infty \|\mathbf{x}(0) \otimes \mathbf{x}(0)\|_1$$

$$\overset{(a)}{\leq} \kappa_s \left\| \mathbf{R} - \mathbf{Q} \right\|_\infty \left\| \mathbf{x}(0) \otimes \mathbf{x}(0) \right\|_1$$

$$\overset{(b)}{=} \kappa_s \left(\max_{1 \leq i \leq n} \{S_i\} \right) \left\| \mathbf{x}(0) \otimes \mathbf{x}(0) \right\|_1$$

$$\overset{(c)}{<} \frac{2 \left(n^2 - 1 \right)}{\prod (1 - \mu_k)} \max_{1 \leq i \leq n} \{S_i\} \left\| \mathbf{x}(0) \otimes \mathbf{x}(0) \right\|_1 ,$$

where we have used the upper bound in (5.83) in step (c). We obtain the statement of the theorem by applying the following standard property of the Kronecker product: $\{\mu_k\}_{k=1}^{n^2} = \{\lambda_i \lambda_j\}_{1 \leq i,j < n}$. □

The following remarks are in order:

Remark 5.5. The bound in (5.84) separates the variance into three multiplicative terms representing each of the following elements (for convenience, we have underlined each of these terms in (5.84)):

(A) The first term exclusively depends on the initial condition as indicated by (5.80).

(B) The second term depends solely on the nodes properties d_i and p_i.

(C) The last term represents the influence from the overall graph structure via the spectral properties of $\mathbb{E}W_k$.

Remark 5.6. Observe that given the sequences of degrees and probabilities, $\{d_i\}_{i=1}^n$ and $\{p_i\}_{i=1}^n$, the influence of the network structure on the variance is given via term (C). Since the eigenvalues of $\mathbb{E}W_k$ are key in this term, it is interesting to briefly describe the homogeneous Markov chain with transition matrix $P \triangleq \mathbb{E}W_k = \Sigma + (I - \Sigma) D_c^{-1} A_c^T$. This Markov chain presents a self-loop in each one of its n states with transition probability $P_{ii} = M_i^{(1)}$, as well as a link from i to j with transition probability $P_{ij} = \left(1 - M_i^{(1)} \right) d_i^{-1}$ for $(i,j) \in \mathbb{E}_c$. From an analysis in [33], we conclude that term (C) is primarily governed by how close the subdominant eigenvalues of $\mathbb{E}W_k$ are from 1. In particular, the further the subdominant eigenvalues of $\mathbb{E}W_k$ are from 1, the lower the upper bound in (5.84).

In what follows, we present several numerical simulations illustrating the developed results by verifying Theorems 5.12 and 5.13.

5.5.3 Numerical example 5.1

In our first simulation we take a graph $\mathbb{G}_c$ composed of 3 stars connected in a chain (see Fig. 5.2). This graph is intended to represent, in a social context, three leaders with a set of followers. In Fig. 5.2, we represent the

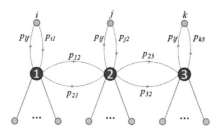

Figure 5.2 A chain of leader nodes, denoted as 1, 2, and 3, with a set of followers.

leaders using large circles marked as 1, 2, and 3, and the followers using smaller circles. We assume that each follower only listens to one leader (the center of a star) and nobody else. In this particular example, the first, second, and third leaders have 4, 8, and 16 followers, respectively. In each time step, a directed random graph $\mathbb{G}_k$ is built by choosing a set of directed communication links from G_c. We have fixed the probability of existence of a directed link coming into followers to be equal to $p_{lf} = 1/2$, for all such links (see Fig. 5.2). Also, the probability of existence of a directed link coming into a leader is inversely proportional to the degree of the leader. More specifically, we have chosen $p_{uv} = 1/d_v$, for $v = 1, 2$, and 3, where p_{uv} represents the existence of a directed communication link from the uth node to the vth leader.

Note in Fig. 5.2 that the connections between leaders are represented as pairs of directed links with existence probabilities p_{uv} and p_{vu}, with $u, v \in \{1, 2, 3\}$. We have also included in this figure the connection of each leader and one of its followers (drawn as a smaller circle on top of each leader). In this case, the directed link incoming the follower (leader) exists with probability p_{lf} (p_{fl}). The rest of the followers are represented under each leader, where we represent each pair of directed links as a single undirected edge for clarity.

In this example, we compute the asymptotic consensus value for 100 realizations of a random consensus algorithm with initial conditions $x_i(0) = i/N$. We represent the histogram of these realizations in Fig. 5.3, where the empirical average of the 100 realizations equals 0.5077. The theoretical expectation for the asymptotic consensus value applying Theorem 5.12 equals $\mathbb{E}(x^*) = \sum_i \rho w_i(i/N) = 0.4595$, which is in accordance with the empirical value. For future reference, we have also computed the empirical standard deviation of the 100 realizations to be 0.0298, and the three eigenvalues of $\mathbb{E}W_k$ closest to 1 are $\{0.9823, 0.9449, 0.75\}$.

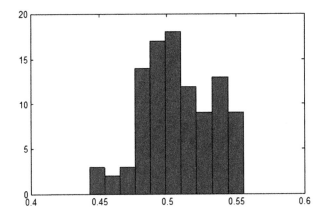

Figure 5.3 Empirical histogram for 100 realizations.

Note in Fig. 5.3 that the realizations arose from the random directed consensus over the graph in Fig. 5.2. The empirical average and the empirical standard deviation of the realizations are 0.4595 and 0.0298, respectively.

5.5.4 Numerical example 5.2

Hereafter, we numerically illustrate some of the implications of Theorem 5.13 in the variance of the asymptotic consensus value. There are some qualitative implications that are consistent throughout our numerical experiments. For example, as we mentioned in Remark 5.6, given a set of initial conditions and node properties, the upper bound in (5.84) is primarily governed by how close the subdominant eigenvalues of $\mathbb{E}W_k$ are from 1. In particular, the further the subdominant eigenvalues of $\mathbb{E}W_k$ are from 1, the lower the upper bound. We illustrate in the following numerical experiments that the lower the upper bound, the lower one should expect the variance to be.

In the first experiment, we slightly modify the network described in the previous subsection and study the influence of this modification on the eigenvalues of $\mathbb{E}W_k$ and the variance of x^*. Our first modification is a change in the probabilities of existence of a directed link from the uth node to the vth leader without changing the network topology. In particular, we choose the new probabilities to be $p_{uv} = 3/d_v$, for $v = 1, 2$, and 3. In the modified network, the eigenvalues of $\mathbb{E}W_k$ closest to 1 are $\{0.9789, 0.9372, 0.75\}$. Hence, since the effect of our modification on the eigenvalues is to move them away from 1, we should expect the variance of x^* to be reduced according to Remark 5.6. In fact, running 100 random

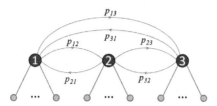

Figure 5.4 Ring of 3 leaders and corresponding followers.

consensus algorithms with the same initial conditions, $x_i(0) = i/N$, using our new probabilities, we obtain a standard deviation 0.0286 (which is less than the 0.0298 obtained before).

In the second experiment, we illustrate that the larger the gap between the eigenvalues of $\mathbb{E}W_k$ and 1, the smaller the variance of x^*. In this case, apart from keeping the new set of probabilities $p_{uv} = 3/d_v$, we convert the 3-chain of leaders into a 3-ring of leaders, as depicted in Fig. 5.4. In this case, the three eigenvalues of EW_k closest to 1 are $\{0.9577, 0.9212, 0.75\}$, which are even further away from 1 than in the second example. Hence, as expected, the standard deviation of 100 random consensus algorithms is even smaller than in the second example, in particular 0.0274. In conclusion, our simulations are consistent with Theorem 5.13 and with the qualitative behavior described in Remark 5.6.

5.6 Notes

The study of cyberphysical systems is involved with the design of resilient and robust control algorithms that can withstand different intruder attacks. Two main forms of attacks in literature are denial-of-service (DoS) and deception attacks. In DoS attacks, the communication link between the interconnected systems is hijacked by an intruder for a short period of time thereby preventing information exchange between the systems. In deception attacks, false information is transmitted by the intruder over the hijacked communication network.

In [17], the problem of distributed consensus control for MAS under DoS attacks was studied. The approach involved using both state-feedback and observer-based controllers to mitigate the effect of an intruder on the network. Further, the authors derived sufficient conditions for consensus to be achieved during the attack period. In [18], secure consensus against deception attacks in synchronous networks was discussed. An algorithm was proposed to guarantee secure consensus and was shown to converge with an exponential rate through matrix analysis.

The relationship between closed-loop stability and the duration of DoS attacks was characterized in [19]. It is concluded that the resilient nature of the control descends from adaptability of its sampling rate to occurrence of DoS. A distributed average consensus problem for linear multiagent systems under DoS attacks was addressed in [20]. The frequency and duration of DoS attacks were analyzed, leading to a distributed event triggered control law for the MAS.

A novel distributed adaptive control architecture was proposed in [21] for networked multiagent systems under the influence of exogenous disturbances and sabotaged sensors and actuators. Also, the proposed controller handles time-varying multiplicative actuator attacks on the followers with indirect link to leaders in the network. The authors claim that the proposed controller guarantees uniform ultimate boundedness of the tracking error for each agent in the mean-square sense.

In [22], distributed tracking control problem was examined for stochastic linear multiagent systems under DoS attacks. The attacks considered were classified as connectivity-broken and connectivity-maintained attacks. Using average dwell time switching between stable and unstable modes based on graph analysis, some conditions were derived for robust mean-square exponential consensus.

Next, we have studied the asymptotic properties of the consensus value in distributed consensus algorithms over switching, directed random graphs. While different aspects of consensus algorithms over random switching networks, such as conditions for convergence and the speed of convergence, have been widely studied, a characterization of the distribution of the asymptotic consensus for general *asymmetric* random consensus algorithms remains an open problem.

We have derived closed-form expressions for the expectation of the asymptotic consensus value as a function of the set of initial conditions, $\{x_u(0)\}_{u \in \mathbb{V}}$, and the set of node properties, $\{(p_u, d_u)\}_{u \in \mathbb{V}}$, as stated in Theorem 5.12. We have also studied the variance of the asymptotic consensus value in terms of several elements that influence it, namely, (*i*) the initial conditions, (*ii*) node properties, and (*iii*) the network topology. In Theorem 5.13, we have derived an upper bound for the variance of the asymptotic consensus value that explicitly describes the influence of each one of these elements. We also provided an interpretation of the influence of the network topology on the variance in terms of the eigenvalues of the expected matrix $\mathbb{E}W_k$. From our analysis we conclude that, in most cases, the variance of x^* is primarily governed by how close the subdominant eigenvalues of $\mathbb{E}W_k$ are from 1.

References

[1] W. Abbas, A. Laszka, X. Koutsoukos, Improving network connectivity and robustness using trusted nodes with application to resilient consensus, IEEE Trans. Control Netw. Syst. 5 (4) (2018) 2036–2048.

[2] M. Franceschelli, A. Giua, A. Pisano, Finite-time consensus on the median value with robustness properties, IEEE Trans. Autom. Control 62 (4) (2017) 1652–1677.

[3] H.J. LeBlanc, H. Zhang, S. Sundaram, X. Koutsoukos, Resilient continuous-time consensus in fractional robust networks, in: Proc. the American Control Conference (ACC), Washington, DC, 2013, pp. 1237–1242.

[4] H. Liu, M. Xu, Y. Wu, N. Zheng, Y. Chen, M.Z.A. Bhuiyan, Resilient bipartite consensus for multi-agent networks with antagonistic interactions, in: Proc. the 17th IEEE Int. Conference on Trust, Security and Privacy in Computing and Communications/12th IEEE International Conference on Big Data Science and Engineering, 2018, pp. 1262–1269.

[5] H.Y. Oksuz, M. Akar, Resilient group consensus in the presence of Byzantine agents, Int. J. Control (2019) 1–16.

[6] O.F. Erkan, O. Cihan, M. Akar, Analysis of distributed consensus protocols with multi-equilibria, J. Franklin Inst. 355 (2018) 332–360.

[7] O.F. Erkan, O. Cihan, M. Akar, Distributed consensus with multi-equilibria in directed networks, in: Proc. the American Control Conference, Seattle, USA, 2017, pp. 4681–4685.

[8] H.J. Leblanc, H. Zhang, X. Khoutsoukos, S. Sundaram, Resilient asymptotic consensus in robust networks, IEEE J. Sel. Areas Commun. 31 (4) (2013) 766–781.

[9] H.J. Leblanc, H. Zhang, X. Khoutsoukos, S. Sundaram, Resilient asymptotic consensus in fractional robust networks, in: Proc. the American Control Conference, IEEE, 2013, pp. 1237–1242.

[10] D. Saldana, A. Prorok, S. Sundaram, M.F. Campos, V. Kumar, Resilient consensus for time-varying networks of dynamic agents, in: Proc. the American Control Conference, IEEE, 2017, pp. 252–258.

[11] H. Zhang, S. Sundaram, A simple median-based resilient consensus algorithm, in: Proc. the Fiftieth Annual Allerton Conference, Allerton House, UIUC, Illinois, 2018, pp. 1734–1741.

[12] S.M. Dibaji, H. Ishii, Resilient consensus of double-integrator multi-agent systems, in: Proc. the American Control Conference, Portland, Oregon, 2014, pp. 5139–5144.

[13] S.M. Dibaji, H. Ishii, Resilient consensus of second-order agent networks: asynchronous update rules with delays, Automatica 81 (2017) 123–132.

[14] J. Huang, Y. Wu, L. Chang, M. Tao, X. He, Resilient consensus with switching networks and heterogeneous agents, Neurocomputing 341 (2019) 70–79.

[15] R.M. Kieckhafer, M.H. Azadmanesh, Reaching approximate agreement with mixed-mode faults, IEEE Trans. Parallel Distrib. Syst. 5 (1) (1994) 53–63.

[16] H.J. LeBlanc, X.D. Koutsoukos, Consensus in networked multiagent systems with adversaries, in: Proc. the 14th Int. Conference on Hybrid Systems: Computation and Control, 2011, pp. 281–290.

[17] A. Lu, G. Yang, Distributed consensus control for multi-agent systems under denial of service, in: Information Services, 2018, pp. 95–107.

[18] C. Zhao, J. He, P. Cheng, J. Chen, Secure consensus against message manipulation attacks in synchronous networks, in: Proc. IFAC Cape Town, South Africa, 2014, pp. 1182–1187.

[19] C. De Persis, P. Tesi, Resilient control under denial-of-service, in: Proc. IFAC Cape Town, South Africa, 2014, pp. 134–139.

[20] Z. Feng, G. Hu, Distributed secure average consensus for linear multi-agent systems under DoS attacks, in: Proc. American Control Conference, 2017, p. 2261.

[21] X. Jin, W. Haddad, An adaptive control architecture for leader-follower multi-agent systems with stochastic disturbances and sensor and actuator attacks, Int. J. Control (2018) 1–9.

[22] Z. Feng, G. Hu, Distributed tracking control for multi-agent systems under two types of attacks, in: Proc. the 19th World Congress of Automatic Control, Cape Town, 2014, pp. 5790–5795.

[23] J.N. Tsitsiklis, Problems in decentralized decision making and computation, Ph.D. dissertation, Massachusetts Institute of Technology, Cambridge, MA, 1984.

[24] A. Jadbabaie, J. Lin, A.S. Morse, Coordination of groups of mobile autonomous agents using nearest neighbor rules, IEEE Trans. Autom. Control 48 (6) (2003) 988–1001.

[25] R. Olfati-Saber, J.A. Fax, R.M. Murray, Consensus and cooperation in networked multi-agent systems, Proc. IEEE 95 (1) (2007).

[26] J. Cortes, S. Martinez, F. Bullo, Analysis and design tools for distributed motion coordination, in: Proceedings of the American Control Conference, Portland, OR, 2005, pp. 1680–1685.

[27] M.H. DeGroot, Reaching a consensus, J. Am. Stat. Assoc. 69 (345) (1974) 118–121.

[28] B. Golub, M.O. Jackson, Naive learning in social networks: convergence, influence, and the wisdom of crowds, 2007, unpublished manuscript.

[29] Y. Hatano, M. Mesbahi, Agreement over random networks, IEEE Trans. Autom. Control 50 (11) (2005) 1867–1872.

[30] S. Boyd, A. Gosh, B. Prabhakar, D. Shah, Randomized gossip algorithms, in: Special Issue, IEEE Trans. Inf. Theory and IEEE/ACM Trans. Netw. 52 (6) (2006) 2508–2530.

[31] A. Tahbaz-Salehi, A. Jadbabaie, Consensus over ergodic stationary graph processes, IEEE Trans. Autom. Control 55 (1) (2010) 225–230.

[32] R.E. Funderlic, C.D. Meyer, Sensitivity of the stationary distribution vector for an ergodic Markov chain, Linear Algebra Appl. 76 (1986) 1–17.

[33] C.D. Meyer, Sensitivity of the stationary distribution of a Markov chain, SIAM J. Matrix Appl. 15 (1994) 715–728.

CHAPTER 6

Consensus on state-dependent fuzzy graphs

6.1 Introduction

The ever-increasing need for automation and autonomous systems, especially those emulating coordination behavior usually observed in natural and biological systems, has driven numerous research interests in coordination among agents over a communication network. Coordination in multiagent systems is either centralized or distributed. In centralized network configurations, agents receive commands from a common control center and communicate their measurements or opinion about a shared variable, called the coordination variable, to the command center. Conversely, distributed networks replace the need for a centralized node with relative information sharing among agents belonging to a common neighborhood set.

Research investigations in multiagent coordination can be categorized under network analysis, synthesis, or algorithms. Network analysis is primarily concerned with issues stemming from coordination over a selected network topology ranging from intruder attacks, network delays, and communication noises. Network synthesis studies deal with synthesis of appropriate communication networks suitable for solving the choice problem. Finally, network algorithm problems investigate centralized or distributed coordination algorithms and protocols leading to agreement (competitive or cooperatives) in multiagent systems. The aforementioned category of problems are not mutually exclusive as the solution of these problems play an integral role in reaching an agreement (or consensus) in multiagent system.

Recent studies in multiagent coordination are mostly centered around designing algorithms which drive the agents to reach an agreement with the underlying assumptions that the network topology is either fixed or switching in a deterministic manner. However, a more realistic approach to designing and modeling true autonomous intelligent systems should consider the interplay between agent's state, graph dynamics, and environmental constraints.

Advanced Distributed Consensus for Multiagent Systems
https://doi.org/10.1016/B978-0-12-821186-1.00014-3

Numerous forms of coordination control problems have been investigated in the algorithmic sense. Coordination control problems are broadly classified under consensus, formation, flocking, swarming, rendezvous, and alignment, to mention a few. In the topological sense, research problems in coordination control can be considered as fixed or switching topology type problems where stability results have been reported for either cases based on Lyapunov and graph-theoretic analysis.

Consensus refers to reaching an agreement on a shared variable of interest often called coordination variable. Every form of coordination problem involves reaching a sort of agreement or consensus. Hence most research interests are centered around consensus-type problems as they find applications in different fields, ranging from biology, chemistry, mathematics, control, and even political science.

In distributed multivehicle coordination, significant theories have been proposed by notable researchers. Some necessary and sufficient conditions were derived in for reaching consensus on fixed topologies. For directed networks, the presence of a spanning tree (or rooted-out branching) is necessary for reaching consensus. This condition is analogous to the connectivity matrix (or Laplacian) having at least one zero eigenvalue with other eigenvalues having positive real parts. For time-varying topologies, a necessary condition is that the union of the set of graphs should have a rooted-out branching [1].

Numerous results have been presented based on the background theories proposed in [1,7]. Significant results have been reported in different branches of consensus, such as group [10–12], bipartite [27–29], leaderless [24], leader–follower [20,30–33], and resilient [16–19] consensus problems. Investigations have been conducted for fixed [24], varying [7,34,35], and random [25,26] topologies, linear and nonlinear systems, as well as networks with phenomenons like cyberphysical attacks [36], and delays [7,21,23].

Human and animal coordination is often modeled by complex social interaction networks which are often dynamic, uncertain, and determined by some local factors influencing each agents' local objective and the overall global objective. Practically, an agent should be able to decide on what its neighborhood set should be at different points in time depending on some local determinants. Network phenomena such as cyberattacks, fading, congestions, delays, and total failures are also practical issues that could pose significant threats to reaching agreements on networks.

Therefore, it is insufficient to represent communication topologies with crisp graphs due to uncertainty, self-interests, and time-dependencies. A more practical approach will be to consider graphs capable of modeling uncertainties and offer flexibility in transiting between topological states depending on the network conditions, system objectives, and states, giving birth to the notion of *state-driven dynamic graphs* [13].

State-driven dynamic graphs, as introduced by [13], are used to refer to transformational mapping from the physical states of agents to the topological states. Some propositions about these graphs in [13] use graph combinatorics, control theory, and algebraic graph theory. Research problems resulting from discussions around state-dependent dynamic graphs are centered around maintaining invariance related to special topological properties (e.g., connectivity) as the graph transition between states.

6.2 Preliminaries

We start by introducing some notations. Let $\vee$ denote maximum or supremum and $\wedge$ denote minimum or infimum. For a nonempty set $\mathbb{X}$, we use $\sim$ to denote an equivalence relation on $\mathbb{V} \times \mathbb{V}$ such that $(x, y) \sim (u, v) \in \mathbb{V} \times \mathbb{V}$ implies $x = u$ and $y = v$. The condition "if and only if" is sometimes abbreviated as iff.

6.2.1 Graph theory

Graph theory is a mathematical framework used in describing interactions among interconnected systems. A graph denoted by $\mathbb{G}(\mathbb{V}, \mathbb{E})$ is a pair consisting of a vertex set $\mathbb{V}$ and edge set $\mathbb{E}$. A graph without edge orientation is called *undirected*, while the presence of edge orientation makes the graph *directed*. Different topological variations of graphs are possible depending on the edge orientation, including simple, complete, signed, balanced, bipartite, and random graphs.

Algebraic graph theory provides spectral analysis for consensus-type problems based on connectivity properties. This spectral analysis is derived based on quantities providing information about the graph, such as degree, adjacency, and Laplacian matrices. For a graph on n vertices and m edges, the degree matrix $\Delta(\mathbb{G}) \in \mathbb{R}^{n \times n}$ is a diagonal matrix, with elements on the diagonal representing the degree $d(v_i)$ of each vertex; $d(v_i)$ is the number of edges incident to the vertex v_i. The adjacency matrix $\mathbb{A}(G)$ is a symmetric $n \times n$ matrix describing the adjacency relationship in $\mathbb{G}$. Each $a_{ij} \in \mathbb{A}(\mathbb{G})$ assumes 1 if $(v_i, v_j) \in \mathbb{E}(\mathbb{G})$ and 0 otherwise. The Laplacian matrix in an undirected graph $\mathbb{L}(\mathbb{G}) = \Delta(\mathbb{G}) - \mathbb{A}(\mathbb{G})$. The incidence matrix $\mathbb{W}$ of a directed graph $\mathbb{D}$ is defined as $\mathbb{W} = [w_{ij}]$, where $w_{ij} = -1$ if v_i is the tail of e_j, $w_{ij} = 1$ if v_i is the head of e_j, and $w_{ij} = 0$ if v_i is not adjacent to e_j. The Laplacian matrix of a directed graph $\mathbb{D}$ is $\mathbb{L}(\mathbb{D}) = \mathbb{W}(\mathbb{D})\mathbb{W}(\mathbb{D})^T$.

6.3 State-dependent graphs

Reaching consensus or agreement in a multiagent setting requires the presence of a communication network. In most consensus analysis, the communication network is allowed to be fixed or varying either stochastically or in a deterministic fashion. However, an important question is how networks are formed in real-life multiagent systems. It is worthy to note that in practical multiagent systems (such as in social networks or biological multiagent systems), connections are formed based on *interest*. This means connections may not be just fixed, deterministic, or random, but rather depend on the states of the each agent participating in the consensus game, thus giving birth to the notion of *state-dependent* graphs. In state-dependent graphs, it is assumed that the order of the graph is fixed, that is, the number of agents in the network is fixed, but the edge-set is a function of the states of agents. Consider an MAS consisting of N agents defined on a graph $\mathbb{G}$, with the following state–evolution dynamics:

$$x_i = f(x_i, u_i) \tag{6.1}$$

where $x_i \in \mathbb{R}^p$ and $u_i \in \mathbb{R}^q$ represent the state and control inputs of agent i, respectively. According to [8], the overall state space of the system assumes the form

$$X := X_1 \times X_2 \times \cdots \times X_N, \tag{6.2}$$

where X_i is the state-space of each agent i. We now proceed to define formally the notion of state-dependent graphs.

Definition 6.1 (State-dependent graphs, [8]). A state-dependent graph is a mapping, g_s, from the distributed system state space, X, to the set of all labeled graphs of order N, $\mathbb{G}_N$, that is,

$$g_s : X \to \mathbb{G}_N \text{ and } g_s(x) = \mathbb{G}. \tag{6.3}$$

Mesbahi [8] used the formalism of graphical equations to provide a means of defining the relation between edge formations and the system states, specifically a subset S_{ij} is defined as

$$S_{ij} \subset X_i \times X_j \quad (i,j) \in \mathbb{E}(g_s(x)) \text{ iff } (x_i, x_j) \in S_{ij}. \tag{6.4}$$

For example, the set S_{ij}, or edge states of vertices i and j, can be formally defined based on nearest neighbor rules:

$$S_{ij} = \{(x_i, x_j) \mid \|x_i - x_j\| \le \rho\}, \quad i, j \in \mathbb{V}, \ i \ne j. \tag{6.5}$$

Thus, based on the edge states S_{ij} for any vertex pair (i, j), a more formal definition of state-dependent graphs is provided as follows:

Definition 6.2 (State-dependent graphs, [8]). Consider $\mathbb{S} = [S_{ij}]$, the state map $g_{\mathbb{S}} : X \to \mathbb{G}_N$, with an image consisting of graphs of order N, having an edge between vertex i and j if and only if $(x_i, x_j) \in S_{ij}$, a state-dependent graph with respect to $\mathbb{S}$. Then the image of the state-dependent graph with respect to $g_{\mathbb{S}}(X)$ is

$$\{\mathbb{G} \mid g_{\mathbb{S}}(x) = \mathbb{G} \text{ for some } x \in X\} = \{\mathbb{G}_x \mid x \in X\}. \tag{6.6}$$

6.3.1 Eigenvalue optimization

In reaching agreement on state-dependent networks, preserving connectivity as the graph transits from one state to another is the main concern. In this regard, some research investigations on state-dependent networks focus on optimization of the second smallest eigenvalue of the Laplacian matrix. We recall some notable results reported by [2]. Consider the MAS on a graph $\mathbb{G}$ with vertex set $\mathbb{V} = \{1, \ldots, n\}$ and edge set $\mathbb{E} = [e_{ij}]$. Each edge e_{ij} is assigned a weight function w_{ij}. The weight function

$$w : \mathbb{R}^3 \times \mathbb{R}^3 \to \mathbb{R}_+ \tag{6.7}$$

can be chosen as function of the Euclidean distance between two adjacent nodes. That is,

$$w_{ij} := w(x_i, x_j) = f(\|x_i - x_j\|). \tag{6.8}$$

The speed of convergence of the MAS depends largely on the magnitude of the second smallest eigenvalue. Thus, in state-dependent consensus problems, the connectivity issue may be posed as follows:

$$\max_{x} \lambda_2(\mathbb{L}(x)). \tag{6.9}$$

In the case of proximity-based graphs, an additional constraint may be posed as

$$d_{ij} := \|x_i - x_j\|^2 \geq \rho_1, \tag{6.10}$$

which ensures that the agents maintain some proximity from each other while maximizing $\lambda_2(\mathbb{G})$.

Corollary 6.1 ([2]). *For the graph Laplacian* $\mathbb{L}$*, the constraint*

$$\lambda_2(\mathbb{L}) > 0 \tag{6.11}$$

is equivalent to

$$\mathbb{P}^T \mathbb{L} \mathbb{P} > 0 \tag{6.12}$$

where $\mathbb{P} = [p_1, p_2, \ldots, p_{n-1}]$ *and the unit vectors* $p_i \in \mathbb{R}^n$ *are chosen such that*

$$p_i^T \mathbf{1}_n = 0 \quad and \quad p_i^T p_j = 0. \tag{6.13}$$

Based on the above corollary, the optimization problem can be restated as

$$\max_x \gamma \tag{6.14}$$

subject to:

$$d_{ij} := \|x_i - x_j\|^2 \geq \rho_1, \tag{6.15}$$
$$\mathbb{P}^T \mathbb{L} \mathbb{P} \geq \gamma_1 I_{n-1}. \tag{6.16}$$

The optimization problem posed above was solved in [2] using the discrete and optimization algorithm based on Euler's first discretization method. By differentiating the constraints,

$$2(\dot{x}_i - \dot{x}_j)^T (\dot{x}_i - \dot{x}_j) = \dot{d}_{ij}. \tag{6.17}$$

Applying Euler's discretization

$$x(t) \to x(k), \quad \frac{x(k+1) - x(k)}{\Delta t}, \tag{6.18}$$
$$2\{x_i(k+1) - x_j(k+1)\}^T \{x_i(k) - x_j(k)\} = d_{ij}(k+1) + d_{ij}(k), \tag{6.19}$$
$$w_{ij}(k+1) = w_{ij}(k) - \epsilon^{(\rho_1 - d_{ij}(k))/(\rho_1 - \rho_2)} (d_{ij}(k+1) - d_{ij}(k)). \tag{6.20}$$

The iterative form of the optimization problem becomes

$$\max_{x(k+1)} \gamma \tag{6.21}$$

subject to:

$$2\{x_i(k+1) - x_j(k+1)\}^T \{x_i(k) - x_j(k)\} = d_{ij}(k+1) + d_{ij}(k), \tag{6.22}$$

$$d_{ij}(k+1) \geq \rho_1, \tag{6.23}$$

$$\mathbb{P}^T \mathbb{L} \mathbb{P} \geq \gamma I_{n-1}. \tag{6.24}$$

The eigenvalue optimization problem can also be posed in a distributed fashion. Consider the following:

$$\lambda_2(\tilde{\mathbb{L}}) z_2^T z_2 \leq z_2^T \tilde{\mathbb{L}} z_2, \tag{6.25}$$

where $\mathbb{L} \neq \tilde{\mathbb{L}}$ and $z_2 \in \mathbb{I}$ is the unit eigenvector of $\mathbb{L}$ corresponding to $\lambda_2(\mathbb{L})$. Expanding the right-hand side gives

$$z_2^T \tilde{\mathbb{L}} z_2 = z_2^T \mathbb{L} z_2 + z_2^T (\tilde{\mathbb{L}} - \mathbb{L}) z_2$$
$$= \lambda_2(\mathbb{L}) + \langle z_2, z_2^T (\tilde{\mathbb{L}} - \mathbb{L}) \rangle. \tag{6.26}$$

Then the optimal Laplacian matrix $\mathbb{L}_\star$ can be obtained as the limit of the subgradient iteration

$$\mathbb{L}_\star^{s+1} = \mathbb{L}_\star^s + \alpha^s \mathbb{Y}^s, \tag{6.27}$$

where $\mathbb{Y} = z_2^T z_2$ is supergradient for $\lambda_2(\mathbb{L})$. Thus the distributed optimization problem is therefore posed as:

$$\min_{x \in \mathbb{R}^d} \left| [\mathbb{L}(x)]_i - [\mathbb{L}_\star^s]_i \right|, \tag{6.28}$$

where $[\mathbb{L}(x)]_i$ denotes the ith row of the Laplacian matrix as a function of the states and $[\mathbb{L}_\star^s]_i$ is the ith row of the optimal Laplacian matrix computed by agent i at the step of the supergradient iteration. The above optimization problem can be solved based on the following potential function:

$$u_i(t) = -\sum_{j \in \mathbb{N}_i} \nabla_{x_i} V_i(t), \tag{6.29}$$

$$V_i(t) = \begin{cases} (||x_i||_2^2 - [\mathbb{L}_\star^s]_i^{-1})^2, & \text{if } ||x_i||_2 \leq \rho_2, \\ (\rho_2 - [\mathbb{L}_\star^s]_i^{-1})^2, & \text{if } ||x_i||_2 > \rho_2. \end{cases} \tag{6.30}$$

6.3.2 Consensus on state-dependent graphs

Reaching agreement on state-dependent networks was studied for systems of the form (6.31) in [5] over an undirected graph $\mathbb{G}$:

$$x_i(t) = f(x_i) + \sum_{j=1}^{n} \omega_{ij}(x_i, x_j) h(x_i, x_j) \tag{6.31}$$

where the $x_i = [x_i^{(1)}, \ldots, x_i^{(m)}]^T$ is an m-dimensional vector and $x = [x_1^T, \ldots, x_n^T]^T$. The state-dependent network was represented with a weighted matrix $W(x) = [w_{ij}(x_i, x_j)] = a_{ij} g_i(x_i, x_j)$. The corresponding weighted Laplacian matrix is a zero row-sum matrix $\mathbb{L}(x)$ such that

$$\ell_{ij}(x_i, x_j) = -w_{ij}(x_i, x_j), \quad \ell_{ii} = \sum_j \ell_{ij}(x_i, x_j).$$

Further, the eigenvalues of the weighted Laplacian are ordered as:

$$0 = \lambda_1(x) \leq \lambda_2(x) \leq \cdots \leq \lambda_N(x). \tag{6.32}$$

The function $g_i(x_i, x_j)$ was chosen in [5] considering the Hebbian learning rule, were the strength of connection depends on some neural activities. Some possible choices for the function $g_i(x_i, x_j)$ at time t are given as follows:

$$g(x_i, x_j) = x_i(t) x_j(t), \tag{6.33}$$

$$g(x_i, x_j) = (V_s - x_i(t))(V_s - x_j(t)), \tag{6.34}$$

$$g^{\text{hebb}}(x_i, x_j) = (V_s + x_i(t))(V_s + x_j(t)). \tag{6.35}$$

In [4], the connectivity-preserving consensus control for MAS based on potential functions was discussed. Consider the MAS described by the following single-integrator dynamics:

$$\dot{x}_i(t) = u_i(t). \tag{6.36}$$

The following distributed consensus protocol was designed for the system (2.9):

$$u_i = \sum_{j \in \mathbb{N}_\sigma(i)} f(x_i - x_j), \tag{6.37}$$

where $\sigma(i, j) = \sigma(j, i) \in \{0, 1\}$ is a symmetric indicator function which determines whether information will be shared over edge (i, j). The neighborhood set $\mathbb{N}_\sigma(i) \subseteq \mathbb{N}_i$ where $\mathbb{N}_i$ represents the neighborhood set of agent i. Now if there are M edges in the graph, the incidence matrix $\mathbf{I}(\mathbb{G}) = [\iota_{ij}]$ is defined as

$$\iota_{ij} = \begin{cases} 1, & \text{if } v_i \text{ is the edge of } e_j, \\ -1, & \text{if } v_i \text{ is the tail of } e_j, \\ 0, & \text{otherwise.} \end{cases} \tag{6.38}$$

The graph Laplacian can be defined as a function of the incidence matrix as

$$\mathbb{L}(\mathbb{G}) = \mathbf{I}(\mathbb{G})\mathbf{I}(\mathbb{G})^T. \tag{6.39}$$

The following control law was considered for the system:

$$\sigma(i, j) = 1, \tag{6.40}$$
$$f(x_i - x_j) = -w(x_i - x_j)(x_i - x_j) \quad \forall (i, j) \in \mathbb{E}(\mathbb{G}). \tag{6.41}$$

Let $x_i = (x_{i,1}, \ldots, x_{i,n})^T$, $i = 1, \ldots, N$ and $x = (x_1^T, \ldots, x_N^T)^T$, a componentwise operator, $c(x, j)$, is defined as

$$c(x, j) = (x_{1,j}, \ldots, x_{N,j})^T \in \mathbb{R}^N, \quad j = 1, \ldots, N. \tag{6.42}$$

Thus it is possible to write

$$\dot{c}(x, j) = -\mathbb{L}(\mathbb{G}) c(x, j), \quad j = 1, \ldots, n. \tag{6.43}$$

The MAS can then be written as

$$\dot{x}_i = -\sum_{j \in \mathbb{N}_G} w(x_i - x_j)(x_i - x_j). \tag{6.44}$$

Note that (6.43) can be rewritten in terms of the incidence matrix as

$$\dot{c}(x, j) = -\mathbb{I}\mathbb{W}(x)\mathbb{I}^T c(x, j), \quad j = 1, \ldots, n, \tag{6.45}$$

where $\mathbb{W}(x) = \text{diag}(w_1(x), \ldots, w_M(x)) \in \mathbb{R}^{M \times M}$. The state dependent weighted graph Laplacian is thus defined as

$$\mathbb{L}_\mathbb{W}(x) = \mathbb{I}\mathbb{W}(x)\mathbb{I}^T. \tag{6.46}$$

The main idea in [4] is for agents within a certain proximity δ to establish connection or exchange information with one another. A region of influence $\mathbb{D}_{\mathbb{G},\delta}^{\epsilon}$ is given by

$$\mathbb{D}_{\mathbb{G},\delta}^{\epsilon} = \{x \in \mathbb{R}^{nN} \mid |\ell_{ij}| \leq (\delta - \epsilon) \; \forall (i,j) \in \mathbb{E}(\mathbb{G})\}. \tag{6.47}$$

An edge-tension V_{ij} can then be defined as

$$\mathbb{V}_{ij}(\delta, x) = \begin{cases} \dfrac{|l_{ij}(x)|^2}{\delta - |\ell_{ij}(x)|}, & \text{if } (i,j) \in \mathbb{E}(\mathbb{G}), \\ 0, & \text{otherwise}, \end{cases} \tag{6.48}$$

$$\frac{\partial V_{ij}}{\partial x_i}(\delta, x) = \begin{cases} \dfrac{2\delta - |\ell_{ij}(x)|}{(\delta - |\ell_{ij}(x)|)^2}(x_i - x_j), & \text{if } (i,j) \in \mathbb{E}(\mathbb{G}), \\ 0, & \text{otherwise}. \end{cases} \tag{6.49}$$

The total tension energy of the graph $\mathbb{G}$ is thus

$$\mathbb{V}(\delta, x) = \frac{1}{2} \sum_{i=1}^{N} \sum_{i=1}^{N} \mathbb{V}_{ij}(\delta, x). \tag{6.50}$$

Lemma 6.1 ([4]). *Consider an initial position $x_0 \in \mathbb{D}_{\mathbb{G},\delta}^{\epsilon}$ for a given $\epsilon \in (0, \delta)$. If the graph $\mathbb{G}$ is connected, then the set $\Omega(\delta, x_0) := \{x \mid \mathbb{V}(\delta, x) \leq \mathbb{V}(\delta, x_0)\}$ is invariant to the system under the following control law:*

$$\dot{x}_i = -\sum_{j \in \mathbb{N}_{\mathbb{G}}} \frac{2\delta - |\ell_{ij}(x)|}{(\delta - |\ell_{ij}(x)|)^2}(x_i - x_j). \tag{6.51}$$

Theorem 6.1 ([4]). *Consider a connected graph $\mathbb{G}$ with initial condition $x_0 \in \mathbb{D}_{\mathbb{G},\delta}^{\epsilon}$. Then the MAS under control law (6.51) converges to the static centroid $\bar{x}(x_0)$ asymptotically.*

Based on $\sigma(i, j)$, the discussion in [4] was extended to the case of dynamic graphs by proposing the following:

$$\sigma(i,j)[t^+] = \begin{cases} 0, & \text{if } \sigma(i,j)[t^-] = 0 \text{ and } |l_{ij}| > \Delta - \epsilon, \\ 1, & \text{otherwise}. \end{cases} \tag{6.52}$$

$$f(x_i - x_j) = \begin{cases} 0, & \text{if } \sigma(i,j)[t^-] = 0 \text{ and } |\ell_{ij}| > \Delta - \epsilon, \\ -\dfrac{\partial V_{ij}(\Delta, x)}{\partial x_i}, & \text{otherwise}. \end{cases} \tag{6.53}$$

Theorem 6.2 ([4]). *Consider an initial position $x_0 \in \mathbb{D}^\epsilon_{g^0, \Delta}$, where ϵ is the switching threshold and $\mathbb{G}^0$ is the initial graph. Assume that the graph $\mathbb{G}^0_\sigma$ is connected, where $\mathbb{G}^0_\sigma$ is the graph induced by the initial indicator function value. Then, by the control law*

$$u_i = -\sum_{j \in \mathbb{N}_\sigma(i)} \frac{\partial V_{ij}(\Delta, x)}{\partial x_i} \tag{6.54}$$

the group of agents converge asymptotically to $\mathrm{span}\{\mathbf{1}\}$.

In [3], the following potential function (6.55), where connectivity violations are treated as an obstacle in the configuration space, was considered:

$$\phi(x) = \log \det(\mathbb{P}^T \mathbb{L}(x)\mathbb{P})^{-1}. \tag{6.55}$$

The feedback controller was then obtained as

$$u = -\nabla_x \phi(x), \tag{6.56}$$

$$u = \frac{1}{\det \mathbb{M}(x)} \begin{bmatrix} \mathrm{trace}[\mathbb{M}^{-1}(x)\dfrac{\partial}{\partial x_1}\mathbb{M}(x)] \\ \vdots \\ \mathrm{trace}[\mathbb{M}^{-1}(x)\dfrac{\partial}{\partial x_n}\mathbb{M}(x)] \end{bmatrix}, \tag{6.57}$$

where $\mathbb{M} = \mathbb{P}^T \mathbb{L}(x)\mathbb{P}$.

6.3.3 Graph controllability

The presence of inherent dynamics dictated by the state evolution of the agents in the network opens up the need to define the notion of *controllability* for dynamic graph processes. Similar to controllability for dynamical systems, the controllability property for graph processes ensures that the graph states are tractable via the state-evolution of each agents; therefore, there exists a mapping from the state-space of agents to the state-space of the graph process. In this section, we recall some results from [8]. Let $\mathbb{G}^\Delta_N$ denote the set of graphs of order N with maximum vertex degree Δ.

Definition 6.3 (Strictly controllable graph process, [8]). A $\mathbb{G}$-process is strictly Δ-controllable if for any $\mathbb{G}_f \in \mathbb{G}^\Delta_N$ and any initial state $x_0 \in \mathbb{X}$, there exists a finite t, and an x-process for which $\mathbb{G}_f \subset g_S(x(t))$. When $\Delta = N - 1$, the Δ-controllable $\mathbb{G}$ processes are called strictly controllable.

Definition 6.4 (Controllable graph process, [8]). A $\mathbb{G}$-process is Δ-controllable if for two graphs $\mathbb{G}_0, \mathbb{G}_f \in \mathbb{G}_N^\Delta$, there exist a finite k and an x-process for which $\mathbb{G}_0 \subset g_s(x(0))$ and $\mathbb{G}_f \subset g_s(x(k))$. When $\Delta = N - 1$, the Δ-controllable $\mathbb{G}$-processes are called simply controllable.

Definition 6.5 (Strictly calm graphs, [8]). A graph $\mathbb{G} \in \mathbb{G}_n^\Delta$ is strictly calm with respect to the controlled x-process if
- for any x_0, x_f for which $\mathbb{G} \subseteq g_s(x_0), g_s(x_f)$, there exists a control sequence that steers x_0 to x_f,
- for all intermediate states, $\mathbb{G} \subseteq g_s(x)$.

Proposition 6.1 ([8]). *If a $\mathbb{G}$-process is Δ-controllable and $g_s(X)$ is calm with respect to the x-process, then the $\mathbb{G}$-process is strictly Δ-controllable in X_Δ.*

Theorem 6.3 ([8]). *A $\mathbb{G}$-process is Δ-controllable if the x-process is controllable and the supergraph $\mathbb{G}(N, \mathbb{S}_{\epsilon,\rho})$ satisfies $(\rho - \epsilon)^\Delta / (1 + \Delta) \geq \epsilon$. On the other hand, the x-process is controllable if the $\mathbb{G}$-process is controllable and $g_s(X)$ is calm with respect to the x-process.*

Definition 6.6. For $\epsilon > 0$, the pair (X_i, X_j) is called ϵ-regular at level ρ if
- $d_{\mathbb{S}}(X_i, X_j) \geq \rho$,
- for every $Y_i \subseteq X_i$ and $Y_j \subseteq X_j$ satisfying $|Y_i| > \epsilon |X_i|$ and $|Y_j| > \epsilon |X_j|$, one has

$$|d_{\mathbb{S}}(X_i, X_j) - d_{\mathbb{S}}(Y_i, Y_j)| < \epsilon. \tag{6.58}$$

6.4 Fuzzy graph theory

Fuzzy graphs are based on the notation of a fuzzy set originally introduced by Lotfi Zadeh in [14]. A fuzzy set $\mathbb{F}$ in $\mathbb{Q}$ is characterized by a function μ which measures the degree of association of members $\mathbb{Q}$ in $\mathbb{F}$. Essentially, fuzzy sets provide mathematical basis for measuring impreciseness or uncertainty and they are used in numerous fields, including communication, biology, social sciences control, and artificial intelligence to model real world phenomena.

6.4.1 Important definitions

In the sequel, we provide some relevant definitions and information pertaining to fuzzy graph theory.

Definition 6.7 (Fuzzy relation, [15]). A fuzzy relation $\mu(v_1, v_2)$ on $\mathbb{V}$ is a fuzzy subset of the set $\mathbb{V} \times \mathbb{V}$ such that $0 \leq \mu(v_1, v_2) \leq 1 \ \forall(v_1, v_2) \in \mathbb{V} \times \mathbb{V}$ and for a fuzzy subset σ, one has the following:

$$\mu(v_1, v_2) \leq \sigma(v_1) \wedge \sigma(v_2). \tag{6.59}$$

Definition 6.8 (Convex fuzzy set, [15]). A fuzzy subset τ of a set $\mathbb{V}$ is convex if for all $\lambda \in [0, 1]$, the following condition is satisfied:

$$\tau(\lambda v_1 + (1 - \lambda)v_2) \geq \tau(v_1) \wedge \tau(v_2) \quad \forall v_1, v_2 \in \mathbb{V}. \tag{6.60}$$

Definition 6.9 (Max–min composition, [15]). Consider fuzzy relations $\mu : \mathbb{S} \times \mathbb{T} \to [0, 1]$ and $\nu : \mathbb{T} \times \mathbb{U} \to [0, 1]$ where μ is a fuzzy relation that maps elements from the fuzzy subset σ of $\mathbb{S}$ into the fuzzy subset τ of $\mathbb{T}$, likewise ν is a fuzzy relation that maps elements from the fuzzy subset ρ of $\mathbb{T}$ into a fuzzy subset η of $\mathbb{U}$, the *max–min composition* of σ and τ is defined as:

$$(\mu \circ \nu)(x, z) = \vee\{\mu(x, y) \wedge \nu(y, z) | y \in \mathbb{T}\} \quad \forall x \in \mathbb{S}, z \in \mathbb{U}. \tag{6.61}$$

Fuzzy graphs rely on the mathematics developed in fuzzy set theory to represent uncertainties in graph topologies. In the field of multivehicle control, it could provide a strong analytical tool for modeling topological uncertainties resulting from varying levels of trust and beliefs among agents, uncertainties resulting from network phenomena such as delays, dropouts, cyberphysical attacks, and physical constraints like line-of-sight. In this section, we reintroduce some basic definitions and results already established in fuzzy graph theory to aid development of our results.

Definition 6.10 (Fuzzy graph, [15]). A fuzzy graph is a triple $((\mathbb{V}, \mathbb{E}), \sigma, \mu)$ denoted by $\mathbb{G}(\sigma, \mu)$ on a nonempty vertex set $\mathbb{V}$ with fuzzy membership function $\sigma : \mathbb{V} \to [0, 1]$ and fuzzy relation $\mu : \mathbb{E} \to [0, 1]$.

Consider the vertex set $\mathbb{V} = \{a, b, c\}$. The fuzzy membership σ on each node in $\mathbb{V}$ is given by $\sigma(a) = 0.5$, $\sigma(b) = 1$, and $\sigma(c) = 0.8$. Let the fuzzy set μ on the edge set $\mathbb{E}$ be defined as $\mu(a, b) = 0.5$, $\mu(b, c) = 0.7$, and $\mu(a, c) = 0.1$. Then the corresponding graph $\mathbb{G}(\sigma, \mu)$ is a fuzzy graph. However, if $\sigma(a) = 0.6$, the graph is not fuzzy.

6.4.2 Connectivity in fuzzy graphs

Similar to crisp graphs, some important notions related to connectivity such as trees, paths, and cycles are also defined for fuzzy graphs. Given a fuzzy graph $\mathbb{G}(\sigma, \mu)$, we recall some of these definitions and theorems from [15]. A sequence of distinct vertices $v_0, v_1, \ldots, v_n$ such that $\mu(v_{i-1}, v_i) > 0$ is called a *path*, $\mathbb{P}$, of the fuzzy graph, and n is the length of the path. The strength of path $\mathbb{P}$, denoted by $s(\mathbb{P})$, is the weight of the weakest edge in the path, that is, $\bigwedge_{i=1}^{n} \mu(v_{i-1}, v_i)$. For any vertex pair, $(v_i, v_j) \in \mathbb{V} \times \mathbb{V}$, the diameter of (v_i, v_j), denoted diam(v_i, v_j), is the length of the longest path between v_i and v_j. The strength of connectedness, $\mu^{\infty}(v_i, v_j)$ between any two vertices $v_i, v_j \in \mathbb{V}$ is the maximum of the strengths of all paths between v_i and v_j. The maximum spanning tree of a connected fuzzy graph $\mathbb{G}(\sigma, \mu)$ is a spanning fuzzy subgraph $\mathbb{T}(\sigma, \mu)$ which is a tree such that $\mu^{\infty}(u, v)$ is the strength of the unique strongest u–v path in $\mathbb{T}$ for all $u, v \in \mathbb{V}$. A fuzzy graph is said to be connected if $\mu^{\infty}(u, v) > 0$ for any pair of vertices $u, v \in \mathbb{V}$. The degree of any node v in $\mathbb{G}(\mu, \sigma)$ is defined as $d(v) = \sum_{u \neq v} \mu(u, v)$. The minimum degree of $\mathbb{G}(\mu, \sigma)$ is $\delta(\mathbb{G}) = \wedge \{d(v) | v \in \mathbb{V}\}$. The maximum degree of $\mathbb{G}(\mu, \sigma)$ is $\Delta(\mathbb{G}) = \vee \{d(v) | v \in \mathbb{V}\}$.

Definition 6.11 (Fuzzy bridge, [15]). Consider a fuzzy graph $\mathbb{G}(\sigma, \mu)$ with two distinct vertices v_i and v_j. If a partial fuzzy subgraph $\mathbb{G}'$ is obtained by deleting the edge (v_i, v_j), that is, $\mathbb{G}'(\sigma, \mu')$, where $\mu'(v_i, v_j) = 0$ and $\mu' = \mu$ for all other edges, then the edge (v_i, v_j) is called a fuzzy bridge if $\mu'^{\infty}(u, v) < \mu^{\infty}(u, v)$, $u, v \subset \mathbb{V}$. A fuzzy bridge is an edge whose deletion reduces the strength of connectedness between any two pair of vertices. Equivalently, (v_i, v_j) is a fuzzy bridge if there exist vertices u, v such that (v_i, v_j) is an edge of every strongest path connecting u to v.

Definition 6.12 (Fuzzy cut vertex, [15]). Given a fuzzy graph $\mathbb{G}(\sigma, \mu)$, let w be any vertex removed from $\mathbb{G}$ resulting in the partial fuzzy subgraph $\mathbb{G}'(\sigma', \mu')$. Then w is a cut vertex if $\mu'^{\infty}(u, v) < \mu^{\infty}(u, v)$ for some $u, v \in \mathbb{V}$, that is, $\sigma'(w) = 0$, $\sigma = \sigma'$, $\mu'(w, z) = 0$, for all vertices $\mu' = \mu$ such that $u \neq v \neq w$.

Theorem 6.4 ([6]). *For a fuzzy graph $\mathbb{G}(\sigma, \mu)$, the following statements are equivalent:*

- (x, y) *is a fuzzy bridge.*
- $\mu'^{\infty}(x, y) < \mu(x, y)$.
- (x, y) *is not the weakest edge of any cycle.*

Corollary 6.2 ([15]). *Every fuzzy graph $\mathbb{G}(\mu, \sigma)$ has at least two vertices which are not fuzzy cut vertices.*

Corollary 6.3 ([15]). *An edge (u, v) of a fuzzy graph $\mathbb{G}(\sigma, \mu)$ is a fuzzy bridge if and only if (u, v) is in every maximum spanning tree of $\mathbb{G}(\mu, \sigma)$.*

Definition 6.13 (Complete fuzzy graph, [15]). A complete fuzzy graph is a fuzzy graph $\mathbb{G}(\sigma, \mu)$ such that $\mu(u, v) = \sigma(u) \wedge \sigma(v) \; \forall u, v \in \mathbb{V}$. If $\mathbb{G}(\sigma, \mu)$ is a complete fuzzy graph, then $\mu^\infty = \mu$ and $\mathbb{G}(\mu, \sigma)$ has no fuzzy cut vertices.

Theorem 6.5. *If* $\mathbb{G}(\sigma, \mu)$ *is a complete fuzzy graph, then for any edge* $(u, v) \in \mathbb{E}$, *then* $\mu^\infty(u, v) = \mu(u, v)$.

Theorem 6.6. *Let* $\mathbb{G}(\sigma, \mu)$ *be a complete fuzzy graph with* $|\mathbb{E}| = n$, *then* $\mathbb{G}(\mu, \sigma)$ *has a fuzzy bridge if and only if there exists an increasing sequence*

$$\{\sigma(u_1), \sigma(u_2), \ldots, \sigma(u_n)\}$$

such that $\sigma(u_{n-2}) < \sigma(u_{n-1}) < \sigma(n)$. *Also, the edge* (u_{n-1}, u_n) *is the fuzzy bridge of* $\mathbb{G}(\sigma, \mu)$.

In the sequel, let $\mathbb{G}(\sigma, \mu)$ be a fuzzy graph containing a cycle.

Theorem 6.7. *Let* $\mathbb{G}(\sigma, \mu)$ *be a fuzzy graph, such that* (σ^*, μ^*) *is a cycle. Then a vertex of* $\mathbb{G}(\sigma, \mu)$ *is a fuzzy cut vertex if and only if it is a common vertex of two fuzzy bridges.*

Theorem 6.8. *If* w *is a common vertex of at least two fuzzy bridges, then* w *is a fuzzy cut vertex.*

Theorem 6.9. *If* (u, v) *is a fuzzy bridge, then* $\mu^\infty(u, v) = \mu(u, v)$.

Definition 6.14 (Strong edge). An edge (x, y) in fuzzy graph $\mathbb{G}(\sigma, \mu)$ is called strong if $\mu(x, y) \geq \mu^\infty(x, y)$.

Definition 6.15. An edge (x, y) in a fuzzy graph $\mathbb{G}(\sigma, \mu)$ is called α-strong if

$$\mu(x, y) > \mu^\infty(x, y).$$

It is called β-strong if $\mu(x, y) = \mu^\infty(x, y)$ and a δ-edge if

$$\mu(xy) < \mu^\infty(x, y).$$

6.5 State-dependent fuzzy graphs

In this section, we propose some new results for reaching consensus over state-dependent fuzzy graphs. First, we introduce the notion of state-dependent fuzzy graphs and proceed to introduce some assumptions and theorems to guarantee convergence of both the graph and states of the system.

6.5.1 Problem statement

In the foregoing discussions, we consider a system of N identical agents where each agent i is defined by the following dynamics:

$$\dot{x}_i(t) = f(x_i(t), u_i(t)), \tag{6.62}$$

where $x_i \in \mathbb{R}^n$ and $u_i \in \mathbb{R}^m$ are the states and control inputs, respectively; f in (6.62) represents any linear or nonlinear state function. In this analysis, we examine the average leaderless consensus problem.

Definition 6.16 (Leaderless asymptotic consensus). A set consisting of N agents described by (6.62) is said to reach average consensus if

$$\lim_{t \to \infty} ||x_i(t) - x_j(t)|| = 0. \tag{6.63}$$

We recall the following lemma for MAS over crisp graphs in [1].

Lemma 6.2. *Consider an MAS defined over the crisp graph* $\mathbb{G} = (\mathbb{V}, \mathbb{E})$, *with agent states* $x = [x_1, \ldots, x_n]$, $x_i \in \mathbb{R}^m$, *and the corresponding Laplacian matrix* $\mathbb{L}_n \in \mathbb{R}^{n \times n}$ *such that*

- $\mathbb{L}_n$ *has a simple zero eigenvalue with an associated eigenvector* $\mathbb{I}_n$ *and all other eigenvalues have positive real parts;*
- $(\mathbb{L}_n \otimes I_m)x = 0$ *implies that* $x_1 = \cdots = x_n$;
- *Consensus is reached asymptotically for the system* $\dot{x} = -(\mathbb{L}_n \otimes I_m)x$;
- *The directed graph of* $\mathbb{L}_n$ *has a directed spanning tree;*
- *The rank of* $\mathbb{L}_n$ *is* $n - 1$.

The conditions for reaching (average) consensus on a crisp graph as given by [1] dictate that the directed graph $\mathbb{D}_n$ is strongly connected and balanced, or simply, the undirected graph $\mathbb{G}_n$ is connected.

Consider the MAS (6.62), let us for simplicity assume that the agents exhibit a uniform single-integrator dynamic behavior, that is,

$$\dot{x}_i(t) = u_i(t), \quad i \in \mathbb{V}, \tag{6.64}$$

where $x_i(t) \in \mathbb{R}$ represents the position of each agent and $u_i(t) \in \mathbb{R}$ describes the control input of the ith agent. For this system, the distributed control protocol is defined as

$$u_i(t) = -\sum_{j \in \mathbb{N}_i} a_{ij}(x_i(t) - x_j(t)), \tag{6.65}$$

where a_{ij} represent the adjacency information between any two neighbors i and j. In classical consensus problems, $a_{ij} = 1$, if there is a link between i and j, and $a_{ij} = 0$ otherwise, with the assumptions that there are no self-loops or multiple edges between any two nodes. Contrary to these definitions, we introduce new definitions based on fuzzy memberships and relations for defining adjacency for distributed multivehicle coordination. We let

$$a_{ij} = \begin{cases} 0, & i = j, \\ \mu_{ij}, & i \neq j, \end{cases} \tag{6.66}$$

where μ_{ij} denotes the fuzzy relation on the set $\mathbb{V} \times \mathbb{V}$ for any pair of neighborhood vertices, (i, j). In very simplistic terms, the fuzzy graph for the distributed multivehicle coordination problem is a weighted graph whose weights are fuzzy in nature and satisfy

$$\mu_{ij} \leq \sigma_i \wedge \sigma_j, \tag{6.67}$$

where σ_i is a fuzzy membership value of node i, and μ_{ij} is a fuzzy relation which measures the degree of relationship between any pair of nodes. Similar to crisp graphs, special matrices such as degree, adjacency and Laplacian are introduced. The degree matrix, $\mathbb{D}$, is a diagonal matrix with entries $d_{ii} = \sum_{j=1}^{N} \mu_{ij}$, the adjacency matrix is defined as $\mathbb{A} = [\mu_{ij}]$. The following fuzzy graph framework can be used to define some important notions for the distributed consensus problem. The membership values can be used to describe the level of trust, or opinions, about the consensus problem. If we consider the consensus objective in (6.63), σ_i can be used to measure the degree by which agent i is close to reaching the distributed objective (6.63).

The fuzzy membership and relation functions applied to the graph theory framework can be used to represent some level of uncertainty in decision making involving coalitions and reaching the global consensus objective. However, in the static form, this fuzzy graph can simply be interpreted as a weighted graph constrained by (6.67). More applications can be derived if we were to introduce state-dependence in the fuzzy graph framework. That is, the fuzzy membership values σ_i and edge relations μ_{ij} are now rewritten as state-dependent functions of the agent states $x(t)$. We now proceed to formally define *state-dependent fuzzy graphs*.

Definition 6.17 (State-dependent fuzzy graph). A state-dependent fuzzy graph $\mathbb{G}(x(t), \sigma(x(t)), \mu(x(t)))$ is a fuzzy graph on a nonempty vertex set $\mathbb{V}$ with state-dependent fuzzy membership function $\sigma(x(t)) : \mathbb{V} \to [0, 1]$

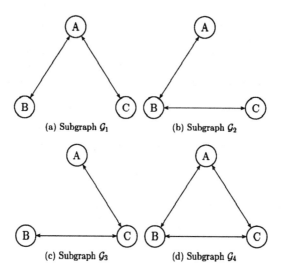

(a) Subgraph $\mathcal{G}_1$ (b) Subgraph $\mathcal{G}_2$

(c) Subgraph $\mathcal{G}_3$ (d) Subgraph $\mathcal{G}_4$

Figure 6.1 Graphs on $n = 3$ vertices.

and state-dependent fuzzy relations $\mu(x(t)) : \mathbb{E} \to [0, 1]$, where $x(t) \in \mathbb{R}^{nm} = [x_1, \dots, x_n]$, $x_i \in \mathbb{R}^m$ represent the states of each node i. That is, the state-dependent fuzzy graph is a mapping from the distributed agent state space X to the set of all possible weighted fuzzy graph topologies of order $|\mathbb{V}|$, whose membership functions and fuzzy relations are dictated by $\sigma(x)$ and $\mu(x)$.

Some important questions are posed by this new approach to defining and reaching multiagent consensus on state-dependent fuzzy graphs:

- How will the state-dependent membership functions $\sigma(x_i(t))$ be defined for each agent?
- How do we define the fuzzy relations $\mu_{ij}(x_i(t), x_j(t))$?
- How will connectivity be preserved?

Consider the MAS (6.64) described by the graph $\mathbb{G}$ consisting of three agents. We consider four possible connected topologies as depicted in Fig. 6.1, with the corresponding Laplacians:

$$\mathbb{L}(\mathbb{G}_1) = \begin{bmatrix} 2 & -1 & -1 \\ -1 & 1 & 0 \\ -1 & 0 & 1 \end{bmatrix}, \quad \mathbb{L}(\mathbb{G}_2) = \begin{bmatrix} 1 & -1 & 0 \\ -1 & 2 & -1 \\ 0 & -1 & 1 \end{bmatrix},$$

$$\mathbb{L}(\mathbb{G}_3) = \begin{bmatrix} 1 & 0 & -1 \\ 0 & 1 & -1 \\ -1 & -1 & 2 \end{bmatrix}, \quad \mathbb{L}(\mathbb{G}_4) = \begin{bmatrix} 1 & 0 & 1 \\ -1 & 2 & -1 \\ -1 & -1 & 2 \end{bmatrix}.$$

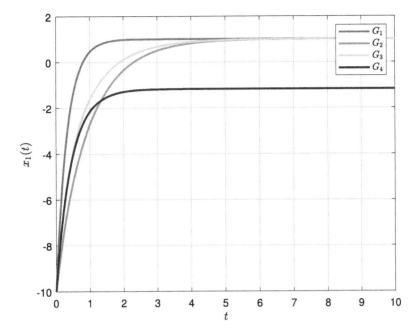

Figure 6.2 State evolution on $\mathbb{G}_1$.

Interestingly, three of the graph structures have the same Laplacian spectrum, that is, $\lambda(\mathbb{L}(\mathbb{G}_1)) - \lambda(\mathbb{L}(\mathbb{G}_2)) = \lambda(\mathbb{L}(\mathbb{G}_3)) = [0, 1, 3]$, due to their structural properties. The Laplacian spectrum of the fourth graph, $\mathbb{G}_4$, is $\lambda(\mathbb{L}(\mathbb{G}_4)) = [0, 2, 3]$. Also $\mathbb{G}_1$, $\mathbb{G}_2$, $\mathbb{G}_3$, are subgraphs of $\mathbb{G}_4$. It is worth-while to note that although the graph topologies, $\mathbb{G}_1$, $\mathbb{G}_2$, $\mathbb{G}_3$, are similar, the behavior of the systems differs significantly due to the difference in the edge dynamics. This means that the edge flow dynamics have some impact, and Figs. 6.2–6.5 show the state response of each agent on different graph topologies while Figs. 6.6– 6.8 compare the state response of each agent on different graph topologies.

6.5.2 Fuzzy membership functions

In other to apply the concept of state-dependent fuzzy graphs to the con-sensus problem, we need to establish a framework to derive membership functions of each node $i \in \mathbb{V}$ based on the current state information $x_i(t)$ on each node. The very first step is to define in clear terms the member-ship set. In this discussion, the membership set refers to the agreement set

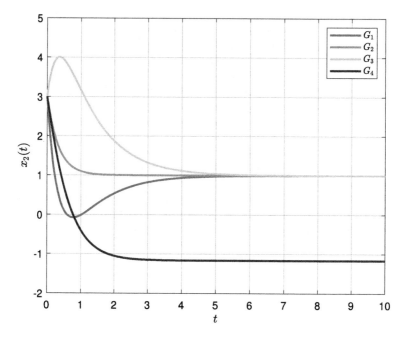

Figure 6.3 State evolution on $\mathbb{G}_2$.

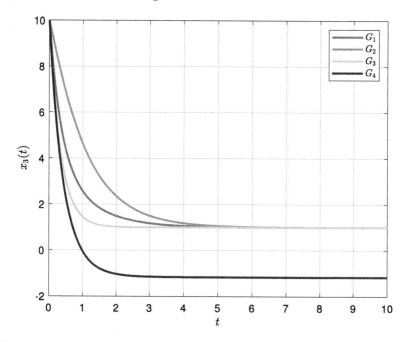

Figure 6.4 State evolution on $\mathbb{G}_3$.

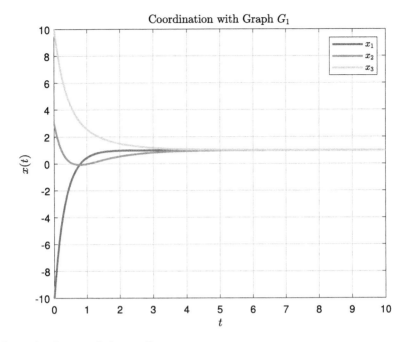

Figure 6.5 State evolution on $\mathbb{G}_4$.

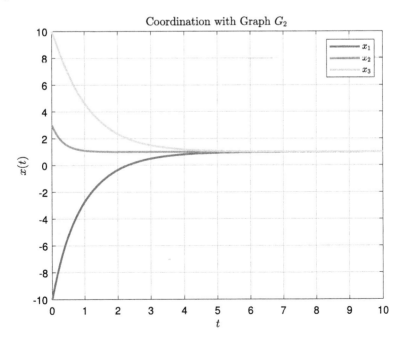

Figure 6.6 Behavior of agent 1 on different topologies.

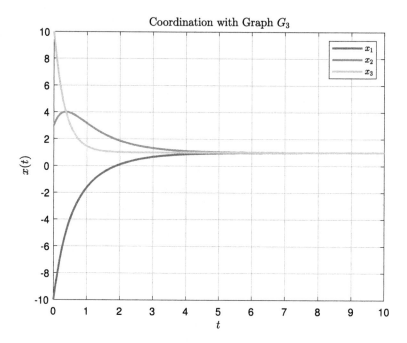

Figure 6.7 Behavior of agent 2.

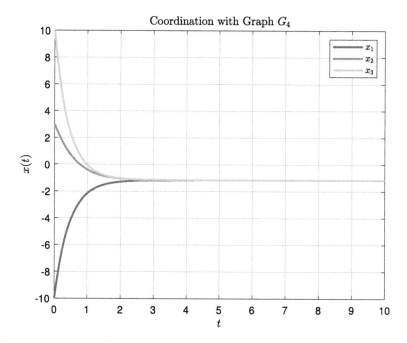

Figure 6.8 Behavior of agent 2.

Table 6.1 Some fuzzy membership functions.

Type	Equation
Triangular	$\sigma_{\mathbb{A}}(x) = \begin{cases} 0, & x \leq a, \\ \dfrac{x-a}{m-a}, & a \leq x \leq m, \\ \dfrac{b-x}{b-m}, & m \leq x < b, \\ 0, & x \geq b \end{cases}$
Trapezoidal	$\sigma_{\mathbb{A}}(x) = \begin{cases} 0, & x < a \text{ or } x > d, \\ \dfrac{x-a}{b-a}, & a \leq x \leq b, \\ 1, & b \leq x \leq c, \\ \dfrac{d-x}{d-c}, & c \leq x < b \end{cases}$
Gaussian	$\sigma_{\mathbb{A}}(x) = e^{\dfrac{(x-m)^2}{2k^2}}$

onto which the states of the agents converge or will converge. We recall, the formal definition of this agreement set as defined in [9].

Definition 6.18 ([9]). The agreement set $\mathbb{A} \subset \mathbb{R}^n$ is the subspace span$\{1\}$, that is,

$$\mathbb{A} = \{x \in \mathbb{R}^n |\ x_i = x_j\ \forall i, j\}. \tag{6.68}$$

Now, we proceed to define some membership relations for the agents which serve as a measure of *belongingness* to the agreement set at each time. Fuzzy membership functions (see Table 6.1) such as triangular, trapezoidal, and Gaussian functions can be used to parametrize the membership of each agent in the agreement set. These functions can be used to *fuzzify* the states of the agents in the agreement space.

In this discussion, we consider the following triangular membership function:

$$\sigma_{\mathbb{A}}(x) = \begin{cases} \epsilon, & x \leq x_{\min}(0), \\ \dfrac{x - x_{\min}(0)}{\bar{x}(t) - x_{\min}(0)}, & x_{\min}(0) \leq x_i(t) \leq \bar{x}(t), \\ \dfrac{x_{\max}(0) - x_i(t)}{x_{\max}(0) - \bar{x}(t)}, & \bar{x}(t) \leq x_i(t) < x_{\max}(0), \\ \epsilon, & x_i(t) \geq x_{\max}(0), \end{cases} \tag{6.69}$$

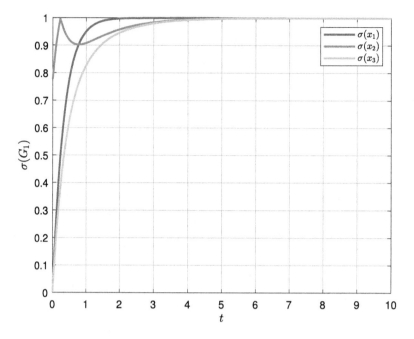

Figure 6.9 Membership functions for agents on $\mathbb{G}_1$.

where $\bar{x} = \dfrac{1}{N} \sum_{j=1}^{N} x_j(t)$. So for an initial condition of $x(0) = [-10; 3; 10]$, the corresponding membership function of each agent becomes

$$\sigma_{\mathbb{G}_1}(x(0)) = \sigma_{\mathbb{G}_2}(x(0)) = \sigma_{\mathbb{G}_3}(x(0)) = \sigma_{\mathbb{G}_4}(x(0))$$
$$= [0.0010; 0.7778; 0.0010].$$

The dynamics of the membership functions for each graph type is depicted in Figs. 6.9–6.12. As shown, the topologies influence the behavior of the membership functions for each state which is made clearer in Figs. 6.13–6.15. This choice of the membership function (6.69) assumes that each agent has global knowledge of the initial conditions $x(0) \in \mathbb{R}^N$ and the states of other agents at any time $t > 0$ when the graph is not complete. This assumption is not quite accurate, a more proper characterization of the membership function would be of a distributed projection of the agreement set $\mathbb{A}$. To establish this membership function, it is necessary to define two important neighborhood sets for the each agent, the reachable and active neighborhood sets.

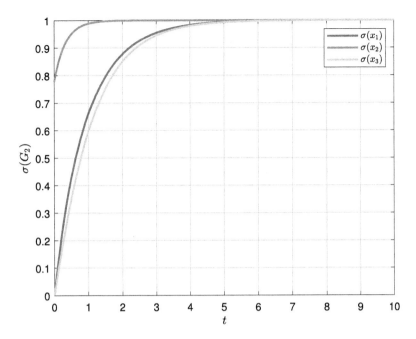

Figure 6.10 Membership functions for agents on $\mathbb{G}_2$.

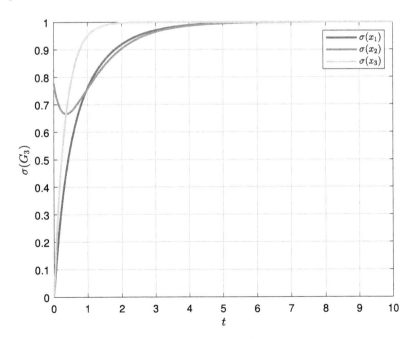

Figure 6.11 Membership functions for agents on $\mathbb{G}_3$.

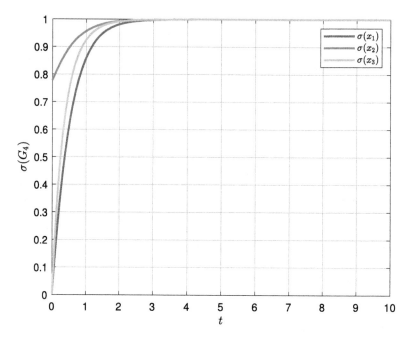

Figure 6.12 Membership functions for agents on $\mathbb{G}_4$.

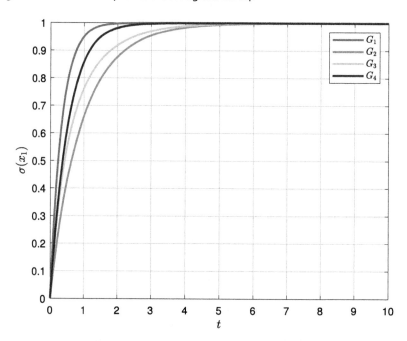

Figure 6.13 Membership functions for agent 1 on different graph topologies.

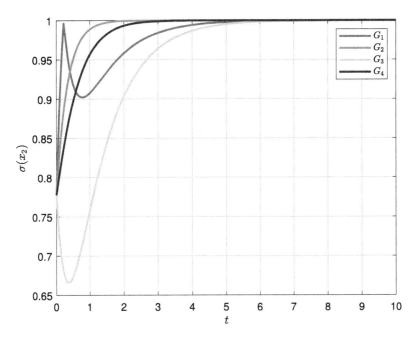

Figure 6.14 Membership functions for agent 2 on different graph topologies.

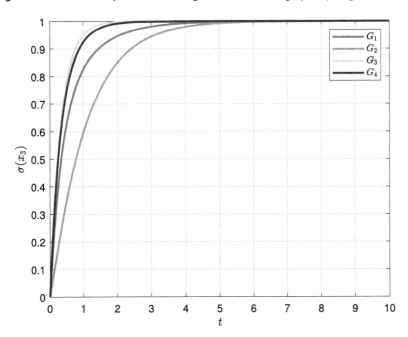

Figure 6.15 Membership functions for agent 3 on different graph topologies.

Definition 6.19 (Reachable neighborhood set). The reachable neighborhood set $\mathbb{N}_i$ for any node i is the set of all agents that agent i could establish communication with at any time t.

Definition 6.20 (Active neighborhood set). The active neighborhood set $\mathbb{N}_i(t)$ for any node i is the set of all agents that agent i exchanges information with at time t. Clearly, $\mathbb{N}_i(t) \subset \mathbb{N}_i$.

Now we proceed to formally define the membership function of each agent based on the reachable neighborhood set:

$$\sigma_\mathbb{A}(x_i) = \begin{cases} \epsilon, & x \leq x_{\min}(0), \ x_{\min}(0) = \min_{\mathbb{N}_i} x(0), \\ \dfrac{x - x_{\min}(0)}{\bar{x}(t) - x_{\min}(0)}, & x_{\min}(0) \leq x_i(t) \leq \bar{x}(t), \\ \dfrac{x_{\max}(0) - x_i(t)}{x_{\max}(0) - \bar{x}(t)}, & \bar{x}(t) \leq x_i(t) < x_{\max}(0), \\ \epsilon, & x_i(t) \geq x_{\max}(0), \ x_{\min}(0) = \max_{\mathbb{N}_i} x(0), \end{cases} \tag{6.70}$$

where $\bar{x}(t) = \sum_{j \in \mathbb{N}_i} x_j(t)$.

6.5.3 Fuzzy relations

Next, we need to establish the basis for establishing weights between agents in the fuzzy graph based on the states and membership relationship of each agent. A necessary condition on fuzzy graphs stipulates that the fuzzy weights satisfy

$$\mu_{ij}(x_i(t), x_j(t)) \leq \sigma(x_i(t)) \wedge \sigma(x_j(t)). \tag{6.71}$$

For simulation purposes, we assume the weights to be

$$\mu_{ij}(x_i(t), x_j(t)) = \sigma(x_i(t)) \wedge \sigma(x_j(t)). \tag{6.72}$$

Based on this approach, the state-response of the system on the fuzzy communication structure is depicted in Fig. 6.16

6.6 Notes

In this chapter, we have discussed state-dependent and fuzzy graphs. We reviewed some important theorems and problems related to connectivity issues on state-dependent graphs. Next we introduced a framework for reaching consensus on state-dependent fuzzy graphs.

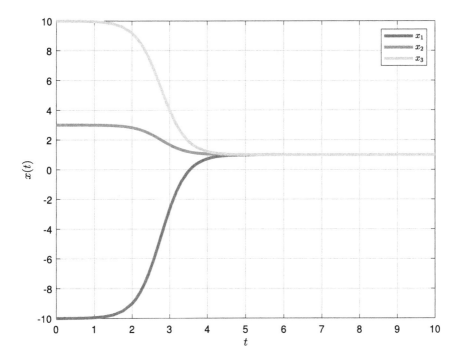

Figure 6.16 Fuzzy consensus.

References

[1] W. Ren, R.W. Beard, Consensus seeking in multiagent systems under dynamically changing interaction topologies, IEEE Trans. Autom. Control 50 (5) (April 2005) 655–661.

[2] Y. Kim, M. Mesbahi, On maximizing the second smallest eigenvalue of the state-dependent graph Laplacian, IEEE Trans. Autom. Control 51 (1) (2006) 116–120.

[3] M.M. Zavlanos, G.J. Pappas, Potential fields for maintaining connectivity of mobile networks, IEEE Trans. Robot. 23 (4) (2007) 812–816.

[4] M. Ji, M. Egerstedt, Distributed coordination control of multiagent systems while preserving connectedness, IEEE Trans. Robot. 23 (4) (2007) 693–703.

[5] A. Bogojeska, M. Mirchev, I. Mishkovski, L. Kocarev, Synchronization and consensus in state-dependent networks, IEEE Trans. Circuits Syst. I, Regul. Pap. 61 (2) (2014) 522–529.

[6] A. Rosenfield, Fuzzy graphs, in: L.A. Zadeh, K.S. Fu, M. Shimura (Eds.), Fuzzy Sets and Their Applications, Academic Press, New York, 1975, pp. 77–95.

[7] R. Olfati-Saber, R.M. Murray, Consensus problems in network of agents with switching topology and time-delays, IEEE Trans. Autom. Control 49 (9) (Sep. 2004) 1520–1533.

[8] M. Mesbahi, On state-dependent dynamic graphs and their controllability properties, IEEE Trans. Autom. Control 50 (3) (March 2005) 387–392.

[9] M. Mesbahi, M. Egerstedt, Graph Theoretic Methods in Multiagent Networks, Princeton Series in Applied Mathematics, 2010.

[10] J. Qin, C. Yu, Cluster consensus control of generic linear multi-agent systems under directed topology with acyclic partition, Automatica 49 (Apr. 2013) 2898–2905.

[11] Y. Feng, J. Lu, S. Xua, Y. Zou, Couple-group consensus for multi-agent networks of agents with discrete-time second-order dynamics, J. Franklin Inst. 350 (2013) 3277–3292.

[12] M.O. Oyedeji, M.S. Mahmoud, Couple-group consensus conditions for general first-order multiagent systems with communication delays, Syst. Control Lett. 117 (2018) 37–44.

[13] M. Mesbahi, On a dynamic extension of the theory of graphs, in: Proc. the American Control Conference, Anchorage, May 2002, pp. 1234–1239.

[14] L.A. Zadeh, Fuzzy sets, Inf. Control 8 (1965) 338–353.

[15] S. Mathew, J.N. Mordeson, D.S. Malik, Graph Theory, Springer, 2018.

[16] Y. Shang, Resilient consensus of switched multi-agent systems, Syst. Control Lett. 122 (2018) 12–18.

[17] J. Usevitch, K. Garg, D. Panagou, Finite-time resilient formation control with bounded inputs, arXiv, 2018.

[18] Y. Wu, X. He, S. Liu, Resilient consensus for multi-agent systems with quantized communication, in: Proc. the American Control Conference, 2016, pp. 5136–5140.

[19] H.J. Leblanc, H. Zhang, X. Koutsoukos, S. Sundaram, Resilient asymptotic consensus in robust networks, IEEE J. Sel. Areas Commun. 31 (4) (2013) 766–781.

[20] J. Shao, W.X. Zheng, T.Z. Huang, A.N. Bishop, On leader–follower consensus with switching topologies: an analysis inspired by pigeon hierarchies, IEEE Trans. Autom. Control 63 (10) (2018) 3588–3593.

[21] C. Tan, G.P. Liu, P. Shi, Consensus of networked multi-agent systems with diverse time-varying communication delays, J. Franklin Inst. 352 (7) (2015) 2934–2950.

[22] Y. Zhang, Y. Sun, X. Wu, D. Sidorov, D. Panasetsky, Economic dispatch in smart grid based on fully distributed consensus algorithm with time delay, in: Chinese Control Conference, CCC, vol. 2018-July, 2018, pp. 2442–2446.

[23] X. Xu, L. Liu, G. Feng, Consensus of discrete-time linear multiagent systems with communication, input and output delays, IEEE Trans. Autom. Control 63 (2) (2018) 492–497.

[24] W. Ren, R.W. Bear, E.M. Atkins, Information consensus in multivehicle cooperative control, IEEE Control Syst. Mag. 27 (2) (2007) 71–82.

[25] P. Ming, J. Liu, S. Tan, S. Li, L. Shang, X. Yu, Consensus stabilization in stochastic multi-agent systems with Markovian switching topology, noises and delay, Neurocomputing 200 (2016) 1–10.

[26] H. Rezaee, F. Abdollahi, Consensus problem in general linear multiagent systems under stochastic topologies, IFAC-PapersOnLine 49 (13) (2016) 13–18.

[27] X. Guo, J. Lu, A. Alsaedi, F.E. Alsaadi, Bipartite consensus for multi-agent systems with antagonistic interactions and communication delays, Phys. A, Stat. Mech. Appl. 495 (2018) 488–497.

[28] J. Shao, L. Shi, Y. Zhang, Y. Cheng, On the asynchronous bipartite consensus for discrete-time second-order multi-agent systems with switching topologies, Neurocomputing 316 (2018) 105–111.

[29] J. Hu, H. Zhu, Adaptive bipartite consensus on coopetition networks, Phys. D: Nonlinear Phenom. 307 (2015) 14–21.

[30] H. Ren, F. Deng, Mean square consensus of leader-following multi-agent systems with measurement noises and time delays, ISA Trans. 71 (2017) 76–83.

[31] W. Zhu, Z.-p. Jiang, Event-based leader-following consensus of multi-agent systems with input time delay, IEEE Trans. Autom. Control 60 (5) (2015) 1362–1367.

[32] T. Xie, X. Liao, H. Li, Leader-following consensus in second-order multi-agent systems with input time delay: an event-triggered sampling approach, Neurocomputing 177 (2016) 130–135.

[33] X. Wang, H. Su, Self-triggered leader-following consensus of multi-agent systems with input time delay, Neurocomputing 330 (2018) 70–77.

[34] X. Lin, Y. Zheng, Finite-time consensus of switched multiagent systems, IEEE Trans. Syst. Man Cybern. Syst. 47 (7) (2017) 1535–1545.

[35] H. Qu, F. Yang, Q.-L. Han, Distributed $\mathcal{H}_\infty$-consensus filtering for a networked time-delay system with switching network topology and packet dropouts, in: Proc. the Australian & New Zealand Control Conference (ANZCC), 2018, pp. 334–339.

[36] Y. Wu, X. He, Secure consensus control for multiagent systems with attacks and communication delays, IEEE/CAA J. Autom. Sin. 4 (1) (2017) 136–142.

CHAPTER 7

Distributed consensus on state-dependent evolutionary graphs

7.1 Introduction

Consensus can be seen as a form of *game* played interactively until the *agents* reach an *agreement* on the shared *variable*. Communication in consensus games typically happen on communication structures called *graphs*. The nature of these graphs can be fixed, switched, or dynamic. In this chapter, we rely on some evolutionary graph theory concepts to describe the *consensus* game on graphs. This chapter offers alternative explanations on how coalitions are formed based on evolutionary games. We consider an MAS where the dynamics of each agent is defined by some discrete or continuous dynamics.

Game theory is a broad interdisciplinary field which provides analytical tools and solution concepts which help in understanding social paradigms and iterative decisions where the participation of players may be cooperative, non-cooperative, or random. Game-theoretic concepts have been applied in different fields, including economics, mathematics, biology, and psychology, offering both theoretical and practical results. Formally, game theory is a study of competition and cooperation in a population using mathematical analysis. Evolutionary game theory (EGT) is a mature field in game theory primarily concerned with the study of population dynamics and evolution of cooperation in biological and social networks. Contrary to classical game theory, evolutionary graph theory offers an approach to selection of strategies based on the process of evolution. Evolutionary games defined by biologists is a tool for predicting behaviors in a population through pairwise interactions [1]. Studies in evolutionary game theory can be broadly classified into evolutionary stability and evolutionary game dynamics. In evolutionary stability, the main concern is the selection of best strategies through evolution such that the population dynamics remain invariant to invasion by a mutant. Evolutionary stability analysis studies primarily the concept of evolutionarily stable strategies (EVSS). In [1], the authors study evolutionary games in wireless networks with respect

Advanced Distributed Consensus for Multiagent Systems
https://doi.org/10.1016/B978-0-12-821186-1.00015-5

to evolutionary stability and evolutionary game dynamics. In this section we provide a brief overview of some concepts related to game theory and evolutionary game theory. We provide formal definition of some concepts related to games and game theory.

7.1.1 Games, payoffs and strategies

This subsection provides a short introduction into the concept of games in game theory. We review some definitions and mathematical concepts directly tied to the notion of games.

Definition 7.1 ([7]). A game is an abstract formulation of an iterative situation with possibly conflicting interests.

Ideally, in a normal strategic form, the following essential elements need to be specified:
- the players of the game,
- feasible actions or moves by each player of the game,
- payoffs or rewards as a consequence of the action chosen by each player of the game.

Consider a game $\mathbb{G}$ consisting of N players. The set of possible strategies available to each player $i \in N$ is denoted by $\mathbb{S}_i = \{e_1^i, \ldots, e_m^i\}$, with $s_i \in \mathbb{S}_i$ representing an arbitrary element of the set of strategies of player i. Now, given a strategy profile of all players in a vector $s = \{s_1, \ldots, s_N\}$, let $u \in \mathbb{R}$ represent the utility or reward resulting from the players adopting a strategy s. Thus, the game can be written as $\mathbb{G} = (\mathbb{S}, \mathbb{U})$, where $\mathbb{S}$ and $\mathbb{U}$ denote the set of strategies and payoffs, respectively.

Definition 7.2 (Utility function, [7]). Given a game $\mathbb{G}$ consisting of N players and $\mathbb{S}$ possible strategies, a utility function u is any function that maps from the state-space of strategies to scalar values defining reward or profit of adopting a particular strategy.

Two-player normal-form games can either be symmetric or asymmetric. Symmetric games, also known as matrix-form games, consists of players which are identical in forms, strategies, and payoffs. That is, the role of the players can be switched, it doesn't matter which player is player 1 or 2. In asymmetric games, the form of the player affects the payoffs and strategies. Typical examples of this can be observed in sender–receiver, predator–prey, male–female, buyer–seller types of games.

Table 7.1 Battle of Bismarck sea game.

	North	South
North	2	2
South	1	3

7.1.2 The battle of the Bismarck Sea

This is a typical zero-sum game. The battle of the Bismarck sea game is a very popular game dating as far back as 1943 during WWII. The game features two admirals, Imamura and Kenney, of Japan and America, respectively. The objective for Imamura is to transport his troops across the Bismark Sea to New Guinea while Kenney wants to bomb this transport. There are two possible strategies or choices for Imamura: to follow a long southern route which will cost him three days or choose a relatively shorter northern route which will cost him two days. Admiral Kenney needs to choose one of the two routes to send his bombers to. However, if he chooses a wrong route, he has the option of calling back his planes and choosing the alternative route which will cost him an extra bombing day. The game can be represented using the payoff in Table 7.1.

Table 7.1 represents the interplay of the game between the two players. Each player is presented with two choices. Assume Kenney chooses a row and Imamura chooses a column, where these choices are assumed to be made independently and simultaneously. That is, if both players choose the northern direction, the payoff for Kenney is 2 and the payoff for Imamura is −2. The convention here is such that the payments in the table are made from player 2 to player 1. In this game, it is safe for Imamura to choose the North, and if Kenney is able to reason the same, he chooses the North, too. Imamura has a dominated choice which is South, no matter what Kenney decides, North is as good as South, and sometimes better. In this game, the (North, North) combination is maximal in its row and minimal in its column. Such a position is known as a saddle point as neither player has an incentive to deviate unilaterally.

7.1.3 Matching pennies

This game is a typical two-player pennies game where each player has a choice of heads or tails. Both players have a coin and simultaneously show heads or tails. If the coins match, player 2 gives his coin to player 1, oth-

Table 7.2 Matching pennies game.

	Heads	Tails
Heads	1	−1
Tails	−1	1

Table 7.3 Prisoner's dilemma game.

	Cooperate	Defect
Cooperate	$(-1,-1)$	$(-10,0)$
Defect	$(0,-10)$	$(-9,-9)$

erwise player 1 gives his coin to player 2. The corresponding payoff matrix for this game is shown in Table 7.2.

In this game, there is no saddle point or dominated choice. Thus, there is no natural way for solving the game. A possible solution is to allow players randomize between their choices, that is, player 1 chooses heads with a probability p and tails with probability $1 - p$. Player 2 chooses heads with probability q and tails with probability $q - 1$. Both players are equally likely of choosing heads or tails. Therefore, the expected payoff for player 1 is given by

$$\frac{1}{2}[q - (1 - q)] + \frac{1}{2}[-q + (1 - q)], \tag{7.1}$$

which is independent of q and equal to zero. The randomized choices of players are usually called *mixed strategies* and are often interpreted as beliefs of the other player about the choice of the player in consideration [10].

7.1.4 Nonzero-sum games

The prisoner's dilemma is a very popular game in game theory texts disclosing a class of nonzero–sum game. The game features two players typically called Bonnie and Clyde who have both committed a crime and are being interrogated by the police department. The punishment for this crime is 10 years imprisonment. Each player is can make two moves, either to cooperate defect. If a player betrays the other player, he gets a reduction in sentence of 1 year. A player that is not betrayed is convicted of a minor crime and spends 1 year in prison. The corresponding payoff matrix for this game is given in Table 7.3.

For both players, cooperation is the dominated choice, defection is better than cooperation for both players.

Table 7.4 Battle of sexes game.

	Opera	Sports
Opera	$(2, 1)$	$(0, 0)$
Sports	$(0, 0)$	$(1, 2)$

Table 7.5 Matching pennies game.

	Heads	Tails
Heads	$(1, -1)$	$(-1, 1)$
Tails	$(-1, 1)$	$(1, -1)$

7.1.5 Battle of sexes

The battle-of-sexes game is considered a two-player asymmetric coordination game. In this game, there are two players, named Bob and Alice. Both players are assumed to have agreed to attending either a sport or opera event but forgotten what the consensus was. Bob prefers sport, and Alice prefers opera. However, they both prefer to be together rather than alone. The payoff matrix for the game is shown in Table 7.4.

In this game, no player has a dominated choice. It is required that the players coordinate without communication. The battle of sexes game is a metaphor for problems requiring coordination.

7.1.6 Matching pennies

Every zero-sum game is a trivial case of a nonzero-sum game. The matching pennies game discussed earlier can be interpreted as a nonzero-sum game as described by the payoff Table 7.5.

Here, no player has a dominated choice, and no strategy is optimal with respect to another.

In the Cournot game, there are two firms (players) who produce a similar product. The market price P of the product is a function of the quantity Q given by $P = \max(1 - Q, 0)$, assuming zero production costs. Each player i chooses a quantity Q_i and makes a profit of $K_i(Q_1, Q_2)$ defined by

$$K_i(Q_1, Q_2) = \begin{cases} 0, & Q_1 + Q_2 \geq 1, \\ Q_i(1 - Q_1 - Q_2), & \text{otherwise.} \end{cases} \tag{7.2}$$

In this game, each player has infinitely many choices and a combination of strategies consists of mutual best replies, resulting in Nash equilibrium.

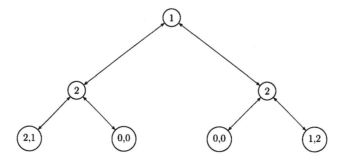

Figure 7.1 The decision tree of the battle of sexes game.

7.1.7 Extensive form games

The structure of games discussed so far has been one-shot in nature. However, in real-life games players move sequentially and observe or partially observe each others' move. These forms of games are called extensive games. Consider the battle of sexes game played in a sequential manner where the man chooses first and the woman can observe the choice of the man.

Player 1 (man) chooses first, and player 2 (woman) observes player 1's choice and then makes her own choice. In Fig. 7.1, the number in each node represents the payoff to each player. The first number is the payoff to player 1, and the second number represents payoff to player 2. The game can be analyzed backwards as follows:

- If player 1 chooses football, then it is optimal for player 2 to choose football.
- If player 1 chooses opera, then it is optimal for player 2 to choose opera.
- Given this sort of response from player 2, and assuming player 1 knows player 2 will behave in this manner, it is optimal for player 1 to choose football.

7.1.8 Cooperative games

In cooperative games, the players focus is on payoffs and coalitions rather than strategies with the implicit assumption that players can make binding agreements. Consider the example in [10] where three cities are required to be connected to a power source. In this game, there are possible transmission links and associated costs between the cities and the power source. The cities can, however, make some binding agreements. If the cities cooperate in hiring the links, they can save on transmission costs, assuming each link

Table 7.6 Three cities game.

S	{1}	{2}	{3}	{1, 2}	{1, 3}	{2, 3}	{1, 2, 3}
$c(S)$	100	140	130	150	130	150	150
$v(S)$	0	0	0	90	100	120	220

has infinite capacity. Let the set of players be denoted by $N = \{1, 2, 3\}$. The player can form coalitions which is based on subset of N. Table 7.6 presents the associated costs and savings for hiring any combination of links.

The expression for the relationship between costs and savings is given by

$$v(S) = \sum_{i \in S} c(\{i\}) - c(S) \text{ for each nonempty } S \subseteq N. \tag{7.3}$$

The cost savings $v(S)$ for any coalition S is the difference in cost corresponding to the situation where all members of S work alone and situation where all members of S work together. The pair (N, v) is termed a cooperative game.

7.2 Evolutionary game theory

Evolutionary game theory studies games in a population of animals or individuals. The main solution strategy here is an evolutionarily stable strategy which guarantees that the population is invariant to a mutant strategy. The concept of evolutionary game theory was birthed originally in biology to explain evolutionary phenomena and rational human behavior in a population.

7.2.1 Evolutionary games: hawk and dove game

The hawk and dove game [5] is a popular game in evolutionary game theory used to describe evolutionary strategies in biological processes. Consider a large population with two kinds of individuals, hawk and dove, where the nomenclature here has been selected to represent strategies employed by biological equivalents. Reproduction in the hawk and dove game is asexual and individuals reproduce over time. The ability of an individual to reproduce is defined by its fitness. The fitness value of an individual represents its ability to reproduce and pass on its genes, that is, an individual with high fitness is more likely to reproduce and pass on its gene than an individual with a low fitness value. Fitness values in the hawk and dove game can be increased by obtaining a resource from the environment. The rules defining the interactions in a hawk and dove game are given as follows:

- When a hawk competes with a dove, the hawk takes the whole resource.
- When a dove competes with a dove, they share the resource equally.
- When a hawk completes with a hawk, they fight for the resource and have equal chances of either obtaining the resource or getting injured.

In mathematical sense, let V denote the value of the resource and C the cost of an injury. The fitness value of an individual increases by V if it fights for a resource and wins, and decreases by C if it fights and loses. Thus the dynamics of the hawk and dove game is described as follows:

- If a hawk encounters a hawk, they have equal chances of increasing their fitness by V or decreasing their fitness by V, on average this results in an increase of fitness by $(V - C)/2$.
- When a dove meets a dove, they share resources equally and each obtains an increase in fitness of $V/2$.
- When hawk meets a dove, the hawk increases its fitness by V, while the dove has no change in fitness.

The replicator dynamics for the hawk and dove game is defined as follows [5]. Let P represent the frequency of hawks in the game, that is, if $P = 0$, there are no hawks in the game, and if $P = 1$, the population is entirely made of hawks. Let $W(H)$ represent the average fitness of hawks defined as

$$W(H) = P((V - C)/2) + ((1 - P)/V). \tag{7.4}$$

The first term in (7.4) denotes the fitness obtained when a hawk competes with another hawk, while the second term denotes the increase in fitness obtained when a dove competes with a dove. The average fitness of doves is given by (7.5) where it is clear that no benefit is obtained by competing with a hawk, while in a competition with a dove resources are shared equally:

$$W(D) = (1 - P)(V/2). \tag{7.5}$$

The average fitness of any individual in the game is defined as

$$W = PW(H) + (1 - P)W(D). \tag{7.6}$$

Thus, the new frequency of hawks in the next iteration is given by

$$P' = P\frac{W(H)}{W} = P\frac{PW(H)}{PW(H) + (1-P)W(D)}. \tag{7.7}$$

At each iteration, each population increases at a rate which is proportional to its fitness. This game continues iteratively and can continue infinitely until the population reaches some equilibrium. Consider a matrix game with a finite set of m strategies labeled $\{e_1, e_2, \ldots, e_m\}$. The payoff matrix is defined by an $m \times m$ matrix P, which has entries $p_{ij} = \pi(e_i, e_j)$. A replicator dynamics can be defined for the game based on the growth rate in the number of individuals using a particular strategy

$$\dot{p}_i = p_i(\pi(e_i, p) - \pi(p, p)). \tag{7.8}$$

7.2.2 Evolutionary stability

In classical evolutionary stability analysis, the main concern is reaching an evolutionarily stable strategy, one which can be adopted by all the members in the population such that no mutant strategy can invade the population by natural selection. We discuss the notion of evolutionary stability based on [2,18]. Consider a population of individuals playing a game with n possible pure strategies. For each strategy i, let s_i denote the proportion of individuals using this strategy, then the state vector $s = (s_1, s_2, \ldots, s_n)$ represents a probability distribution of the population. The payoff or fitness $F(i|s)$ is the fitness of the strategy i in s. The more fit a strategy is at the current instant, the more likely it is that it is adopted in the next instant as agents tend to switch from worst to best strategy. Thus is a dynamic game, it is imperative to study the evolution of strategies, look for equilibrium states, and examine their stability.

If $\phi = \{p : \sum p_i = 1, p_i \geq 0\}$ denotes the state space of the population, a state p is said to be an equilibrium state if the fitnesses $F(i|p)$ are equal for all pure strategies i actually used by individuals in a population in state p. An equilibrium state is said to be stable if it continues to persist when undisturbed and when disturbed, it returns to the same equilibrium state.

Definition 7.3. A state p is called an evolutionarily stable state if for every state $q \neq p$, if we let $\bar{p} = (1-\epsilon)p + \epsilon q$, then $F(q|\bar{p}) < F(p|\bar{p})$ for sufficiently small $\epsilon > 0$.

Definition 7.4. An equilibrium state p is a regular EVSS if $F(i|p) < F(p|p)$ whenever $i \notin \text{supp}(p)$, and $xAx < 0$ whenever $\text{supp}(x) \subset \text{supp}(p)$, $x \neq 0$, and $\sum x_i = 0$.

7.3 Evolutionary graph theory

Evolutionary graph theory provides a framework for studying how the population structure affects evolutionary dynamics. As originally introduced by Liberman [3], an evolutionary graph consists of a finite population N defined by a graph $\mathbb{G}(\mathbb{V}, \mathbb{E})$, however, the edge set is formed based on the outcome of an evolutionary game.

7.3.1 Moran process

Given a population consisting of N individuals, the Moran process provides a stochastic approach for modeling evolution in a population. In the typical Moran process, in each iteration, an individual is selected at random to reproduce, then another individual is selected to die and be replaced by the duplicate of the reproducing individual. A population with a lower fixation probability is considered evolutionarily stable and more resistant to invasion by a mutant,

$$\rho_1 = \frac{1 - 1/r}{1 - 1/r^N}. \tag{7.9}$$

7.3.2 Evolutionary graph

An evolutionary graph (EG) is a graph defined on $\mathbb{V}$ vertices and edges $\mathbb{E}$ where $\mathbb{V}$ denotes the set of individual in the population and the edges of the graph are defined by a weight matrix $\mathbb{W} = [w_{ij}]$, where w_{ij} specifies the weight of the directed edge between any two pair of vertices v_i and v_j; w_{ij} typically specifies the probability that if v_i is selected to reproduce then it replaces v_j. Thus for any given vertex v_i, the following relationship is satisfied: $\sum_j w_{ij} = 1$. Consider an undirected graph $\mathbb{G} = (\mathbb{V}, \mathbb{E})$, where $\mathbb{V}$ is the number of vertices and $\mathbb{E}$ is the number of edges; the graph is assumed to be simple, that is, no vertex has self-loops and there are no parallel edges. Evolutionary dynamics studies the possibility that a mutant introduced into the population is able to replace all the agents in the initial population. Consider the evolutionary game described in [3], where a vertex i is chosen at random to be replaced by a mutant with fitness r and all other vertices have fitness 1. At each discrete step, the population dynamics is updated based on the following algorithm [4]:
- A base vertex is selected based on its fitness.
- A neighbor of the base vertex is selected at random.
- The base vertex places a copy of itself in the neighboring vertex.

If f_i denotes the fitness of an individual $i \in V$, $f_i \in \{1, r\}$, where $f_i = r$ means that the individual is a mutant. Any individual can be selected at random with a probability s_i defined by

$$s_i = \frac{f_i}{\sum_{v \in V} f_v}. \tag{7.10}$$

Let $\mathbb{W} = [w_{ij}]$ represent the structure of the graph, where w_{ij} is the weight between each node i and j and specifies the probability of vertex i replacing vertex j by itself, provided vertex i was selected for reproduction; w_{ij} is defined as

$$w_{ij} = \begin{cases} \dfrac{1}{e_i}, & \text{if } i \text{ and } j \text{ are connected,} \\ 0, & \text{otherwise,} \end{cases} \tag{7.11}$$

where e_i is the number of edges incident to vertex i. Let the set of mutants be defined as $C \subset V$, and P_C denote the probability of mutant fixation given in [4] as

$$P_c = \frac{\sum_{i \in C} \sum_{j \notin C} r w_{ij} P_{C \cup \{j\}} + w_{ji} P_{C \setminus \{i\}}}{\sum_{i \in C} \sum_{j \notin C} r w_{ij} + w_{ji}}, \tag{7.12}$$

where $P_\emptyset = 0$ and $P_V = 1$.

7.3.3 Fixation probability

We start with the following definition:

Definition 7.5 (Fixation probability). Consider a population of size N each with fitness 1, into which a mutant is introduced with fitness r. The probability that the mutant replaces the population is known as the fixation probability.

Some studies have been carried out for computation of fixation probabilities on graphs of certain topologies. In [3], some studies were conducted on some special graphs to illustrate how fixation probability can be modified based on the structure of the graph. For a one-rooted graph such as a directed tree, the fixation probability is given for any value of r by

$$\rho = \frac{1}{N}. \tag{7.13}$$

Some studies focused on computation of the fixation probability given that a certain percentage of the population are mutants. If the mutant set is denoted by $C \subset V$, the following relationship exists between the fixation probability ρ and the probability P that a vertex v_i is picked:

$$N\rho = \sum_i P_{v_i}. \tag{7.14}$$

A superstar graph $\mathbb{G}_{L,M}^K$ [3] is a graph structure consisting of a central vertex v_c surrounded by L leaves. A leaf l contains M reservoir vertices $r_{l,m}$ and $K - 2$ ordered chain vertices $c_{l,1}, \ldots, c_{l,K-2}$. The directed edges in $\mathbb{G}_{L,M}^K$ are of the form $(r_{l,m}, c_{l,1})$, $(c_{l,w}, c_{l,w+1})$, $(r_{l,k-2}, v_c)$, and $(v_c, r_{l,m})$. The fixation probability for a superstar graph $\mathbb{G}_{L,M}^K$ is given in [3] as

$$\lim_{L,M \to \infty} \rho(\mathbb{G}_{L,M}) = \frac{1 - r^K}{1 - 1/r^{KN}}, \tag{7.15}$$

where K denotes the amplification parameter. In an undirected evolutionary graph, the weight w_{ij} is defined as [4]

$$w_{ij} = \begin{cases} d_i^{-1}, & \text{if } (v_i, v_j) \in \mathbb{E}, \\ 0, & \text{otherwise.} \end{cases} \tag{7.16}$$

In the case of an undirected EG, where the mutants occupy the set $C \subset V$, and $r = 1$, the fixation probability is computed as

$$P_c \frac{\sum_{v_i \in C} d_i^{-1}}{\sum_{v_i \in V} d_i^{-1}}, \tag{7.17}$$

and the probability of a vertex v_i being selected for reproduction is

$$P_{v_i} = \frac{r}{r + \sum_{v_j \in V \setminus \{v_i\}} w_{ji}}. \tag{7.18}$$

An exhaustive study was conducted in [12] to compute the relative mutant advantage

$$\frac{P_{v_i}}{P_{v_j}} = \left(\frac{d_j}{d_i}\right)^2. \tag{7.19}$$

It was demonstrated in [12] that, as the relative fitness of the mutant increases, the fixation probabilities increase more rapidly for mutant placed into vertices with higher degrees [2]. The fixation probability for an undirected star graph is given in [11] as

$$
\rho_{undir-star} = \frac{L\dfrac{rL}{r(L+1)} + \dfrac{r}{r+L}}{(L+1)\left(1 + \dfrac{L}{L+r}\sum_{j=1}^{L-1}\left(\dfrac{L+r}{r(rL+1)}\right)^j\right)}, \tag{7.20}
$$

$$
\lim_{L\to\infty}\rho_{undir-star} = \frac{1 - \dfrac{1}{r^2}}{1 - \dfrac{1}{r^{2L}}}. \tag{7.21}
$$

Fixation probability can also be computed using dynamical equations based on the variation of the well-known Kimura diffusion equation [2]. This approach was developed in [13] for well-mixed populations and extended to graphs in [14]. This approach is typically based on the master equations governing a continuous-time birth–death process in a community of fixed population size N, where the probability of observing k events in an infinitesimal time dt is proportional to Poissonian events dt^k. The transition probabilities χ derived from the probability density of for the system to change its size from n to m individuals during the infinitesimal period of time as $dt \to 0$ becomes the weights of the master equation which is summarized as:

$$
\chi(n \to n+1) = \chi^+(n), \tag{7.22}
$$
$$
\chi(n \to n-1) = \chi^-(n), \tag{7.23}
$$
$$
\chi(n \to n+k) = 0 \quad \text{if } |k| > 1. \tag{7.24}
$$

Thus, the equation governing the dynamics of the population based on the probability $P(n, t)$ of observing n individuals at time t is given by

$$
\frac{\partial P(n, t)}{\partial t} = \chi^+(n-1)P(n-1) - \chi^+(n)P(n)
$$
$$
+ \chi^-(n-1)P(n-1) - \chi^-(n)P(n). \tag{7.25}
$$

The problem can be generalized to a continuous-time Moran process for haploid populations [2].

7.3.4 Time to fixation

Another important aspect of research in evolutionary graph theory is computing time for fixation to occur in an evolutionary graph. For undirected evolutionary graphs, some studies [12,15,16] have discovered that structures promoting fixation also increase time to fixation. The mean time to fixation is defined as

$$t_i = \frac{1}{P_i} \sum_{t=0}^{\infty} t P_i^{(t-1)}. \tag{7.26}$$

The following recursive equations holds for $P_i^{(t-1)}$:

$$P_i^{(t)} = \mu_i P_{i-1}^{(t-1)} + (1 - \mu_i - \lambda_i) P_i^{(t-1)} + \lambda_i P_{i+1}^{(t-1)}, \tag{7.27}$$

$$P_i t_i = \mu_i P_{i-1}(t_{i-1} + 1) + (1 - \mu_i - \lambda_i) P_i(t_i + 1) + \lambda_i P_{i+1}(t_{i+1} + 1), \tag{7.28}$$

where μ_i and λ_i represent the transition probability from state i to state $i-1$ and from i to $i+1$. Using some algebraic manipulations, the mean time to fixation is

$$t_1 = \sum_{n=1}^{N-1} \frac{s^{0,n-1} s_{n,N-1}}{\lambda_n q_n s_{0,N-1}}, \tag{7.29}$$

where $s_{n,m} = \sum_{k=n}^{m} q_k$ and $q_i = \prod_{j=1}^{i} \frac{\mu_j}{\lambda_j}$. For large isothermal structures, the mean time to fixation is [17]

$$t_1 = \sum_{n=1}^{N-1} \frac{(r^n - 1)(r^{N-n} - 1)(1 + r)}{(r^N - 1)(r - 1)}, \tag{7.30}$$

and for large N and a k-star structure, the mean time to fixation is

$$t_1 = \sum_{n=1}^{N-1} \frac{(r^{nK} - 1)(r^{(N-n)K} - 1)(1 + r^K)}{(r^{NK} - 1)(r^K - 1)}. \tag{7.31}$$

7.4 Games, coalitions, evolution, and consensus

7.4.1 Games, graphs, and payoffs

Consider a finite set of N agents on a graph $\mathbb{G} = (\mathbb{V}, \mathbb{E})$ with $|\mathbb{V}| = N$. Each member of the set participates in a game to get a reward or payoff determined by a payoff function v that determines the worth of the coalition $\mathbb{S} \subset \mathbb{V}$. Different kinds of games on graphs have been studied by earlier researchers. In [19], the following infinite two-person game was considered, where the game starts at a vertex $a_0 \in \mathbb{V}$. The first player chooses an edge

$e_1 = (a_0, a_1)$, and the second player chooses an edge $e_2 = (a_1, a_2) \in \mathbb{E}$ and the game continues indefinitely. The objective of the first player is to maximize (7.32) while the second player wants to minimize (7.33):

$$\liminf_{n\to\infty} \frac{1}{n} \sum_{i=1}^{n} w(e_i), \tag{7.32}$$

$$\limsup_{n\to\infty} \frac{1}{n} \sum_{i=1}^{n} w(e_i). \tag{7.33}$$

Definition 7.6 (Mean payoff games). A mean payoff game is played over a finite graph whose edges are labeled by integer weights. The interaction of the two players, called Eve and Adam, describe a path in the graph. The goal of Eve is to ensure that the (lower) limit of the weights' average is nonnegative.

It was demonstrated by [19], that it is possible to derive a payoff function v such that the first player can obtain a strategy that allows

$$\liminf_{n\to\infty} \frac{1}{n} \sum_{i=1}^{n} w(e_i) \geq v$$

and the second player obtains a strategy that allows

$$\liminf_{n\to\infty} \frac{1}{n} \sum_{i=1}^{n} w(e_i) \leq v.$$

These strategies for both players can be obtained using a positional strategy [19], that is, a strategy in which the next move depends only on the current node. These sort of games are known as *mean payoff games*. In mean payoff games, the graph is assumed to be bipartite. In the finite version of the game, the game begins at a specific vertex in $\mathbb{G} = (\mathbb{V}, \mathbb{E})$, which is a bipartite graph and continues until a cycle is formed. The outcome of the game is the mean weight of the edges on this cycle. Player 1 tries to maximize this value, and player 2 tries to minimize it. In [19], it was demonstrated that both players have positional optimal strategies for the finite version of the game and, more importantly, that the value v for the finite game is the same for the infinite game. Another class of mean payoff games on graphs was considered in [20], termed cyclic games, where a game is played on a graph which is not necessarily bipartite, starting at a vertex $a_0 \in \mathbb{V}$. The vertex set on which the game is played is divided into two vertex sets, $\mathbb{V}_1$

and $\mathbb{V}_2$. If the endpoint of the path formed belongs to the set $\mathbb{V}_1$, the first player chooses the next edge, while if the endpoint belongs to the set $\mathbb{V}_2$, the second player chooses the next edge. The strategy for player 1 is such that the long-term average weight of the path formed is maximized, while player 2 seeks to minimize it.

In [21], the computational complexity of mean payoff games was analyzed for a family of games with perfect information. The payoff value v for the mean payoff game of two players was given in [21] for a path of length $k \geq 1$ as

$$v_k(a) = \begin{cases} \max\limits_{(a,b) \in \mathbb{E}} \{w(a,b) + v_{k-1}(b)\} & \text{if } a \in \mathbb{V}_1, \\ \min\limits_{(a,b) \in \mathbb{E}} \{w(a,b) + v_{k-1}(b)\} & \text{if } a \in \mathbb{V}_2. \end{cases} \tag{7.34}$$

In the mean payoff game, a positional strategy for player 1 is a mapping $\psi : V_1 \to V$ such that $(a_1, \psi_1(a_1)) \in \mathbb{E}$ for every $a_1 \in \mathbb{V}_1$,

Theorem 7.1 ([21]). *The values $v_k(a)$ for every $a \in V$ can be computed in* $\mathbf{O}(k|\mathbb{E}|)$.

Theorem 7.2 ([21]). *For every $s \in V$, we have*

$$kv(a) - 2nW \leq v_k(a) \leq kv(a) + 2nW. \tag{7.35}$$

In contrast, a discounted payoff game (DPG) is a five-tuple ($\mathbb{V}_{\max}, \mathbb{V}_{\min}, \mathbb{E}, w, \lambda$) defined as:

- $\mathbb{V}_{\max}$ and $\mathbb{V}_{\min}$ are disjoint sets of vertices belonging to the players max and min, respectively; $\mathbb{V} = \{v_1, \dots, v_n\}$ denotes $\mathbb{V}_{\max} \cup \mathbb{V}_{\min}$.
- $\mathbb{E} = \{e_1, \dots, e_m\}$ is a set of directed edges between vertices in $\mathbb{V}$, such that each vertex has at least one outgoing edge. There is the possibility of multiple edges between the same ordered pair of vertices, and we denote the set of edges from u to v by $\mathbb{E}(u, v)$; we denote the set of outgoing edges from v by $\mathbb{E}(v)$; $\mathbb{E}_p = \bigcup_{v \in \mathbb{V}_p} \mathbb{E}(v)$ for $p \in \{\max, \min\}$.
- $w : \mathbb{E} \to \mathbb{Q}$ is a weight function.
- $\lambda : \mathbb{E} \to \{x \in Q : 0 < x < 1\}$ is a discount function.

In the discounted payoff game, a token is placed at some initial vertex $v_0 \in \mathbb{V}$. Then the following steps are repeated indefinitely: the owner of the vertex where the token is currently placed chooses an outgoing edge from this vertex and then moves the token to the head of the chosen edge. The objective of the max and min players is to maximize and minimize, respectively, the value function:

$$\mu(\pi) = w(e_{i0}) + \lambda(e_{i0})(w(e_{i1}) + \lambda(e_{i1})(\ldots)) = \sum_{j=0}^{\infty} \left(w(e_{ij}) \prod_{0 \leq k \leq j} \lambda(e_{ik}) \right).$$

(7.36)

A pure positional strategy σ for a player $p \in \{\max, \min\}$ is a selection of exactly one outgoing edge from each vertex owned by p, that is, an element of the set $\prod_{v \in \mathbb{V}_p} \mathbb{E}(v)$ which we denote by $\mathbb{P}_p$. A play where p only uses edges in σ is said to be consistent with σ.

Definition 7.7 (Cooperative or coalitional game). A coalition game with transferable utility is expressed by (N, v) where
- N is the set of agents and $|N|$ is the size of the grand coalition,
- $v : 2^N \to \mathbb{R}$ is a function that maps each group of agents $S \subset N$ to a real-valued payoff.

Definition 7.8 ([24]). An allocation rule or the value of a game is a function $f : \mathbb{G}_n \to \mathbb{R}^n$ which allocates the worth $v(N)$ to all players in N, where $f_i(N, v)$ signifies the payoff for each player $i \in N$.

The Shapley value [25] is one of such allocation rule:

$$f_i(N, v) = \sum_{S \subset N \setminus \{i\}} \frac{s!(n-s-1)!}{n!} [v(S \cup \{i\}) - v(S)] \quad \text{for all } i \in N. \quad (7.37)$$

Definition 7.9 (Graph game, [24]). A graph game on a set N is a triplet $(N, v, \mathbb{L})$ such that (N, v) is a cooperation game with transferable utility and $(N, \mathbb{L})$ is a communication graph.

Allocation rules can be characterized by some properties:
- **(Efficiency)** An allocation rule f on $\mathbb{G}^N$ is efficient if $\sum_{i \in N} f_i(N, v) = v(N)$ for each $v \in \mathbb{G}^N$;
- **(Additivity)** An allocation rule f on $\mathbb{G}^N$ is additive if

$$f(N, v + w) = f(N, v) + f(N, w) \quad \text{for each } v, w \in \mathbb{G}^N;$$

- **(Symmetry)** An allocation rule f on $\mathbb{G}^N$ is symmetric if for each $v \in \mathbb{G}^N$ and any two $i, j \in N$ with $v(S \cup \{i\}) = v(S \cup \{j\})$ for all $S \subseteq N \setminus \{i, j\}$, then $f_i(N, v) = f_j(N, v)$;
- **(Null-player property)** An allocation rule f on $\mathbb{G}^N$ satisfies the null-player property if for each $v \in \mathbb{G}^N$ and any $i \in N$ with $v(S) = v(S \cup \{i\})$ for all $S \subseteq N$, then $f_i(N, v) = 0$;

- (**Component efficiency**) An allocation rule f is component efficient for any graph graph $(N, v, \mathbb{L}) \in \mathbb{G}_C^S$ and any $C \in N\backslash\mathbb{L}$ where $\mathbb{G}_C^S$ is the graph associated with a graph subset S, as expressed by

$$\sum_{i \in C} f_i(N, v, \mathbb{L}) = v(C);$$

- (**Fairness**) An allocation rule f is fair if for any graph game $(N, v, \mathbb{L}) \in C$ and any $ij \in \mathbb{L}$, it holds that

$$f_i(N, v, \mathbb{L}) - f_i(N, v, \mathbb{L} \setminus ij) = f_j(N, v, \mathbb{L}) - f_j(N, v, \mathbb{L} \setminus ij);$$

- (**Balanced contributions**) An allocation rule f has balanced contributions if for any graph game $(N, v, \mathbb{L}) \in C$ and any $ij \in \mathbb{L}$, it holds that $f_i(N, v, \mathbb{L}) - f_i(N, v, \mathbb{L}\backslash\mathbb{L}_j) = f_j(N, v, \mathbb{L}) - f_j(N, v, \mathbb{L}\backslash\mathbb{L}_i)$, where $\mathbb{L}_k = \{e \in \mathbb{L} : k \in e\}$.

Definition 7.10 (Myerson value). The Myerson value $\mu(N, v, L)$ is the Shapley value $f^{\text{sh}}(N, v^L)$ of the graph restricted game v^L defined by:

$$v^L(S) = \sum_{T \in S/L_S} v(T) \text{ for all } S \subseteq N. \tag{7.38}$$

Definition 7.11 (Digraph game). [24] A digraph game on a set N is a triplet (N, v, D) such that $(N, v) \in \mathbb{G}_N$ is a cooperation game with transferable utility and $(N, D) \in \mathbb{D}_N$ is a communication graph.

The digraph restricted game (N, v^D) associated with the digraph game is defined by [24]:

$$v^D(S) = \sum_{T \in S/D_S} v(T) \text{ for all } S \subseteq N \tag{7.39}$$

Similarly, the Myerson value $\mu(N, v, D)$ is an allocation rule which assigns to every digraph game (N, v, D) the Shapley value of v^D, that is,

$$\mu(N, v, D) = f^{\text{sh}}(N, v^D). \tag{7.40}$$

In [24], some properties were used to characterize the Myerson values for digraph games such as:

- (**Strong component efficiency**) f is called strong component efficient if

$$\sum_{i \in T} f_i(N, v, D) = v(T), \tag{7.41}$$

for any $(N, v, D) \in \mathbb{G}_S^D$ and any strong component $T \in N/D$;

- (**Fairness**) f is called fair if for any $(N, v, D) \in \mathbb{G}_S^D$ and any $(i, j) \in D$, then

$$f_i(N, v, D) - f_i(N, v, D \backslash ij) = f_j(N, v, D) - f_j(N, v, D \backslash ij); \qquad (7.42)$$

- (**Bi-fairness**) f is called fair if for any $(N, v, D) \in \mathbb{G}_S^D$ and any $(i, j) \in D$,

$$f_i(N, v, D) - f_i(N, v, D \backslash ij) = f_j(N, v, D) - f_j(N, v, D \backslash ij). \qquad (7.43)$$

7.4.2 Replicator dynamics for games on graphs

In this subsection, we introduce the concept of replicator equations for evolutionary games played on a graph as presented in [26]. In an evolutionary game with n strategies, let $A \in \mathbb{R}^{n \times n}$ represent a payoff matrix where each a_{ij} is a payoff when strategy i is played against strategy j. The popularity of each strategy is denoted by x_i, which may be interpreted as the number of times the strategy is played. The fitness of each strategy is popularity (or the number of times its played, that is, x_i) determined by $f_i = \sum_{j=1}^{n} x_j a_{ij}$. The average fitness of the population is given by $\phi = \sum_{i=1}^{n} x_i f_i$. Thus, the replicator dynamics is given as

$$\dot{x}_i(t) = x_i(f_i - \phi), \quad i = 1, \dots, n. \qquad (7.44)$$

The replicator dynamics is defined on an invariant set S_n, that is, every trajectory which begins in the set remains there for all time. In the above replicator dynamics, we do not consider the effect of mutation, and every corner of the invariant set represents an equilibrium. A fixed point of the system is an asymptotically stable equilibrium which corresponds to a corner point of the invariant set, or equivalently, an evolutionarily stable strategy, or a strict Nash equilibrium. In the evolutionary game on graphs, the fitness of each individual is determined locally based on the interactions of an agent with its surrounding neighbors, the replicator dynamics in (7.44) assumes that the graph is complete (see Fig. 7.2).

Typical update rules in evolutionary dynamics are classified under birth–death, death–birth, and imitation. In birth–death updating, an individual is selected for reproduction based on its fitness, the offspring from this reproduction then replaces a randomly chosen neighbor. Conversely in death–birth updating, a random individual is chosen to die and it is replaced by its neighbor based on its fitness. In imitation-based update rules, an individual is chosen at random, and it has two choices based on fitness; either to keep its strategy or adopt a neighboring strategy. For regular

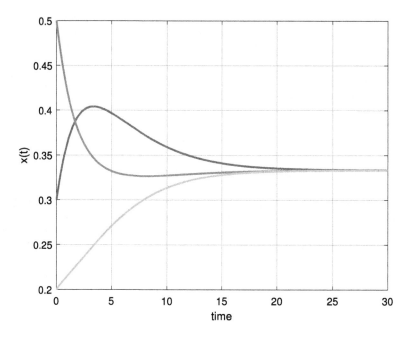

Figure 7.2 Simulation of the replicator dynamics (7.44).

graphs of degree k, let B represent a matrix for the updating rule [26],

$$b_{ij} = \frac{a_{ii} + a_{ij} - a_{ji} - a_{jj}}{k - 2}, \tag{7.45}$$

$$b_{ij} = \frac{(k+1)a_{ii} + a_{ij} - a_{ji} - (k+1)a_{jj}}{(k-1)(k-2)}, \tag{7.46}$$

$$b_{ij} = \frac{(k+3)a_{ii} - 3a_{ij} - 3a_{ji} - (k+3)a_{jj}}{(k+3)(k-2)}. \tag{7.47}$$

The following modified replicator equation on graphs was defined in [26]:

$$\dot{x}_i(t) = x_i(f_i + g_i - \phi), \quad i = 1, \ldots, n, \tag{7.48}$$

where g_i is defined based on the update mechanism as follows:

$$g_i = \sum_{j=1}^{n} x_j b_{ij}. \tag{7.49}$$

7.5 Consensus in evolutionary graphs

We start by presenting the problem formulation.

7.5.1 Problem formulation

An earlier study has been conducted on consensus on evolutionary graphs in [8]. In this study, the authors considered an evolutionary game where payoff function for each agent was based on a continuous-iterative prisoners' dilemma game. Consider the MAS on a graph $\mathbb{G}$ described by the following single integrator dynamics:

$$\dot{x}_i(t) = u_i(t). \tag{7.50}$$

The network graph $\mathbb{G}$ is a dynamic graph with fixed and predetermined vertex set and a time-varying edge-set $\mathbb{E}$, which depends on the state of the agents. In [8], the edge-set $\mathbb{E} = \mathbb{E}(x, t, \omega) \subseteq \mathbb{V} \times \mathbb{V}$ is assumed to be a state-dependent random set in a probability space $(\omega, \mathbb{F}, \mathbb{P})$ where ω is the sample space, $\mathbb{F}$ is a σ-algebra on ω, and $\mathbb{P}$ is a probability measure. Likewise, the adjacency matrix $\mathbb{A} = \mathbb{A}(x, t, \omega)$ and Laplacian matrix $\mathbb{L} = \mathbb{L}(x, t, \omega)$ are state- or time-dependent.

There are two possible neighborhood sets for each agent $i \in \mathbb{V}$, the set $\mathbb{N}_i$ denotes the set of all feasible neighbors, that is, the set of neighbors i could establish connection with; $\mathbb{N}_i^1(t) \subset \mathbb{N}_i$ denotes the set of active neighbors while $\mathbb{N}_i^2(t) \subset \mathbb{N}_i$ denotes the set of inactive neighbors, or the set of nodes i is not connected with at time t. The degree of each node i at each time t is denoted as $\mathbb{N}_i^1(t)$.

Assumption 7.5.1 ([8]). The graph $\mathbb{G}' = (\mathbb{V}, \mathbb{E}')$ with $\mathbb{E}' := \bigcup_{i \in \mathbb{V}} \bigcup_{j \in \mathbb{N}_i} \{(i, j)\}$ is connected.

The following distributed consensus protocol was considered in [8]:

$$u_i(t) = \sum_{j \in \mathbb{N}_i^1(t)} a_{ij}^m(t)(x_j(t) - x_i(t)) + \sum_{j \in \mathbb{N}_i^1(t)} a_{ij}^c(t)(x_j(t) - x_i(t)) \tag{7.51}$$

where $a_{ij}^m(t), a_{ij}^c(t) \in [0, 1]$ are time-varying variables that indicate either to maintain or create a link in the active and inactive neighborhood set of agent i, respectively. In [8], the decision to maintain, drop, or create a new link was based on a Bernoulli random process which is dependent on success probabilities $0 \le w_{ij}^m(t) \le 1$ and $0 \le w_{ij}^c(t) \le 1$. It is worthy to note that these success probabilities are dependent on the states of the agent, that is, $w_{ij}^m(t) = w_{ij}^m(x_i(t), x_j(t))$ and $w_{ij}^c(t) = w_{ij}^c(x_i(t), x_j(t))$. Thus the definition of the corresponding Laplacian matrix is $\mathbb{L}(t) = [l_{ij}(t)]$, where

$$
l_{ij} = \begin{cases} -\{1\}_{j\in\mathbb{N}_i^1} a_{ij}^m(t) - \{1\}_{j\in\mathbb{N}_i^2} a_{ij}^c(t), & i \neq j, \\ -\sum_{j\neq i} l_{ij}, & i = j. \end{cases}
\tag{7.52}
$$

Thus, the MAS becomes

$$
\dot{x}(t) = \mathbb{L}(t)x(t),
\tag{7.53}
$$

which is described as a continuous Markov decision process. The rules for obtaining the weights $w_{ij}^m(x_i(t), x_j(t))$ and $w_{ij}^c(x_i(t), x_j(t))$ are based on a continuous actions of an iterative prisoners dilemma (CAIPD) [9] game. The reward function for each agent is defined as

$$
f_i(x) = b \sum_{j\in\mathbb{N}_i^1} x_j - c|\mathbb{N}_i^1|x_i,
\tag{7.54}
$$

where the constants are chosen to satisfy $b > c > 0$ based on the assumption that the gain-per-unit of coordination from cooperating is higher than the per-unit loss. Now, the change in fitness of agent i when it creates or drops a link are denoted by $\tilde{f}_{ij}^c$ and $\tilde{f}_{ij}^d$, respectively. The mathematical definitions for $\tilde{f}_{ij}^c$ and $\tilde{f}_{ij}^d$ are:

$$
\tilde{f}_{ij}^c = b \sum_{j\in\mathbb{N}_i^1\cup\{i\}} x_j - c|\mathbb{N}_i^1 + 1|x_i - \left(b \sum_{j\in\mathbb{N}_i^1} x_j - c|\mathbb{N}_i^1|x_i \right),
\tag{7.55}
$$

$$
\tilde{f}_{ij}^d = b \sum_{j\in\mathbb{N}_i^1\setminus\{i\}} x_j - c|\mathbb{N}_i^1 - 1|x_i - \left(b \sum_{j\in\mathbb{N}_i^1} x_j - c|\mathbb{N}_i^1|x_i \right).
\tag{7.56}
$$

Based on the evolutionary game, the agents can decide either to drop or create a link if coordination is mutually beneficial, that is, if the creation or dropping of a link either increases or decreases the agents fitness function. If both agents benefit from keeping/creating a link, the corresponding probability should be higher than in the case when only one agent benefits. The weights corresponding to these probabilities are given by $w_{ij}^m(t)$ and $w_{ij}^c(t)$ which are defined mathematically as:

$$
\begin{aligned}
w_{ij}^m &= \frac{1}{2} - \frac{1}{2} \tanh(\tilde{f}_{ij}^d(x) + \tilde{f}_{ij}^c(x)) \\
&= \frac{1}{2} - \frac{1}{2} \tanh((c - b)(x_i + x_j)),
\end{aligned}
\tag{7.57}
$$

$$w_{ij}^m = \frac{1}{2} + \frac{1}{2} \tanh(\tilde{f}_{ij}^d(x) + \tilde{f}_{ij}^c(x))$$
$$= \frac{1}{2} + \frac{1}{2} \tanh((b - c)(x_i + x_j)). \tag{7.58}$$

Finally, the weighted Laplacian of the network is defined as

$$l_{ij} = \begin{cases} -\{1\}_{j \in \mathbb{N}_i^1} w_{ij}^m(t) - \{1\}_{j \in \mathbb{N}_i^2} w_{ij}^f(t), & i \neq j, \\ -\sum_{j \neq i} w_{ij}, & i = j. \end{cases} \tag{7.59}$$

7.5.2 Consensus: a coalition game

In classical multiagent consensus problems, the objective is for the participating agent to reach an agreement on the consensus variable. For example, consider a multiagent system consisting of N agents modeled by a single integrator dynamics

$$\dot{x}_i(t) = u_i(t), \tag{7.60}$$

where $x_i \in \mathbb{R}$ and $u_i \in \mathbb{R}$ represent position and velocity, respectively. Suppose the agents start at some arbitrary initial condition $x_0 \in \mathbb{R}^n$. In this case we assume the agents are playing the average consensus game and are required to converge to a position which is the average of the agents' initial positions, that is,

$$\bar{x} = \frac{1}{N} \sum_{i=1}^{N} x_i(0). \tag{7.61}$$

In order for agents to reach an agreement, they need to share information over a communication network. This communication network is modeled by a graph $\mathbb{G}$, which may be directed or undirected, as well as weighted or unweighted graph. In classical consensus problems, a popular distributed consensus protocol is designed for the each agent based on relative state information as

$$u_i(t) = -\sum_{j=1}^{N} a_{ij}(x_i(t) - x_j(t)), \tag{7.62}$$

where $a_{ij} \in \{0, 1\}$ determines if there's a communication path between agents i and j. In an unweighted graph, if $a_{ij} = 1$, there's a communication path between agents i and j, and no information is exchanged between agents i and j when $a_{ij} = 0$. In directed graphs or digraphs, a_{ij} may not be equal to a_{ji}, that is, there is some edge orientation in the communication path. The choice of communication structure may influence the information flow in the multiagent network and thus impact the outcome of the consensus game. Some notable studies such as [22,23] have pointed out that it a necessary condition on the graph for average consensus is that the detailed balance condition must be satisfied. Also, in general some equivalent conditions were listed by Ren in [22] involving spanning trees and eigenvalues of the Laplacian of the graph formed by the consensus network. Now, there are well-established results in consensus theory for reaching consensus on graphs with fixed and switching topologies.

Consider an alternative approach to establishing consensus (or agreement) on graphs. Suppose the agents are able to form coalitions (or define the communication structures) suitable to the coordination objective based on the outcome of an evolutionary game played iteratively until the consensus objective is achieved. That is, let us consider the case where the each pair of agents play a game at each time instant to determine the weights of the adjacency matrix $a_{ij} \in \{0, 1\}$. For this consideration, we consider time-varying network version of (7.63) as

$$u_i(t) = -\sum_{j=1}^{N} \phi_{ij}(t)(x_i(t) - x_j(t)). \tag{7.63}$$

The weights of the adjacency matrix here are determined by the outcome (payoff) of an evolutionary game played between adjacent agents. The evolutionary game in turn depends on the states of the agent, $x(t)$, and the nature of the consensus game. Thus, we need to design a time-varying payoff function $v(t, x(t))$ that will determine the weight $\phi_{ij}(t)$. Note that $\phi_{ij}(t)$ can be an interval-valued function in $\{0, 1\}$ which may be discrete or continuous. Alternatively, the connection weight $\phi_{ij}(t)$ can be defined as

$$\phi_{ij}(t) = a_{ij}w_{ij}(t, x_i(t), x_j(t)). \tag{7.64}$$

Eq. (7.64) models the case where the communication topology $\mathbb{G}$ is a fixed interaction topology, with the adjacency information being described by the network weights a_{ij}. However, the factors $w_{ij}(t, x_i(t), x_j(t))$ are state-dependent weights which are outcomes of an evolutionary game played by agents i and j; $w_{ij}(t)$ is a real-valued function in the interval $[0, 1]$, which reflects the opinion or trust by agent i about the information from agent j.

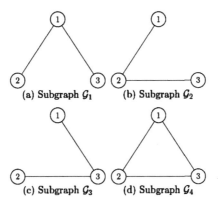

Figure 7.3 Connected graphs on $N = 3$ vertices.

The trust (or opinion) level about the information from neighboring agent is dependent on the states of the trusting agent and the trusted agent. In an undirected graph, the state-dependent weights $w_{ij}(t)$ lack orientation, that is, $w_{ij}(t) = w_{ji}(t)$, this can be interpreted as mutual trust or mutual cooperation between the agents.

Now, consider an average consensus game on $N = 3$ vertices. Fig. 7.3 shows the possible topologies for this game, where we assume that the edge set is undirected. Let us assume the case of the complete graph, that is, $\mathbb{G}_4$. The strategy for each player is determined by the replicator dynamics in (7.60) (see Figs. 7.4, 7.5, 7.6, and 7.7). Here, the payoff for each player of the game is determined the Shapley function

$$\phi_i(v) = \sum_{S \subseteq N, i \in S} \frac{(n-s)!(s-1)!}{n!} [v(S) - v(S \setminus \{i\})]. \qquad (7.65)$$

We assume here that the game is played in a distributed fashion, that is, each node i has its own neighborhood set on which a consensus game is played with its neighbors. Thus the Shapley function (7.65) is distributed over each neighborhood set and is thus formally defined:

$$\phi_j(v, \mathbb{N}_i) = \sum_{S \subseteq \mathbb{N}_i, j \in S} \frac{(n-s)!(s-1)!}{n!} [v(S) - v(S \setminus \{j\})], \qquad (7.66)$$

where $\phi_j(v, \mathbb{N}_i)$ is interpreted as the payoff obtained by agent j, by playing a game in the neighborhood set $\mathbb{N}_i$. Since we are interested in average

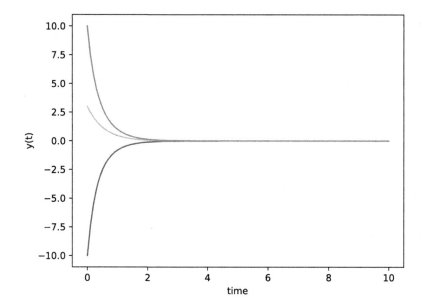

Figure 7.4 Simulation of the replicator dynamics (7.44).

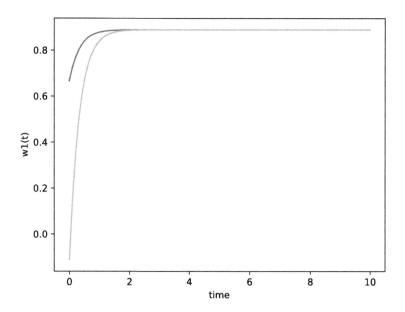

Figure 7.5 Weights distribution at node 1.

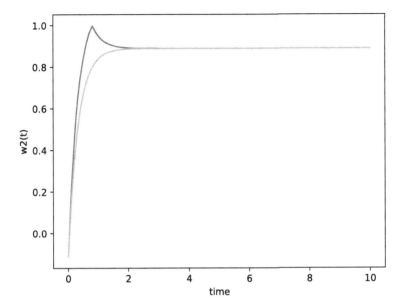

Figure 7.6 Weights distribution at node 2.

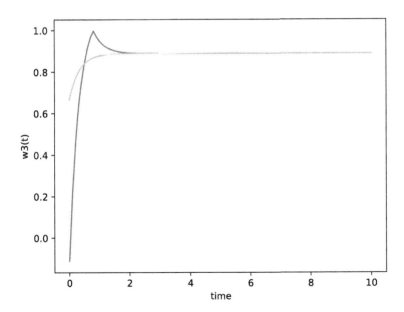

Figure 7.7 Weights distribution at node 3.

consensus, the worth v of the game is defined by

$$v(M, x) = \frac{1}{M} \sum_{j \in M} x_j(t). \tag{7.67}$$

It is worthy to mention that the worth function should be reflective of the consensus objectives. Also, we note that the Shapely function used in this context is a decreasing function which approaches zero as the distance between the agents tends to zero. Using this, we can choose the weights related to the adjacency matrix (7.64) as

$$w_{ij}(t) = 1 - \phi_j(v, \mathbb{N}_i). \tag{7.68}$$

Thus the weight matrix becomes

$$W(t, v(N, x(t))) = \begin{bmatrix} 0 & 1 - \phi_2(v, \mathbb{N}_1) & \dots & 1 - \phi_N(v, \mathbb{N}_1) \\ 1 - \phi_1(v, \mathbb{N}_2) & 0 & \dots & 1 - \phi_N(v, \mathbb{N}_2) \\ \vdots & \vdots & \ddots & \vdots \\ 1 - \phi_1(v, \mathbb{N}_N) & 1 - \phi_2(v, \mathbb{N}_i) & \dots & 0 \end{bmatrix}. \tag{7.69}$$

Definition 7.12 (State-dependent evolutionary graph). A state-dependent evolutionary graph $\mathbb{G} = (\mathbb{V}, \mathbb{E})$ is a graph on $N = |\mathbb{V}|$ vertices, which has time-varying edge set $\mathbb{E}(t, x(t)) = [e_{ij}(t)]$ where the weights of adjacent edges $e_{ij}(t) = e_{ij}(t, x_i(t), x_j(t))$ are based on the outcome of an evolutionary game played by neighboring agents i and j whose strategy is defined by the corresponding states x_i and x_j.

7.5.3 Numerical example 7.1

Consider a multiagent system consisting of $N = 4$ agents described by the topology in Fig. 7.8 with the following dynamics:

$$\dot{x}_i(t) = -x_i(t) + u_i(t). \tag{7.70}$$

The states of the system are assumed to be at the following initial conditions $x(0) = [-3, 2.5, 0, 4]$. The static adjacency and Laplacian matrices describing the multiagent network are:

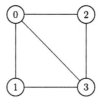

Figure 7.8 Connected graphs on $N = 4$ vertices.

$$L = \begin{bmatrix} 3 & -1 & -1 & -1 \\ -1 & 2 & 0 & -1 \\ -1 & 0 & 2 & -1 \\ -1 & -1 & -1 & 3 \end{bmatrix}, \quad A = \begin{bmatrix} 0 & 1 & 1 & 1 \\ 1 & 0 & 0 & 1 \\ 1 & 0 & 0 & 1 \\ 1 & 1 & 1 & 0 \end{bmatrix}. \tag{7.71}$$

We consider the following distributed consensus protocol for the MAS:

$$u_i(t) = \sum_{j \in \mathbb{N}_i} a_{ij}(t)(x_j(t) - x_i(t)), \tag{7.72}$$

where $a_{ij}(t)$ is determined by the outcome of an evolutionary game on a graph. The state response for the MAS (7.70) on graph with topology in (7.8) under a static version of the control protocol (7.72) is shown in Fig. 7.9.

The allocation rule for the game will be determined by the Shapley function (7.66). Suppose the weights on each adjacent link are determined by the outcome of an evolutionary game whose payoff function is defined by Shapley function. We briefly illustrate here the computation of the Shapley value for the multiagent game on $N = 4$ vertices with the initial conditions $x(0) = [-3, 2.5, 0, 4]$ at time $t = 0$. Now, the neighborhood set of each agent is defined based on the topology as follows: $\mathbb{N}_1 = \{2, 3, 4\}$, $\mathbb{N}_2 = \{1, 4\}$, $\mathbb{N}_3 = \{1, 4\}$, $\mathbb{N}_4 = \{1, 2, 4\}$. For $\mathbb{N}_1 = \{2, 3, 4\}$, agent 1 can form the following coalitions with its neighbors: $\mathbb{S}_1^1 = \{1, 2\}$, $\mathbb{S}_2^1 = \{1, 3\}$, $\mathbb{S}_3^1 = \{1, 4\}$, $\mathbb{S}_4^1 = \{1, 2, 3\}$, $\mathbb{S}_5^1 = \{1, 2, 4\}$, $\mathbb{S}_6^1 = \{1, 3, 4\}$, $\mathbb{S}_7^1 = \{1, 2, 3, 4\}$, which means the agent has 7 possible choices of connections to make with its neighbors. Table 7.7 shows the possible coalitions that can be formed by each agent at the initial time $t = 0$.

Next, we compute the worth of each coalition for this average consensus game using the worth function (see Table 7.8)

$$v(M, x) = \frac{1}{|M|} \sum_{j \in M} x_j(t). \tag{7.73}$$

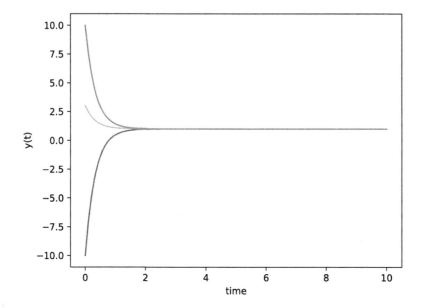

Figure 7.9 Initial state response of MAS described by $\mathbb{G}$.

Table 7.7 Possible neighborhood coalitions.

Agent 1	Agent 2	Agent 3	Agent 4
$\mathbb{N}_1$	$\mathbb{N}_2$	$\mathbb{N}_3$	$\mathbb{N}_4$
$\{1, 2\}$	$\{2, 1\}$	$\{3, 1\}$	$\{4, 1\}$
$\{1, 3\}$	$\{2, 4\}$	$\{3, 4\}$	$\{4, 2\}$
$\{1, 4\}$	$\{2, 1, 4\}$	$\{2, 1, 4\}$	$\{4, 3\}$
$\{1, 2, 3\}$			$\{4, 1, 2\}$
$\{1, 2, 4\}$			$\{4, 1, 3\}$
$\{1, 3, 4\}$			$\{4, 2, 3\}$
$\{1, 2, 3, 4\}$			$\{4, 1, 2, 3\}$

Based on $\Phi(\mathbb{N}, x(0))$, each agent can make informed decisions on the neighborhood coalitions to form, where

$$\Phi(\mathbb{N}, x(0)) = \begin{bmatrix} 0 & -0.263 & 0.218 & 0.9945 \\ -0.124 & 0 & 0 & 0.4575 \\ -0.364 & 0 & 0 & 0.3192 \\ -0.752 & -0.665 & -0.169 & 0 \end{bmatrix}. \tag{7.74}$$

Table **7.8** Worth of each neighborhood coalitions.

Coalition	Worth
$v(\{1, 2\})$	-2.75
$v(\{1, 3\})$	-1.5
$v(\{1, 4\})$	0.5
$v(\{2, 4\})$	0.75
$v(\{3, 4\})$	2
$v(\{1, 2, 3\})$	-1.833
$v(\{1, 2, 4\})$	-0.5
$v(\{1, 3, 4\})$	0.33
$v(\{1, 2, 3, 4\})$	-0.375

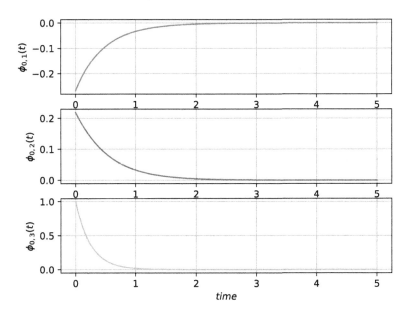

Shapley values at neighborhood 0

Figure 7.10 Shapley value at node 0.

Just for the analytic purposes, we evaluate the Shapley function (7.66) for the MAS based on the state data generated from the fixed topology case via the static Laplacian. This Shapley function is evaluated at each instant of time for each neighborhood i as shown in Figs. 7.10–7.13. The figures can be interpreted as the worth or importance which each agent attaches

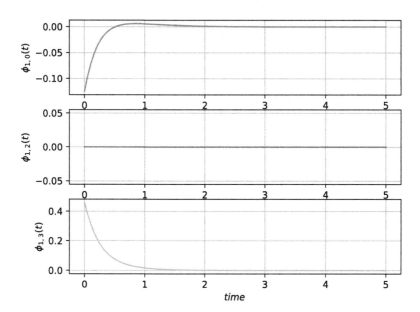

Figure 7.11 Shapley value at node 1.

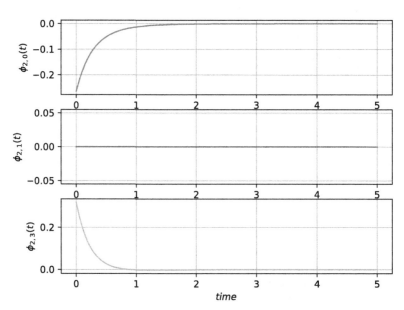

Figure 7.12 Shapley value at node 2.

Shapley values at neighborhood 3

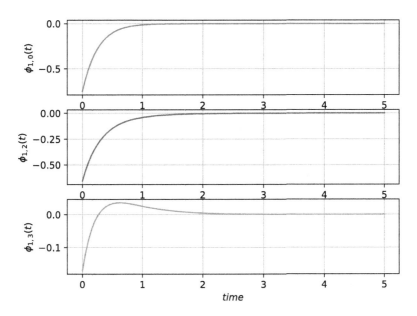

Figure 7.13 Shapley value at node 3.

to its neighbors in seeking an evolutionarily stable state, that is, in this case the average consensus value. As demonstrated by the figures, as each node approaches the equilibrium or the consensus state, the Shapley value for node j at each neighborhood i strictly approaches zero. The interpretation of this is that as an agent approaches the consensus state, it sees absolutely no need to maintain connection with its neighbors.

Next, we use the following intuition to design a state-dependent consensus protocol based on (7.72) where the weights of the adjacency matrix will be determined based on the Shapley-value function mentioned earlier. Fig. 7.14 shows the state response of the MAS where the network weights are determined by the Shapley function. Figs. 7.15 and 7.16 compare the state responses for the static graph and state-dependent graph states. A quick evaluation shows that with the state-dependent graph, the agents converge to a consensus state $x = -0.803$, while in the case of the fixed topology the agents converge directly to the average of the initial conditions of the system.

State response on state-dependent evolutionary graphs

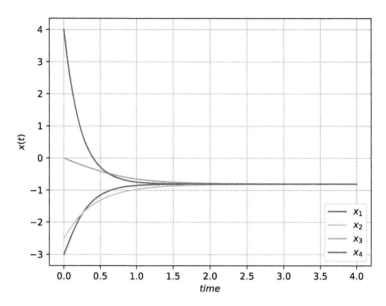

Figure 7.14 State response on state-dependent graphs.

Comparison between static graph and state-dependent graph response

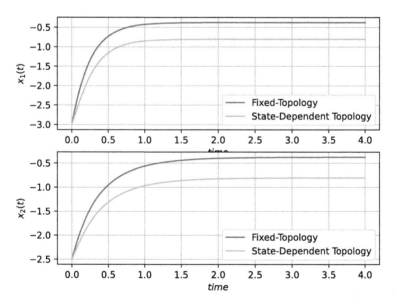

Figure 7.15 Comparison of state response (x_1 and x_2) on fixed and state-dependent graphs.

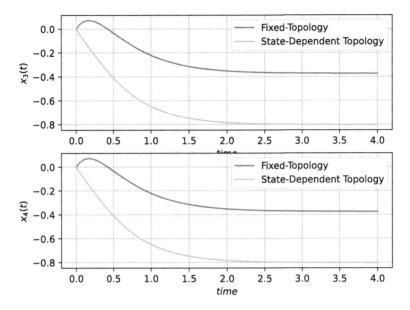

Figure 7.16 Comparison of state response (x_3 and x_4) on fixed and state-dependent graphs.

7.6 Notes

In summary, we have studied the average consensus problem via state-dependent evolutionary graphs. The chapter began by reviewing some important concepts related to game theory, evolutionary games, and graph theory. We proceeded to establish the consensus as a game on a graph problem. We introduced the Shapley function for computing the state-dependent weights on the edges of the graph.

Some future considerations for investigation will be to examine value functions other than the Shapely function for determining the state dependent weights. In our analysis, we have assumed that the weights of the edges are computed continuously in time. A better approach may be to update the weights via an event-triggered approach using an appropriate switching topology. Also, the function used here for computing the worth of each coalition is simplistic. Better worth function can be investigated not just depending on the consensus objective but also inclusive of environmental constraints for each agent in the network. The game and graph theory considered in this discussion may also prove effective in designing better resilient consensus algorithms for the MAS.

References

[1] H. Tembine, E. Altman, R. El-Azouzi, Y. Hayel, Evolutionary games in wireless networks, IEEE Trans. Syst. Man Cybern., Part B, Cybern. 40 (3) (2010) 634–646.

[2] P. Shakarian, P. Roos, A. Johnson, A review of evolutionary graph theory with applications to game theory, Biosystems 107 (3) (2012) 66–80.

[3] E. Lieberman, C. Hauert, M.A. Nowak, Evolutionary dynamics on graphs, Nature 433 (7023) (2005) 312–316.

[4] M. Broom, J. Rychtar, B. Stadler, Evolutionary dynamics on small-order graphs, J. Interdiscip. Math. 12 (2) (2009) 129–140.

[5] J.M. Smith, Evolution and the Theory of Games, vol. 12, Cambridge University Press, 2009.

[6] S. Phelps, M. Woolridge, Game theory and evolution, IEEE Intell. Syst. 28 (7/8) (2013) 76–81.

[7] G. Szabo, G. Fath, Evolutionary games on graphs, arXiv:cond-mat/0607344v3, Phys. Rep. 12 (2) (2009) 129–140.

[8] M. Smrynakis, N.M. Ferris, H. Tembine, Consensus over evolutionary graphs, in: Proc. European Control Conference, Limassol, Cyprus, 2018, pp. 2218–2223.

[9] B. Ranjbar-Sahraei, H. Bou Ammar, D. Bloembergen, K. Tuyls, G. Weiss, Evolution of cooperation in arbitrary complex networks, in: Proc. the 2014 Int. Conference Autonomous Agents and Multiagent Systems, 2014, pp. 667–684.

[10] H. Peters, Game Theory: A Multi-Leveled Approach, Springer Texts in Business and Economics, 2015.

[11] M. Broom, J. Rychtar, An analysis of the fixation probability of a mutant on special classes of non-directed graphs, Proc. R. Soc. A 464 (2008) 2609–2627.

[12] M. Broom, J. Rychtar, B. Stadler, Evolutionary dynamics on graphs – the effect of graph structure and initial placement on mutant spread, J. Stat. Theory Pract. 5 (3) (2011) 369–381.

[13] B. Houchmandzadeh, M. Vallade, Alternative to the diffusion equation in population genetics, Phys. Rev. E 82 (2010).

[14] B. Houchmandzadeh, M. Vallade, The fixation probability of a beneficial mutation in a geographically structured population, New J. Phys. 13 (2011).

[15] C.J. Paley, S.N. Taraskin, S.R. Elliott, Temporal and dimensional effects in evolutionary graph theory, Phys. Rev. Lett. 98 (2007).

[16] P.A. Whigham, G. Dick, Evolutionary dynamics for the spatial Moran process, Genet. Program. Evol. Mach. 9 (2) (2008) 157–170.

[17] P.Y. Nie, P.A. Zhang, Fixation time for evolutionary graphs, Int. J. Mod. Phys. B 24 (2010) 5285–5293.

[18] P.D. Taylor, L.B. Jonker, Evolutionarily stable strategies and game dynamics, Math. Biosci. 40 (1978) 145–156.

[19] A. Ehrenfeucht, J. Mycielski, Positional strategies for mean payoff games, Int. J. Game Theory 8 (1979) 109–113.

[20] V.A. Gurvich, A.V. Kamanov, L.G. Khachiyan, Cyclic games and an algorithm to find minimax cycle means in directed graphs, USSR Comput. Math. Math. Phys. 28 (1988) 85–91.

[21] U. Zwick, M. Patterson, The complexity of mean payoff games on graphs, Math. Games 158 (1996) 343–359.

[22] W. Ren, R.W. Bear, E.M. Atkins, Information consensus in multivehicle cooperative control, IEEE Control Syst. Mag. 27 (2) (2007) 71–82.

[23] M. Mesbahi, M. Egerstedt, Graph Theoretic Methods in Multiagent Networks, Princeton Series in Applied Mathematics, 2010.

[24] D.L. Li, E. Shan, The Myerson value for directed graph games, Oper. Res. Lett. 48 (2010) 142–146.

[25] L.S. Shapley, A value for n-person games, in: H. Kuhn, A.W. Tucker (Eds.), Contribution to the Theory of Games, vol. II, Princeton University Press, Princeton, 1953, pp. 307–317.
[26] H. Ohtsuki, M.A. Nowak, The replicator equation on graphs, J. Theor. Biol. 243 (1) (2006) 86–97.

CHAPTER 8

Multivehicle cooperative control

8.1 Introduction

Research studies in distributed multivehicle coordination is primarily motivated by the need to emulate distributed and social behaviors exhibited by nature-inspired swarms. The core of any multiagent (or multivehicle) system comprises primarily three features which include communication, computation, and coordination, featuring different algorithms at each level. However, the objectives for these systems vary depending on the nature of the problem being investigated. Some research studies have been conducted for instance in path-planning, collision and obstacle avoidance, and evasion techniques for these kind of systems.

As we continue to evolve into the age of robotics and automation, more advances are being made in distributed robotics with several industrial applications. Robotic systems are being deployed in industries to perform tasks which cannot be otherwise performed easily or efficiently by human beings. The past few years alone witnessed some significant advances in self-driving vehicles, distributed energy systems involving microgrids and humanoid robots. Some desirable properties of any multiagent system include [1]:

- **Autonomous.** The agents participating in the MAS framework should possess some autonomous properties in their interaction with neighboring agents and their environment.
- **Homogeneous or heterogeneous.** The agents in the MAS network may exhibit homogeneous or heterogeneous dynamics. The coordination variables are however similar. For example, position, velocity, acceleration, and so on.
- **Decentralization.** Cooperation in an MAS can be achieved either through centralized or decentralized (or distributed) control architectures. Recently, decentralized information structures became desirable in cooperative control due to inherent robust and redundant properties.
- **Local sensing and communication.** Decentralization (or distributed) control architectures help save costs for long-range communication equipment. Agents are required to communicate in short ranges with their neighbors to complete a global objective.

Advanced Distributed Consensus for Multiagent Systems
https://doi.org/10.1016/B978-0-12-821186-1.00016-7

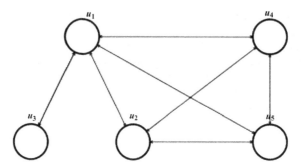

Figure 8.1 A representative of multiagent systems [1].

- **Flexibility and adaptability.** It is desired that MAS swarm architectures are flexible in switching between possible network topologies
  and strategies to cooperatively solve the global problem in a distributed
  fashion. MAS networks are capable of performing several minitasks
  which aids convergence to the global solution. Flexibility also ensures that the behavior of the agents is not localized to one single
  environment. Agents are able to show versatility even in unknown environments.

In this chapter, we present an overview of some of these problems in
multirobot coordination in general with more emphasis on multivehicle
coordination. We present typical models for the agents in the multiagent
coordination, as well as discuss different possible coordination objectives
for the multiagent system (see Fig. 8.1).

8.2 Multivehicle architecture

In this section, we examine the modeling of components pertaining to
multivehicles.

8.2.1 Agent models

Every MAS design begins with the elementary model of the agents participating in the MAS framework. In multivehicle applications, models define
elementary actions such as movement in either two- or three-dimensional
space. Models representing each agent are either continuous or discrete dynamical models. The dynamics representing agents may be homogeneous
or heterogeneous where agents behave differently from each other. However, the coordination variable which forms the basis for which the agents

exchange information should be uniform for all the agents. The dynamics of each vehicle in the multivehicle framework should capture dynamics for motion, observation, and estimation of either neighboring agents or the environment.

8.2.2 Attitude dynamics and kinematics

Two common modeling approaches are empirical and axiomatic approaches [2]. Empirical models are based on identification methods which can be used to develop input–output mathematical models, usually termed *black-box* models. Axiomatic approaches develop mathematical models based on first principles physics models. Mechanical components that make up vehicle systems can be described as multibody, finite element, or continuous systems.

The bodies and particles are interconnected with each other and with the environment by springs, dampers, and actuators, as well as by bearings, joints, and servomechanisms, all of them without mass. An important hypothesis in vehicle dynamics is the assumption of rigid bodies for the modeling of low frequency motions. Rigid bodies are characterized by constant distances between all material points resulting in six degrees of freedom for each body. However, the number of degrees of freedom of a multibody system is reduced by the bearings. Due to the compliant elements acting at a discrete number of node points on the rigid bodies, the stiffness distribution within a multibody system is inhomogeneous. Furthermore, there are no restrictions with respect to the geometry, resulting in the high adaptability of multibody systems. The motion behavior is completely described by generalized coordinates the number of which is equal to the number f of degrees of freedom. The general dynamic model based on multibody dynamics is given by

$$\mathbf{M}\ddot{y}(t) + \mathbf{D}\dot{y}(t) + \mathbf{K}y(t) = f(t), \qquad (8.1)$$

where $\mathbf{M}$, $\mathbf{D}$, $\mathbf{K}$ specify the inertia, damping, and stiffness behavior of the component under consideration; $y(t)$ denoted the generalized coordinates of the body; and f is the number of degrees of freedom of the body.

Finite element systems (FESs) consist of simple, nonrigid elements, like rods, beams, plates, etc., with a finite number of degrees of freedom. At node points there are acting discrete loads, and there may be attached particles, springs, bearings, etc. Further, the boundary conditions are taken into account by motions or forces, respectively, at the node points. The linear independent motions of the node points represent the degrees of freedom of the total FES. The number of degrees of freedom is finite due to the finite number of node points, however, the numbers are much higher than for comparable multibody systems due to the nonrigid elements. The deformations within the finite elements between the node points are identified by shape functions. The modeling procedure with FES is very flexible, there are no restrictions with respect to the geometric system design. Modeling is very efficient by using uniform or similar elements, and by collecting groups of elements in superelements or substructures, respectively.

The inertial frame is rigidly attached to the surface of the earth and is assumed to be flat with orthonormal basis given by the following set of axis $\{\hat{e}_1, \hat{e}_2, \hat{e}_3\}$, where $\hat{e}_1$ is orientated towards the North, $\hat{e}_2$ is orientated towards the East, and $\hat{e}_1$ points towards the center of the earth. The body frame $\mathbb{B}_i, i \in 1, \ldots, N$ is attached to the center of gravity of the rigid body with the associated basis $\{\hat{e}_{i1}, \hat{e}_{i2}, \hat{e}_{i3}\}$, where $\hat{e}_{i1}$ is orientated towards the front of the vehicle, $\hat{e}_{i2}$ points to right, and $\hat{e}_{i3}$ is directed downwards. The attitude of the rigid body can be described using different representations including rotation matrix, Euler angles, axis–angle parameterization, unit quaternion, and Rodrigues parameterization. The rotation matrix denoted by $\mathbf{R}$ is a representation of the orientation of axes of frame $\mathcal{F}_1$ in frame $\mathcal{F}_2$. The column vectors of $\mathbf{R}$ describe the axis of frame $\mathcal{F}_1$ in frame $\mathcal{F}_2$ and contain only cosines. The axes of the rotation matrices are orthonormal and the rotation matrix is orthogonal and belongs to the following set:

$$SO(3) = \{\mathbf{R} \in \mathbb{R}^{3 \times 3} \mid \det(\mathbf{R}) = 1, \mathbf{R}\mathbf{R}^T = \mathbf{R}^T\mathbf{R} = \mathbf{I}_3\}. \qquad (8.2)$$

Another representation commonly used for describing attitude dynamics is the Euler angles. The orientation of the body-fixed frame can then be obtained by the composition of these three rotations, provided that adjacent rotations are not made about parallel axes. As a result, 12 distinct sets of Euler angles are possible. Common Euler angles are the roll (ϕ), pitch (θ), and yaw (ψ) angles. Relative orientation of two reference frames can be expressed by a single rotation about a given normalized vector by a given rotation angle. Consider the unit vector $\hat{\kappa} \in \mathbb{R}^3$, where v denotes the angle of rotation about $\hat{\kappa}$. Based on the angle–axis representation, the corresponding rotation matrix:

$$\mathbf{R}(v, \hat{\kappa}) = \mathbf{I}_3 - \sin(v)\mathbf{S}(\hat{\kappa}) + (1 - \cos(\phi))\mathbf{S}(\hat{\kappa})^2, \qquad (8.3)$$

where $\mathbf{S}(x)$ is a skew-symmetric matrix associated with a vector $\mathbf{x} = [x_1\ x_2\ x_3]$,

$$\mathbf{S}(x) = \begin{bmatrix} 0 & -x_3 & x_2 \\ x_3 & 0 & -x_1 \\ -x_2 & x_1 & 0 \end{bmatrix}, \tag{8.4}$$

satisfying $\mathbf{S}(x)y := x \times y$ where $x, y \in \mathbb{R}^3$.

The unit quaternion is another possible representation for attitude of a rigid body which provides a four-element representation denoted by

$$\mathbf{Q} = \begin{bmatrix} \mathbf{q} \\ \eta \end{bmatrix} \in \mathbb{Q}, \tag{8.5}$$

where $\mathbf{q} \in \mathbb{R}^3$ and $\eta \in \mathbb{R}$; $\mathbb{Q}$ is the set of unit quaternions defined by

$$\mathbb{Q} = \{\mathbf{Q} \in \mathbb{R}^4 \mid |Q| = 1\}. \tag{8.6}$$

The unit quaternion is considered as an angle–axis representation. In terms of the unit quaternion, it is possible to define a rotation by an angle v about a unit length vector as $\hat{\kappa} \in \mathbb{R}^3$ as

$$\mathbf{Q} = \begin{bmatrix} \hat{\kappa} \sin(v/2) \\ \cos(v/2) \end{bmatrix}. \tag{8.7}$$

The Rodriques formula can be used to obtain a transformation that provides the rotation matrix associated with the unit quaternion as

$$\mathbf{R}(\mathbf{Q}) = (\eta^2 - \mathbf{q}^T\mathbf{q})\mathbf{I}_3 + 2\mathbf{q}\mathbf{q}^T - 2\eta\mathbf{S}(\mathbf{q}), \tag{8.8}$$

where $\mathbf{S}$ is a skew-symmetric matrix. The rotation matrix associated with the unit quaternion has the following properties:
- $\mathbf{R}(\mathbf{Q}) = \mathbf{R}(-\mathbf{Q})$;
- Given the quaternions $\mathbf{Q}_1 = [\mathbf{q}_1\ \eta_1]^T$ and $\mathbf{Q}_2 = [\mathbf{q}_2\ \eta_2]^T$ such that

$$\mathbf{Q}_3 = \mathbf{Q}_1 \cdot \mathbf{Q}_2 = \begin{bmatrix} \eta_1\mathbf{q}_2 + \eta_2\mathbf{q}_1 + \mathbf{S}(\mathbf{q}_1)\mathbf{q}_2 \\ \eta_1\eta_2 - \mathbf{q}_1^T\mathbf{q}_2 \end{bmatrix}, \tag{8.9}$$

the rotation matrix is $\mathbf{R}(\mathbf{Q}_3) = \mathbf{R}(\mathbf{Q}_2)\mathbf{R}(\mathbf{Q}_1)$;
- Given the quaternions $\mathbf{Q} = [\mathbf{q}\ \eta]$ and $\mathbf{Q}^{-1} = [-\mathbf{q}\ \eta]$, $\mathbf{Q} \cdot \mathbf{Q}^{-1} = \mathbf{Q}^{-1} \cdot \mathbf{Q} = \mathbf{Q}_I$, where $\mathbf{Q}_I = [\mathbf{O}_3\ 1]$;
- $\mathbf{R}(\mathbf{Q}^{-1}) = \mathbf{R}(\mathbf{Q})^T$.

8.2.3 Information exchange

Control protocols for achieving coordination are either centralized or distributed. Distributed information structures are increasingly being adopted for multivehicle coordination problems. In both structures, the information exchange is defined by a control protocol which is suited to the global objective to be reached. For example, in formation control problems, agents are required to maintain some particular geometric formation while performing some tasks which may range from surveillance to object manipulation in two- or three-dimensional space. Information exchange structures are mostly defined by mathematical structures called graphs which may be fixed, switching, or dynamic. Communication media in MAS networks can be through direct communication, environment, and through sensing. In direct communications, agent communicate through wireless information networks such as wireless ethernet or bluetooth communication networks. Agents can also communicate through the environment by mimicking pheromone-type information exchange as observed in biological agents. In this case, agents leave trails for their neighbors to observe and follow. Agents can also interact with their environments and distinguish between neighbors and obstacles using sensor nodes.

8.2.4 Behavior module

The behavioral module of the multivehicle architecture is designed to allow the agents in the network perform some functions directly related to the global objective. For example, consider a multi-UAV network where the agents are tasked to move some objects from one location to another. Behavioral functions for this objective may include algorithms which enables the agent to compute the shortest distance between source and destination. Behavioral algorithms are usually adaptive in nature and may in some cases feature AI learning algorithms.

8.3 Aerial vehicle systems

Aerial vehicle systems find practical applications in both civilian and military applications. Practical applications of the aerial vehicle systems such as helicopters, miniaircrafts, and quadrotors include surveillance, load transportation, and data acquisition in areas that may be inaccessible. Common types of aerial vehicle systems include the quadrotor (quadcopter) and helicopter system. Aerial vehicle systems can be categorized based on the type of lift and landing mode. Vertical take-off and landing (VTOL) aerial vehicle systems are more often discussed due to their durability and maneuverability in real-life applications. The quadrotor system is a typical example of a VTOL aircraft.

8.3.1 Vertical take-off landing aircrafts

Consider a VTOL aircraft with position $\mathbf{p}$ and velocity $\mathbf{v}$ of its center of gravity defined in inertial frame $\mathbb{F}_0$. Let the unit quaternion $\mathbf{Q}$ represent the orientation of its body fixed frame, and ω denote the body-referenced angular velocity of the aircraft; $\mathbf{R}(\mathbf{Q})$ denotes the rotation matrix that transforms the orientation from inertial frame to the body fixed frame; $\mathbf{J} \in \mathbb{R}^{3 \times 3}$ represents the symmetric positive-definite constant inertia matrix associated with the body fixed frame $\mathbb{F}_b$. The nominal model for a VTOL can be expressed in the translational and rotational dynamics given by:

$$\dot{\mathbf{p}} = \mathbf{v}, \tag{8.10}$$

$$\dot{\mathbf{v}} = g\hat{e}_3 - \frac{\mathcal{T}}{m}\mathbf{R}(\mathbf{Q})^T\hat{e}_3, \tag{8.11}$$

$$\dot{\mathbf{Q}} = \frac{1}{2}\mathbf{T}(\mathbf{Q})\omega, \tag{8.12}$$

$$\mathbf{J}\dot{\omega} = \mathbf{\Gamma} - \mathbf{S}(\omega)\mathbf{J}\omega, \tag{8.13}$$

where $\hat{e}_3 := (0, 0, 1)^T$ and $\mathcal{T}$ and $\mathbf{\Gamma}$ denote the thrust and torque inputs applied to the aircraft, respectively. The nominal model given here assumes zero or negligible aerodynamic and external disturbances.

In the quadrotor aircraft system, there are four independent rotors, namely the front and rear rotors which rotate counterclockwise about the positive z-axis and the left and right rotors which rotate in the clockwise direction (see Fig. 8.2). The equations of motion of the quadrotor aircraft can be derived as [17]:

$$\dot{\mathbf{p}} = \mathbf{v}, \tag{8.14}$$

$$m\dot{\mathbf{v}} = mg\hat{e}_3 - \mathcal{T}\mathbf{R}(\mathbf{Q})^T\hat{e}_3, \tag{8.15}$$

$$\dot{\mathbf{Q}} = \frac{1}{2}\mathbf{T}(\mathbf{Q})\omega, \tag{8.16}$$

$$\mathbf{J}\dot{\omega} = \mathbf{\Gamma} - \mathbf{S}(\omega)\mathbf{J}\omega - G_a. \tag{8.17}$$

In addition the dynamics of each rotor is defined as

$$\mathbf{J}_r\dot{\omega}_i = \mathbf{v}_i - W_i, \quad i \in \{1, 2, 3, 4\}, \tag{8.18}$$

Figure 8.2 Quadrotor system [3].

where $\mathbf{J}_r$ and $\bar{\omega}_i$ represent the moment of inertia and rotational speed of each rotor, respectively; $\hat{e}_3 := (0, 0, 1)^T$ is the unit vector in the inertial frame $\mathbb{F}_0$; $\mathcal{T}$ denotes the total thrust applied to the airframe by the four rotors in the direction of $\hat{e}_{3b}$; $\mathbf{\Gamma} \in \mathbb{R}^3$ is the external torque applied to the airframe by the four rotors in the body frame $\mathbb{F}_b$; G_a denotes the total gyroscopic torques due to the combination of the airframe and the four rotors; and $W_i = \kappa \bar{\omega}_i^2$ is the reactive torque on the airframe due to the rotor drag generated by the ith rotor

$$G_a = \sum_{i=1}^{4} (-1)^{i+1} \mathbf{J}_r \mathbf{S}(\mathbf{w}) \hat{e}_3 \bar{\omega}_i. \tag{8.19}$$

The thrust $\mathcal{T}$, torque $\mathbf{\Gamma}$, and speeds $\mathbf{w}$ are related by the expression

$$\begin{bmatrix} \mathcal{T} \\ \mathbf{\Gamma} \end{bmatrix} = \mathbf{M} \begin{bmatrix} \bar{\omega}_1^2 \\ \bar{\omega}_2^2 \\ \bar{\omega}_3^2 \\ \bar{\omega}_4^2 \end{bmatrix}, \tag{8.20}$$

where $\mathbf{M}$ is defined as

$$\mathbf{M} = \begin{bmatrix} b & b & b & b \\ 0 & bd & 0 & -bd \\ -bd & 0 & bd & 0 \\ \kappa & -\kappa & \kappa & -\kappa \end{bmatrix} \tag{8.21}$$

with κ and b being positive constants which depend on physical quantities such as the density of air, size, shape, and pitch angle of the blades;

Table 8.1 6-DOF quadrotor system.

Symbol	Description
p, q, r	body angular rates in the body frame
I_x, I_y, I_z	moment of inertia
L	distance from the center of mass to the rotors
m	mass of the quadrotor
J_r	moment of inertia of the rotor blade
$\delta_{collective}$	collective input command
δ_{pitch}	pitch input command
δ_{roll}	roll input command
δ_{yaw}	yaw input command
ϕ	roll angle
θ	pitch angle
ψ	yaw angle

d represents the distance between the rotor to the COG of the aircraft. The matrix **M** is nonsingular. Control of the aircraft is achieved by designing appropriate thrusts and torque inputs.

The dynamic model of a 6-DOF quadrotor was described in [21] as:

$$\dot{p} = qr\left(\frac{I_y - I_z}{I_x}\right) - \frac{J_r}{I_x}qd + \frac{L}{I_x}\delta_{roll}, \tag{8.22}$$

$$\dot{q} = pr\left(\frac{I_z - I_x}{I_y}\right) - \frac{J_r}{I_y}rd + \frac{L}{I_x}\delta_{pitch}, \tag{8.23}$$

$$\dot{r} = pq\left(\frac{I_x - I_y}{I_z}\right) - \frac{1}{I_z}\delta_{yaw}, \tag{8.24}$$

$$\dot{\phi} = p + \tan\theta(q\sin\phi + r\cos\phi), \tag{8.25}$$

$$\dot{q} = q\cos\phi + r\sin\phi, \tag{8.26}$$

$$\dot{r} = (q\sin\phi + r\cos\phi)\sec\theta. \tag{8.27}$$

The parameters describing the quadrotor system are listed in Table 8.1.

The VTOL ducted-fan aircraft is another interesting aerial vehicle system which provides increased thrust efficiency at low velocities. The system is usually actuated with one or two coaxial propellers which generate the thrust and ailerons which control the orientation of the UAV. This system has been proposed as an experimental platform in [18–20]. Consider the dynamical model of a ducted-fan VTOL with two coaxial propellers and four servo-actuated ailerons:

Figure 8.3 Helicopter system.

$$\dot{\mathbf{p}} = \mathbf{v}, \tag{8.28}$$

$$m\dot{\mathbf{v}} = mg\hat{e}_3 - \frac{1}{l}\mathbf{R}(\mathbf{Q})^T\mathbf{S}(\hat{e}_3)\mathbf{\Gamma} + \mathbf{R}(\mathbf{Q})^T\mathbf{F}_d, \tag{8.29}$$

$$\dot{\mathbf{Q}} = \frac{1}{2}\mathbf{T}(\mathbf{Q})\omega, \tag{8.30}$$

$$\mathbf{J}\dot{\omega} = -\mathbf{S}(\omega)\mathbf{J}\omega + \mathbf{\Gamma} - \epsilon\mathbf{S}(\hat{e}_3)\mathbf{J}\mathbf{F}_d. \tag{8.31}$$

The dynamic model of the quadrotor system with respect to the body frame is [22]:

$$\ddot{x} = (\sin\psi\sin\phi + \cos\psi\sin\theta\cos\phi)\frac{U_1}{m}, \tag{8.32}$$

$$\ddot{y} = (-\cos\psi\sin\phi + \sin\psi\sin\theta\cos\theta)\frac{U_1}{m}, \tag{8.33}$$

$$\ddot{z} = -g + (\cos\theta\cos\phi)\frac{U_1}{m}, \tag{8.34}$$

$$\dot{p} = \frac{I_{yy} - I_{zz}}{I_{xx}}qr - \frac{J_{TP}}{I_{xx}} + \frac{U_2}{I_{xx}}, \tag{8.35}$$

$$\dot{q} = \frac{I_{zz} - I_{xx}}{I_{yy}}pr - \frac{J_{TP}}{I_{YY}}p\Omega + \frac{U_3}{I_{yy}}, \tag{8.36}$$

$$\dot{r} = \frac{I_{xx} - I_{yy}}{I_{zz}}pq + \frac{U_4}{I_{zz}}. \tag{8.37}$$

Another interesting aircraft system worthy of note in UAV literature is the helicopter system (see Fig. 8.3). The equations of motion defining the dynamical behavior of the helicopter system are given by [22] as:

$$\dot{p}^I = v^I, \tag{8.38}$$

$$\dot{v}^I = \frac{1}{M} R f^B, \tag{8.39}$$

$$\dot{R} = R \hat{w}^B, \tag{8.40}$$

$$I \dot{\omega}^B = -\omega \times (I \omega^B) + \tau^B, \tag{8.41}$$

$$\tau_f \dot{a} = -a - \tau_f q + A_b b + A_{\text{lon}} \delta_{\text{lon}}, \tag{8.42}$$

$$\tau_f \dot{b} = -b - \tau_f p + B_b a + B_{\text{lat}} \delta_{\text{lat}}, \tag{8.43}$$

$R(\Theta) =$

$$\begin{bmatrix} \cos(\psi)\cos(\theta) & \cos(\psi)\sin(\phi)\sin(\theta) - \cos(\phi)\sin(\psi) & \sin(\phi)\sin(\psi) + \cos(\phi)\cos(\psi)\sin(\theta) \\ \cos(\theta)\sin(\psi) & \cos\phi\cos\psi + \sin(\phi)\sin(\psi)\sin(\theta) & \cos(\theta)\sin(\phi)\sin(\theta) - \cos(\psi)\sin(\phi) \\ -\sin(\theta) & \cos(\theta)\sin(\phi) & \cos(\phi)\cos(\theta) \end{bmatrix}.$$

The dynamic model of the helicopter system with respect to the body frame is [22]:

$$\dot{u} = rv - qw + R_{31}g + \frac{X}{m}, \tag{8.44}$$

$$\dot{v} = pw - ru + R_{32}g + \frac{Y}{m}, \tag{8.45}$$

$$\dot{w} = qu - pv + R_{33}g + \frac{Z}{m}, \tag{8.46}$$

$$\dot{p} = \frac{qr(J_{yy} - J_{zz})}{J_{xx}} + \frac{L}{J_{xx}}, \tag{8.47}$$

$$\dot{q} = \frac{pr(J_{zz} - J_{xx})}{J_{yy}} + \frac{M}{J_{yy}}, \tag{8.48}$$

$$\dot{r} = \frac{qp(J_{xx} - J_{yy})}{J_{zz}} + \frac{N}{J_{zz}}. \tag{8.49}$$

8.4 Ground vehicle systems

Ground vehicle systems generally refer to vehicles moving on the ground terrain (or environment) which may be smooth or rough, as well as deterministic or nondeterministic. Models used for representing each ground vehicle agent in a multiagent framework considers the following subsystems:

- the vehicle body;
- the propulsion;
- the guidance and suspension devices; and
- the road.

Figure 8.4 Military ground vehicle.

Figure 8.5 Unmanned ground vehicle.

Autonomous ground vehicle systems can be used in cooperative frameworks to deliver payload between two locations, team surveillance missions, or navigating and mapping complex terrains (see Figs. 8.4 and 8.5). Models describing each ground vehicle system should define to some level the task to be carried out by the agents. Therefore results derived based on theoretical analysis depend largely on the degree of sophistication of the choice model. Models describing multivehicle dynamics can be decomposed into interacting subsystems as shown in Fig. 8.6. Models for ground vehicle systems can range from simple to complex models based on the intended analysis. We summarize some vehicle models here as obtained in the literature.

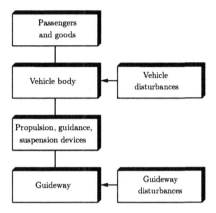

Figure 8.6 Components of a ground vehicle system.

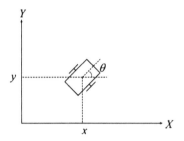

Figure 8.7 Differential drive vehicle.

8.4.1 Differential drive vehicle

The differential drive vehicle is a simple vehicle system in 2D frame defined by the coordinate frame $q = [x, y, \theta]$, where x and y represent the position of the vehicle in 2D and θ defines the orientation (see Fig. 8.7). The Lagrangian equation defining the motion of the vehicle in 2D plane is given as

$$L = \frac{1}{2}m(\dot{x}^2 + \dot{y}^2) + \frac{1}{2}J\dot{\theta}^2, \tag{8.50}$$

where m is the mass of the vehicle and J is the inertia of the vehicle with respect to the vertical axis. If it is assumed that the vehicle rolls without side slipping, the following nonholonomic constraint is defined:

$$\dot{x}\sin\theta - \dot{y}\cos\theta = 0. \tag{8.51}$$

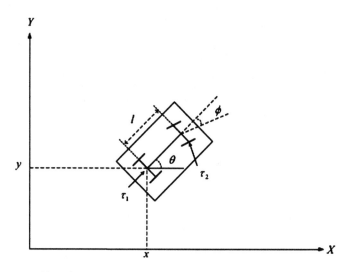

Figure 8.8 Car-like robot.

The kinematic model of the vehicle is thus defined as:

$$\dot{x} = (\cos\theta)u_1, \tag{8.52}$$

$$\dot{y} = (\sin\theta)u_1, \tag{8.53}$$

$$\dot{\theta} = u_2, \tag{8.54}$$

where u_1 and u_2 represent the driving and steering velocity, respectively, and are defined as:

$$u_1 = \frac{\rho}{2}(\omega_r + \omega_l), \tag{8.55}$$

$$u_2 = \frac{\rho}{2}(\omega_r + \omega_l). \tag{8.56}$$

8.4.2 Car-like vehicle

A car-like vehicle is basically a four-wheel vehicle system described in a two-dimensional space (see Fig. 8.8). The front wheels of the vehicle steer it, and its rear wheels have a fixed orientation with respect to the body. The system is defined in a two-dimensional space with the following generalized coordinate, $q = [x\,y\,\theta\,\phi]$; x and y describe the position of the body in two-dimensional Cartesian coordinates; ϕ is the steering angle; and θ is the orientation of the body with respect to the x-axis:

$$\dot{x} = (\cos\theta)u_1, \tag{8.57}$$

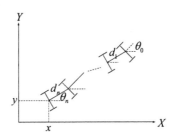

Figure 8.9 Trailer systems.

$$\dot{y} = (\sin\theta)u_1, \tag{8.58}$$

$$\dot{\theta} = \frac{1}{l}(\tan\phi)u_1, \tag{8.59}$$

$$\dot{\phi} = u_2, \tag{8.60}$$

where u_1 and u_2 are the linear velocity and the steering rate of the body and can be defined mathematically as

$$u_1 = \rho\omega_1 \quad u_2 = \omega_2, \tag{8.61}$$

where ρ is the radius of the back wheels, ω_1 is the angular velocity of the back wheels, and ω_2 is the steering rate of the front wheels. In terms of ω_1 and ω_2, the system model becomes:

$$\dot{x} = (\rho\cos\theta)\omega_1, \tag{8.62}$$
$$\dot{y} = (\rho\sin\theta)\omega_1, \tag{8.63}$$
$$\dot{\theta} = \frac{\rho}{l}(\tan\phi)\omega_1, \tag{8.64}$$
$$\dot{\phi} = \omega_2. \tag{8.65}$$

8.4.3 Tractor-trailer systems

The tractor-trailer systems refer generally to linked vehicles connected at the middle axle, while the first and third axle are allowed to be steered (see Fig. 8.9). The generalized coordinate is defined as $q = [x\ y\ \phi_1\ \theta_1\ \phi_2\ \theta_2]$; x and y define the position of the guide-point, ϕ_1 is the steering angle of the front wheel, θ_1 is the orientation of the tractor, ϕ_2 is the steering angle of the third axle; and θ_2 is the orientation of the trailer:

$$\dot{x} = (\cos\theta_1)u_1, \tag{8.66}$$
$$\dot{y} = (\sin\theta_1)u_1, \tag{8.67}$$

Figure 8.10 Off-road light autonomous vehicle (OLAV) [13].

$$\dot{\phi}_1 = u_2, \tag{8.68}$$

$$\dot{\theta}_1 = \frac{1}{l_f}(\tan\phi_1)u_1, \tag{8.69}$$

$$\dot{\phi}_2 = u_3, \tag{8.70}$$

$$\dot{\theta}_2 = -\frac{1}{l_b}(\sec\phi_2)\sin(\phi_2 - \theta_1 + \theta_2)u_1. \tag{8.71}$$

8.4.4 Off-road light autonomous vehicle

The dynamic model of an off-road light autonomous vehicle (OLAV) was studied by [13]. The OLAV UGV has 13 states and 5 control inputs defined as follows (see Fig. 8.10 and Tables 8.2, 8.3, and 8.4):

$$x = [\omega_e \ \dot{\phi}_p \ \phi_p \ \rho_b \ \omega_{fw} \ \omega_{rw} \ F_{fw,x} \ F_{rw,x} \ \omega_v \ \theta_v \ v_x \ r_x \ r_z],$$

$$u = [\alpha \ \beta \ G\theta_g \ \frac{d\theta_g}{r_x}].$$

Table 8.2 UGV states; definitions and units [13].

Symbol	Definition	Unit
ω_e	Angular velocity of engine	rad s^{-1}
$\dot{\phi}_p$	Displacement speed of CVT Pulley	
ϕ_p	Displacement of CVT pulleys, proportional to the speed ratio of the CVT	
ρ_b	Delayed brake input	rad s^{-1}
ω_{fw}	Angular velocity of front wheels	rad s^{-1}
ω_{rw}	Angular velocity of rear wheels	rad s^{-1}
$F_{fw,x}$	Tractive force of front wheels	N
$F_{rw,x}$	Tractive force of rear wheels	N
ω_v	Angular velocity of vehicle body around pitch axis	rad s^{-1}
θ_v	Pitch angle of vehicle in inertia frame	rad
v_x	Vehicle speed along the ground	ms^{-1}
r_x	Vehicle horizontal position in inertia frame	m
r_z	Vehicle vertical position in inertia frame	m

The dynamics of the system is defined by:

$$\dot{x} = \begin{bmatrix} c\dfrac{1}{I_e}[T_e(\alpha, \omega_e) - T_b] \\[2mm] \dfrac{k_{cvt}}{u_{cvt}}\omega_e - \dfrac{k_{cvt}}{u_{cvt}}(R_{cvt}T_b) - b_{cvt}\dot{\phi}_p - \Phi_{k,cvt}(\phi_p) \\[2mm] \dot{\phi}_p \\[2mm] \dfrac{1}{\tau_b}[\beta - \rho_p] \\[2mm] \dfrac{1}{I_{fw}}[B_{bk}T_{bk}(\omega_{fw}, \rho_{bk}) + T_{rr}(\omega_{fw}, F_{fw,z}) - r_{dyn}F_{fw,x}] \\[2mm] \dfrac{1}{I_{ra}}[R_{cvt}R_t(G)T_b + (1 - B_b k)T_{bk}(\omega_{rw}\rho_{bk}) \\ \qquad + T_{rr}(\omega_{rw}, F_{rw,z})] \\[2mm] \dfrac{1}{\tau_t}[F_{fw,x,\text{stat}} - F_{rw,x}] \\[2mm] \dfrac{1}{\tau_t}[F_{rw,x,\text{stat}} - F_{fw,x}] \\[2mm] \dfrac{1}{I_{v,\theta}}[(F_{fw,x} + F_{rw,z})r_c g + k_s(\theta_g + \theta_0 - \theta_v) \\ \qquad + b_s(\omega_g - \omega_v)] \\[2mm] \omega_v \\[2mm] \dfrac{1}{m_v}[F_{g,x} + F_d + F_{fw,x} + F_{rw,x}] \\[2mm] v_x\cos(\theta) \\[2mm] v_x\sin(\theta) \end{bmatrix} \qquad (8.72)$$

Table 8.3 UGV parameters [13].

Symbol	Definition
$m_{v,e}$	Total mass of vehicle without any occupants or additional equipment
$I_{v,\theta}$	Moment of inertia of the vehicle around the pitch axis
k_s	Spring coefficient of suspension
b_s	Damper coefficient of suspension
$\theta_{v,0}$	Pitch of vehicle relative to ground at steady state
r_f	Distance along longitudinal axis from center of gravity to rear wheel contact patch
r_r	Distance along longitudinal axis from center of gravity to front wheel contact patch
r_{cg}	Vertical distance from ground to center of gravity
b_{ar}	Drag coefficient
Ie	Moment of inertia of engine including the primary CVT pulley
$P_{e,\max}$	Peak power of engine at full throttle
b_e	Viscous friction coefficient of engine
$\omega_{pe,\max}$	Angular velocity of engine at peak power
$\omega_{pe,\mathrm{idle}}$	Engine idle speed
$\omega_{e,\max}$	Angular velocity for rev limiter
$p_{e,\mathrm{idle}}$	Proportional gain of idle controller
$v_{\max}$	Speed limiter set point
ω_{be}	Engine speed needed to engage belt
u_{cvt}	Coefficient controlling the relationship between primary pulley angular velocity and shifting speed.
d_{cvt}	Coefficient controlling the relationship between torque on the secondary pulley and shifting speed
k_{cvt}	Shifting spring coefficient
b_{cvt}	Viscous friction coefficient of shifting
$R_{cvt,U}$	Speed ratio of CVT at full up-shift
$R_{cvt,D}$	Speed ratio of CVT at full down-shift
$\mu_{b,\max}$	Maximum friction between belt and the primary pulley
$s_{b,\max}$	Slip between belt and primary pulley at maximum friction
I_t	Moment of inertia of the gear box including secondary pulley
b_t	Transmission friction
R_H	Speed ratio of transmission in high range
R_L	Speed ratio of transmission in low range
R_R	Speed ratio of transmission in reverse
I_{fw}	Moment of inertia of forward wheels, drive shafts, brake disks, and differential
I_{rw}	Moment of inertia of rear wheels, drive shafts, brake disks, and differential
b_{rr}	Rolling resistance coefficient
$\mu_{t,\max}$	Maximum friction produced by the tires

Table 8.4 UGV parameters (continued) [13].

Symbol	Definition
$s_{\mu_t,\max}$	Tire slip producing maximum friction
τ_t	Time constant for dynamic tire forces
r_{eff}	Effective tire radius
r_{dyn}	Static tire radius
$T_{b_k,\max}$	Maximum braking torque
τ_{b_k}	Time constant for brake system
B_{b_k}	Front to rear brake torque distribution

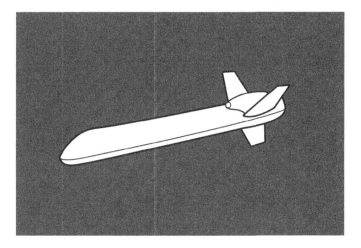

Figure 8.11 Underwater vehicle system.

8.5 Underwater vehicle systems

The kinematic equations defining the motion of an underwater vehicle system cruising at a constant linear speed can be expressed as (see Fig. 8.11):

$$\dot{x} = \cos\theta \cos\psi v_x, \tag{8.73}$$

$$\dot{y} = \sin\psi \cos\theta v_x, \tag{8.74}$$

$$\dot{z} = -\sin\theta, \tag{8.75}$$

$$\dot{\phi} = \omega_x + \sin\phi \tan\theta \omega_y + \cos\phi \tan\theta \omega_z, \tag{8.76}$$

$$\dot{\theta} = \cos\phi \omega_y - \sin\phi \omega_z, \tag{8.77}$$

$$\dot{\psi} = \sin\phi \sec\theta \omega_y - \cos\phi \sec\theta \omega_z, \tag{8.78}$$

$$\dot{x} = \cos(\phi)\cos(\theta)u - \sin(\phi)v + \sin(\theta)\cos(\phi)w, \tag{8.79}$$

$$\dot{y} = \sin(\phi)\cos(\theta)u + \cos(\phi)v + \sin(\theta)\sin(\phi)w, \tag{8.80}$$

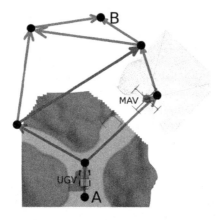

Figure 8.12 Aerial-assisted vehicular system [3].

$$\dot{z} = -\sin(\theta)u + \cos(\theta)w, \tag{8.81}$$

$$\dot{\theta} = q, \tag{8.82}$$

$$\dot{\phi} = \frac{r}{\cos\theta}, \tag{8.83}$$

$$\dot{u} = \frac{m_{22}}{m_{11}}vr + \frac{m_{33}}{m_{11}}wq - \frac{d_{11}}{m_{11}}u + \frac{1}{m_{11}}\tau_u, \tag{8.84}$$

$$\dot{v} = -\frac{m_{11}}{m_{22}}ur - \frac{d_{22}}{m_{22}}v, \tag{8.85}$$

$$\dot{w} = \frac{m_{11}}{m_{33}}uq - \frac{d_{33}}{m_{33}}w, \tag{8.86}$$

$$\dot{q} = \frac{m_{33} - m_{11}}{m_{55}}uw - \frac{d_{55}}{m_{55}}q - \frac{\rho g \nabla G \bar{M}_L \sin\theta}{m_{55}} + \tau_q \frac{1}{m_{55}}\tau_u, \tag{8.87}$$

$$\dot{r} = \frac{m_{11} - m_{22}}{m_{66}} - \frac{d_{66}}{m_{66}}r + \frac{1}{m_{66}}\tau_r. \tag{8.88}$$

8.6 Drone-assisted vehicular networks

In some applications, aerial vehicles are used to provide support to ground vehicles in performing tasks such as navigation (see Fig. 8.12). One of such problems was posed by [3], where the authors considered cooperative surveillance and navigation problem involving a microaerial vehicle (MAV) and unmanned ground vehicle (UGV). The approach is such that the ground vehicle directs the activity of the aerial vehicle by being its eye in the sky. Therefore, the UGV is able to survey and estimate terrains in the environment to determine which are suitable and which terrains to avoid completely. The model described in this setting is a single-agent model. The environment and possibly the state of the vehicle are assumed to be partially observable, therefore the states are described using probability density functions (pdf). At each instant of time, the UGV module is required to make a decision $a_t \subset \mathcal{A}$, where $\mathcal{A} = \mathcal{A}_m \cup \mathcal{A}_s$ represents a space of decisions representing movement actions $\mathcal{A}_m$ or sensing actions $\mathcal{A}_s$. The control actions are defined based on a stochastic transition function $p(s_{t+1}|s_t, a_t)$. For a given sensing action, an observation $z_{t+1} \in \mathbb{Z}$ is fed back based on a probabilistic observation model $p(z_{t+1}|s_{t+1}, a_t)$. Consequently, $b_{t+1} \equiv \tau(b_t, a_t, z_{t+1})$ denotes a Bayesian-filtering operation which estimates new belief state recursively based on current belief, action, and observation. Suppose the objective is to reach a destination while minimizing the cumulative cost of traversal $C(s, a)$. The following cost function was formulated as a recursive dynamic programming problem in [3]:

$$Q_{t+1}(b, a) = \mathbb{E}\left[C(s, a) + \gamma \sum_{z \in Z} p(z|b, a) J_t(\tau(b, a, z)) | b \right], \tag{8.89}$$

$$J_t(b) = \min_{a \in \mathcal{A}} Q_t(b, a), \tag{8.90}$$

where $C(s_t, a_t)$ is the cumulative sum of costs accrued along the way; $\gamma \in [0, 1)$ is a discount factor which determines the relative importance of future and immediate costs; $p(z|b, a)$ is the *a priori* probability of observing z; J_t is the cost functional which determines the minimum expected cost over t horizons. The solution to this problem is a set of control policies, $\pi^* = [\pi_1^*, \dots, \pi_T^*]^T$, for some desired time horizon T which may be infinite.

8.7 Load transportation

Load transportation is one of the real-life applications of autonomous vehicles. Autonomous vehicles may find potential applications in moving complex objects in and out of environments which may be harmful or simply inaccessible to humans. Examples of these environments include radioactive areas, collapsed buildings, or high-rise buildings. Load transportation problem is mostly investigated for UAV/MAVs and sometimes hybrid systems involving a cooperative combination of both ground and aerial vehicle systems. Here, we describe some models for load transportation problem as described in the literature ranging from single to multiple vehicle systems.

Table 8.5 Single quadrotor with slung load.

Symbol	Description
$m_Q \in \mathbb{R}$	Mass of quadrotor
$m_L \in \mathbb{R}$	Mass of slung load
$l \in \mathbb{R}$	Length of the string
$g \in \mathbb{R}$	Gravitational constant
$p_Q \in \mathbb{R}^3$	Position of the quadrotor
$p_L \in \mathbb{R}^3$	Position of load
$v_Q \in \mathbb{R}^3$	Velocity of the quadrotor
$v_L \in \mathbb{R}^3$	Velocity of the load
$r_L \in \mathbb{S}^2$	Unit vector from quadrotor to load
$\Omega_L \in \mathbb{R}^3$	String angular velocity
$T_L \in \mathbb{R}$	String tension
$\tau_L \in \mathbb{R}^3$	Angular acceleration of string
$R_Q \in \mathbb{SO}(3)$	Rotation matrix from $\{B\}$ to $\{I\}$
$w_Q \in \mathbb{R}^3$	Quadrotor angular velocity in $\{B\}$
$T_Q \in \mathbb{R}$	Quadrotor thrust force in $\{B\}$
$n_Q \in \mathbb{R}^3$	Quadrotor torque force in $\{B\}$
$e_1, e_2, e_3 \in \mathbb{R}^3$	Unit vectors along the reference frame

8.7.1 Single quadrotor with slung load

The coordinate-free model of the quadrotor system with slung load described by parameters in Table 8.5 is given by [5] as:

$$\dot{\mathbf{p}}_L = \mathbf{v}_L, \tag{8.91}$$

$$\dot{\mathbf{v}}_L = -\frac{1}{m_Q} T_L r_L + ge, \tag{8.92}$$

$$\dot{r}_L = S(\Omega_L) r_L, \tag{8.93}$$

$$\dot{\Omega}_L = \Pi_{r_L} \tau_L, \tag{8.94}$$

$$\dot{R}_Q = R_Q S(w_Q), \tag{8.95}$$

$$\dot{w}_Q = \frac{1}{J}(n_Q - S(\omega)J_\omega), \tag{8.96}$$

where the string tension and angular acceleration are

$$T_L = \frac{m_L}{m_Q + m_L} r_L^T T_Q r_Q + \frac{m_L m_Q}{m_Q + m_L} ||\omega_L||^2, \tag{8.97}$$

$$\tau_L = \frac{1}{m_Q L} S(r_L) T_Q r_Q. \tag{8.98}$$

The relationship between quadrotor and load position is given by

$$\mathbf{p}_Q = \mathbf{p}_L - Lr. \tag{8.99}$$

This model generally captures the dynamics of three coupled subsystems, namely the load, string, and quadrotor attitude kinematics and dynamics. The generalized coordinate-free state vector for the system is $x = [p_L\ v_L\ r_L\ \Omega_L\ R_Q\ \omega_Q]$ with control inputs $u = [T_Q\ n_Q]$. The system satisfies the differential flatness property as defined by [16].

Definition 8.1 (Differential flatness, [16]). A system $\dot{x} = f(x, u)$, $x \in \mathbb{R}^n$, $u \in \mathbb{R}^n$ satisfies the differential flatness property of there exist outputs $y \in \mathbb{R}^m$ of the form $y = y(x, u, \dot{u}, \ldots, u^{(p)})$ such that the states and the inputs can be expressed as $x = x(y, \dot{y}, \ldots, y^{(q)})$, $u = u(y, \dot{y}, \ldots, y^{(q)})$, where p, q are finite integers. Simply stated, a system is said to be differentially flat if there exists a set of outputs such that all the states and inputs can be expressed as functions of that output and its time derivatives.

The differential flatness property has been applied by different researchers to the slung-load transportation problem to address objectives such as motion planning and tracking control. The authors of [12] show that differential flatness property is satisfied for a system comprising a single quadrotor and a load connected by an inelastic massless cable, and they also extend the definition of differential flatness for a full-hybrid system which may result from the case where the cable link is not rigid. This work was extended for a quadrotor array consisting of three or more robots in [14]. The differential flatness property is also an indicator that a modified backstepping procedure can be used to obtain a feedback controller for the system [5]. The dynamics of the system in free-flight mode is given as:

$$\dot{\mathbf{p}} = \mathbf{v}, \tag{8.100}$$

$$\dot{\mathbf{v}} = -\frac{1}{m} TR\mathbf{e}_3 + g\mathbf{e}_3, \tag{8.101}$$

$$\dot{R} = RS(\omega), \tag{8.102}$$

$$\dot{\omega} = \tau. \tag{8.103}$$

Factoring out the yaw movement and defining the following variable changes:

$$\Omega = R\Pi_{e_3}\omega, \tag{8.104}$$

$$\tau' = R\Pi_{e_3}\tau, \tag{8.105}$$

the resulting reduced system dynamics becomes:

$$\dot{\mathbf{p}} = \mathbf{v}, \tag{8.106}$$

$$\dot{\mathbf{v}} = -\frac{1}{m} T r_3 + g e_3, \tag{8.107}$$

$$\dot{r}_3 = S(\Omega) r_3, \tag{8.108}$$

$$\dot{\Omega} = \Pi_{r_3} \tau', \tag{8.109}$$

where $\mathbf{r}_3 = R\mathbf{e}_3$. Thus in free-flight mode, the quadrotor dynamics can be written in the following compact form:

$$\dot{\mathbf{x}} = f(\mathbf{x}) + G(\mathbf{x}) \begin{bmatrix} T \\ \tau' \end{bmatrix}, \tag{8.110}$$

with $f(\mathbf{x}) : \mathbb{R}^{12} \to \mathbb{R}^{12}$ and $G(\mathbf{x}) : \mathbb{R}^3 \to \mathbb{R}^{12}$; T_l and τ_L are affine functions of $T_Q r_Q$ given as

$$\begin{bmatrix} T_L \\ \tau_L \end{bmatrix} = H(x_L) T_Q r_Q + h(x_L). \tag{8.111}$$

The generalized coordinate-free model of the quadrotor–slung-load system is thus written in compact form as:

$$\dot{x}_L = f_L(x_L) + G_L(x_L) T_Q^* r_Q^* + G_L x_L (T_Q r_Q - T_Q^* r_Q^*), \tag{8.112}$$

$$\dot{R}_Q = R_Q S(\omega_Q), \tag{8.113}$$

$$\dot{\omega}_Q = \tau_Q, \tag{8.114}$$

where T_Q^* and r_Q^* are virtual inputs for the string–load subsystem, and $f_L(x_L) = f(x_L) + G(x_L)h(x_L)$ and $G_L(x_L) = G(x_L)H(x_L)$. If the quadrotor attitude dynamics are ignored, the virtual inputs T_Q^* and r_Q^* are desired stabilizing inputs for the string–load subsystem which can be obtained as

$$T_Q^* r_Q^* = \left(\frac{m_L + m_Q}{m_L} T_L^*(z_L, p_d^{(4)}) - m_Q l \|\omega_L\|^2 \right) r_L + m_Q l S(r_L) \tau^*(z_L, p_d^{(4)}). \tag{8.115}$$

Nonlinear Lyapunov theory was proposed in [5] for trajectory tracking in the case of a single quadrotor and suspended load via backstepping approach which guarantees asymptotic stabilization with large region of attraction.

The full model of a single quadrotor with slung load was analyzed in [11] for domains where the inputs and angular velocity remain taut with the quadrotor's thrust.

Cable–load behavior was described by a mass–spring system in [6] to account for elasticity behaviors, leading to nondifferentially flat system output. Quadrotor with slung load system where the spring is modeled by a spring–damper system was presented in [6] as:

$$\dot{\mathbf{p}}_L = \mathbf{v}_L, \tag{8.116}$$

$$(m_Q + m_L)(\dot{v}_L + ge_3) = (m_Q \ddot{l} + (r_L.fRe_3) - m_Q l(\dot{r}_L \cdot \dot{r}_L))r_L, \tag{8.117}$$

$$\dot{r}_L = \omega \times r_L, \tag{8.118}$$

$$m_Q l\dot{w} = -(r_L \times fRe_3) - 2m_Q \dot{l}\omega, \tag{8.119}$$

$$\dot{R} = R\hat{\Omega}, \tag{8.120}$$

$$J\dot{\Omega} = M - (\Omega \times J\Omega), \tag{8.121}$$

$$\ddot{l} = l(\dot{r}_L \cdot \dot{r}_L) + \frac{m_Q + m_L}{m_Q m_L} k(L - l) - \frac{m_Q + m_L}{m_Q m_L} c\dot{l}$$
$$- \frac{1}{m_Q}(r_L \cdot fRe_3), \tag{8.122}$$

where L represents fixed length of the spring, M represents moment vector of the quadrotor in body-fixed frame, and k is the spring constant of the suspended spring.

Remark 8.1 ([6]). Although the quadrotor with a (inelastic) cable-suspended load is shown to be a differentially flat system in [17], the quadrotor with elastic-cable-suspended load is not differentially flat. This is due to the additional degree of underactuation caused by the presence of the spring.

8.7.2 Dual quadrotor with slung load

In [4], the load transportation problem was investigated for unmanned aerial vehicles (UAV) considering two quadrotors with masses m_1 and m_2, joined by massless rigid links with lengths l_1 and l_2, respectively, connected to a point-mass load with mass l_1. An inertial reference frame I and two body reference frames, B_1 and B_2, were considered, each fixed to center of mass of each quadrotor. The direction of each cable was described by a unit vector $q_i \in \mathbb{S}^2$, where $\mathbb{S}^2 = \{q_i \in \mathbb{R}^3 || q_i| = 1\}$ was expressed in the inertial frame and centered at the origin of the body frame B_i, $i = 1, 2$. The position of the

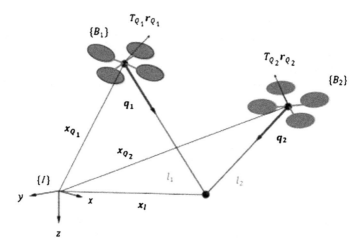

Figure 8.13 Dual quadrotor system with point load [3].

Table 8.6 Dual-quadrotor system with slung load.

Symbol	Description
$m_{Q_i} \in \mathbb{R}$	Mass of quadrotor i
$m_L \in \mathbb{R}$	Mass of slung load
$l_i \in \mathbb{R}$	Length of the string i
$g \in \mathbb{R}$	Gravitational constant
$p_{Q_i} \in \mathbb{R}^3$	Position of the quadrotor i
$p_L \in \mathbb{R}^3$	Position of load
$v_{Q_i} \in \mathbb{R}^3$	Velocity of the quadrotor i
$v_L \in \mathbb{R}^3$	Velocity of the load
$r_i \in \mathbb{S}^2$	Unit vector from quadrotor i to load
$T_L \in \mathbb{R}$	String tension
$\tau_L \in \mathbb{R}^3$	Angular acceleration of string
$R_Q \in \mathbb{SO}(3)$	Rotation matrix from $\{B\}$ *to* $\{I\}$
$w_Q \in \mathbb{R}^3$	Quadrotor angular velocity in $\{B\}$
$T_{Q_i} \in \mathbb{R}$	Quadrotor i thrust force in $\{B\}$
$e_1, e_2, e_3 \in \mathbb{R}^3$	Unit vectors along the reference frame

load was defined in the inertial frame I by the state x_l and the orientation of the quadrotor i, or specifically, the rotation matrix from the body reference frame B_i to the inertial reference frame I, was defined as $R_{Q_i} \in \mathbb{SO}$, where $\mathbb{SO}(3) = \{RR^T = I | \det(R) = I\}$ denotes the special orthogonal group of order 3 (see Fig. 8.13 and Table 8.6).

The mathematical model of the system as given by [4] was:

$$m_l\ddot{x}_l = -T_1 q_1 - T_2 q_2 + m_l g e_3 = -T_{L_t} q_t + m_l g e_3,$$ (8.123)

$$m_i\ddot{x}_i = T_i q_i - T_{Q_i} r_{Q_i} + m_{Q_i} g e_3,$$ (8.124)

where $\ddot{x}_l$ denotes the load linear acceleration, $\ddot{x}_{Q_i}$ represents the acceleration of quadrotor i, g is the gravitational acceleration, T_i is the tension applied to link i, T_{Q_i} is the thrust applied by quadrotor Q_i, r_{Q_i} is the direction of the thrust which coincides with the z-axis of the vehicle; T_{L_t} and $q_t \in \mathbb{S}^2$ define the total tension norm and direction, respectively, and are related by $T_{L_t} q_t = T_1 q_1 + T_2 q_2$. Assuming that each quadrotor has an inner loop controller which provides tracking of angular velocity and considering the kinematics of only r_{Q_i}, the total model of the system is written as:

$$\dot{p}_l = v_l,$$ (8.125)

$$\dot{v}_l = g e_3 - M_q^{-1} \sum_{i=1}^{n} \alpha_i q_i,$$ (8.126)

$$\ddot{q}_i = -|\dot{q}_i|^2 q_i + \frac{1}{l_i} \Pi_{q_i}\left(\frac{1}{m_{Q_i}} T_{Q_i} r_{Q_i} - M_q^{-1} \sum_{j=1}^{n} \alpha_j q_j\right),$$ (8.127)

$$\dot{r}_{Q_i} = u_{Q_i},$$ (8.128)

where
- v_l is the velocity of the load,
- M_q is a positive-definite symmetric matrix given by $M_q = M_l I + \sum_{i=1}^{n} m_{Q_i} q_i q_i^T$,
- α_i is an auxiliary variable given by $\alpha_i = q_i^T T_{Q_i} r_{Q_i} + m_{Q_i} l_i |\dot{q}_i|^2$,
- u_{Q_i} is the simplified quadrotor angular velocity input.

8.7.3 Multiquadrotor systems with slung load

A geometric control approach was proposed in [9] for an arbitrary number of quadrotors for slung-load transportation problem based on inner–outer loop control structure, while the work in [8] extends the results to a rigid-body load. That study considered an n quadrotor UAV connected to a point mass via massless links (see Fig. 8.14). The inertia reference frame $\{\vec{e}_1, \vec{e}_2, \vec{e}_3\}$ and fixed-body reference frame $\{b_{i1}, b_{i2}, b_{i3}\}$ were used for describing the dynamics for each $i \in \{1, \dots n\}$ quadrotor system:

$$\dot{x}_L = f_L(x_L) + G_L(x_L) T_Q^* r_Q^* + G_{Lx_L}(T_Q r_Q - T_Q^* r_Q^*),$$ (8.129)

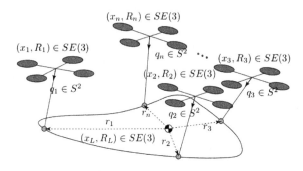

Figure 8.14 Multiquadrotor system.

$$\dot{R}_Q = R_Q S(\omega_Q), \tag{8.130}$$

$$\dot{\omega}_Q = \tau_Q. \tag{8.131}$$

8.8 Notes

In this chapter, we have covered some important aspects of multivehicle co-operative control framework. Different agent models have been discussed by highlighting some important control problems pertaining to multivehicle cooperative control.

References

[1] Y. Tan, Z. Zheng, Research advance in swarm robotics, Defense Technol. 9 (2013) 18–39.
[2] K. Popp, W. Schiehlen, Ground Vehicle Dynamics, Springer-Verlag Berlin Heidelberg, 2009.
[3] J. Melin, M. Lauri, A. Kolu, J. Koljonen, R. Ritala, Cooperative sensing and planning in a multi-vehicle environment, IFAC-PapersOnLine 48 (9) (2013) 198–203.
[4] T. Valentim, R. Cunha, P. Oliviera, D. Cabecinhas, C. Silvestre, Multivehicle cooperative control for load transportation, IFAC-PapersOnLine 52 (19) (2019) 358–363.
[5] D. Cabecinhas, R. Cunha, C. Silvestre, A trajectory tracking control law for a quadrotor with slung load, Automatica 106 (2019) 384–389.
[6] P. Kotaru, G. Wu, K. Sreenath, Dynamics and control of a quadrotor with a payload suspended through an elastic cable, in: 2017 American Control Conference (ACC), 2017, pp. 3906–3913.
[7] T. Lee, Optimal hybrid controls for global exponential tracking on the two-sphere, in: Proc. the 55th IEEE Conference on Decision and Control, 2016, pp. 3331–3337.
[8] T. Lee, Geometric control of quadrotor UAVs transporting a cable-suspended rigid body, IEEE Trans. Control Syst. Technol. 26 (1) (2018) 255–264.
[9] T. Lee, K. Sreenath, V. Kumar, Geometric control of cooperating multiple quadrotor UAVs with a suspended payload, in: Proc. the IEEE Conference on Decision and Control, 2013, pp. 5510–5515.

[10] P.O. Pereira, D.V. Dimarogonas, Control framework for slung load transportation with two aerial vehicles, in: Proc. the 56th IEEE Annual Conference on Decision and Control (CDC), 2017, pp. 4254–4259.

[11] P.O. Pereira, M. Herzog, D.V. Dimarogonas, Slung load transportation with a single aerial vehicle and disturbance removal, in: Proc. the 24th Mediterranean Conference on Control and Automation (MED), 2016, pp. 671–676.

[12] K. Sreenath, V. Kumar, Dynamics, control and planning for cooperative manipulation of paylods suspended by cables from multiple quadrotor robots, in: Robotics: Science and Systems, 2013, pp. 671–676.

[13] A.P. Magnus, Modeling and Parameter Estimation of an Unmanned Ground Vehicle with Continuously Variable Transmission, Master's Thesis, Department of Physics, University of Oslo, 2016.

[14] K. Sreenath, N. Micheal, V. Kumar, Trajectory generation and control of a quadrotor with a cable suspended load—a differentially-flat hybrid system, in: Proc. the IEEE Int. Conference on Robotics and Automation (ICRA), 2013, pp. 4888–4895.

[15] D.K.D. Villa, A.S. Brandao, M. Sarcinelli-Filho, Load transportation using quadrotors: a survey of experimental results, in: Proc. the Int. Conference on Unmanned Aircraft Systems, 2013, pp. 84–93.

[16] R.M. Murray, Trajectory generation for a towed cable system using differential flatness, in: IFAC World Congress, San Francisco, CA, 1996.

[17] A. Tayebi, S. McGilvray, Attitude stabilization of a quadrotor aircraft, IEEE Trans. Control Syst. Technol. 14 (2006) 562–571.

[18] J.M. Pflimlin, P. Soures, T. Hamel, Position control of a ducted fan VTOL UAV in crosswind, Int. J. Control 80 (5) (2006) 666–683.

[19] A. Roberts, Attitude estimation and control of a ducted fan VTOL UAV, Master's Thesis, Lakehead University, 2007.

[20] A. Roberts, A. Tayebi, Adaptive position tracking of VTOL-UAVs, IEEE Trans. Robot. 27 (1) (2011) 129–142.

[21] B.L. Stevens, F.L. Lewis, Aircraft Control and Simulation, 2nd ed., Wiley, Hoboken, NJ, 2003.

[22] J. Alvarenga, N.I. Vitzilaios, K.P. Valavanis, M.J. Rutherford, Survey of unmanned helicopter model-based navigation and control techniques, J. Intell. Robot. Syst. 80 (2015) 87–138.

CHAPTER 9

Path planning in autonomous ground vehicles

9.1 Introduction

One of the most important tasks desired of autonomous vehicles is navigation and/or surveillance. At any point in time, an autonomous vehicle will be required to navigate between two or more locations. To facilitate this navigation, the vehicle needs to be able to plan its movement between any two points in the space. Path planning is a subproblem in autonomous vehicle control which deals with finding optimal and feasible paths for an autonomous vehicle in a complex environment. Path planning problems appear in wide range of real-life applications which involve mostly, but are not limited to, autonomous vehicles. Path planning is considered the most studied problem in robotics. Deterministic planning techniques based on exact roadmap methods such as Voronoi diagrams, Delaunay triangulation, and adaptive roadmaps are primarily interested in capturing connectivity in the robots configuration search space. Grid-based algorithms such as Dikstra, A* and D* approach the problem by discretizing the search area into small cells using some cell decomposition methods. These classes of algorithms are plagued with the curse of dimensionality and are generally considered inefficient when dealing with robots with high degrees of freedom or large search spaces. Generally speaking, path planning problems require that some feasible path be established between a source and a target location subject to some environmental constraint.

Definition 9.1 (Feasible path planning). Consider a path planning problem defined by a triple $(\mathcal{C}, q_{\text{init}}, q_{\text{goal}})$, which means finding a path $\pi \in \Pi_{\text{free}}$ such that $\pi(0) = q_{\text{init}}$ and $\pi(1) = q_{\text{goal}}$, if one exists or reporting the solution as infeasible otherwise.

Definition 9.2 (Probabilistic completeness). An algorithm A is probabilistically complete if for any robustly feasible path planning problem $(\mathcal{C}, q_{\text{init}}, q_{\text{goal}})$, the probability that A fails to return a solution when one exists decays to zero as the running time of A approaches infinity.

Advanced Distributed Consensus for Multiagent Systems
https://doi.org/10.1016/B978-0-12-821186-1.00017-9

Definition 9.3 (Cost-space path planning). A cost-space path planning problem is defined by a quadruplet $(\mathcal{C}, q_{\text{init}}, q_{\text{goal}}, c)$, where $c : \mathcal{C} \to \mathbb{R}_+$ is a scalar differentiable cost function associating to each configuration of the $\mathcal{C}$-space a positive value. The solution to this problem involves finding a feasible path planning problem defined by the triple $(\mathcal{C}, q_{\text{init}}, q_{\text{goal}})$ while taking the cost function c into account in the exploration of the $\mathcal{C}$-space.

The approach to the path planning problem can be viewed in two ways. On the one hand, research studies are devoted to developing algorithms that are concerned with finding feasible quality paths. On the other hand, some studies are concerned with finding low-cost paths. Consider a positive scalar function $c_p : \Pi_{\text{free}} \to \mathbb{R}_+$ which associates to every feasible path a positive cost based on the configuration cost $c : \mathcal{C} \to \mathbb{R}_+$. This function measures the quality of a path and can be defined in several ways. Possible criteria to measure the quality of a path include the integral cost along the path and the mechanical work of a path. The integral cost of a path is defined by (9.1) which gives a discrete approximation of the integral cost with a constant step size of $\delta = \dfrac{1}{n}$:

$$c_p(\pi) = \frac{\text{length}(\pi)}{n} \sum_{k=1}^{n} c\left[\pi\left(\frac{k}{n}\right)\right]. \tag{9.1}$$

The mechanical cost measures the sum of positive cost variations in a bid to determine the effort exerted by a vehicle in navigating the path. A discrete approximation of the mechanical work along a path with a constant step size $\delta = \dfrac{1}{n}$ is given by

$$c_p(\pi) = \sum_{k=1}^{n} \max\left[0, c\left(\pi\left(\frac{k}{n}\right)\right) - c\left(\pi\left(\frac{k-1}{n}\right)\right)\right]. \tag{9.2}$$

Definition 9.4 (Optimal path planning). An optimal path planning problem is denoted as quintuplet $(\mathcal{C}, q_{\text{init}}, q_{\text{goal}}, c, c_p)$ which admits an optimal path π^*. That is, given a path planning problem $(\mathcal{C}, q_{\text{init}}, q_{\text{goal}})$, a configuration cost function $c : \mathcal{C} \to \mathbb{R}_+$, and a monotonic bounded path-quality criterion $c_p : \Pi_{\text{free}} \to \mathbb{R}_+$, it is required to find a path $\pi^* \in \Pi_{\text{feas}}$ such that $c_p(\pi^*) = \min\{c_p(\pi)|\pi \in \Pi_{\text{feas}}\}$.

Definition 9.5 (Asymptotic optimality). An algorithm $\mathcal{A}$ is asymptotically optimal if for any optimal path planning problem $(\mathcal{C}, q_{\text{init}}, q_{\text{goal}}, c, c_p)$ admitting a robustly optimal solution path with a finite cost $c^* \in \mathbb{R}_+$, the cost of

Figure 9.1 A car-like vehicle.

the current solution path that can be returned by $\mathcal{A}$ converges towards c^* as the running time of $\mathcal{A}$ approaches infinity.

Important criteria considered in developing path planning algorithms include the following:
- shortest length between source and destination,
- minimal energy consumption in navigation of the path,
- path smoothness,
- adaptability to dynamic environments, and
- minimal information requests about environments.

9.2 Vehicle model

Consider a car-like vehicle, popularly known as the *Ackermann* vehicle, or the bicycle model. As shown in Fig. 9.1, the vehicle has a rear-wheel fixed to its body while the front wheel of the vehicle rotates around the vertical axis for steering purposes. Let the pose of the vehicle be described in coordinate frame $\{\mathcal{V}\}$, where the vehicle forward movement is defined in the axis of $\{V\}$ and the origin of the vehicle is fixed at its rear axles. The vehicle configuration is defined in the generalized coordinates by $q = (x, y, \theta) \in \mathbb{C}$, $\mathbb{C} \in SE(2)$, where x, y, and θ define position and orientation of the body in a plane. The overall kinematic model of the vehicle is defined by

$$\dot{x} = v\cos(\theta), \tag{9.3}$$

$$\dot{y} = v\sin(\theta), \tag{9.4}$$

$$\dot{\theta} = \frac{v}{L}\tan(\phi), \tag{9.5}$$

where v is the velocity of the body, ϕ is the steering angle, and L is the length of the wheel base. The steering angle is limited mechanically and is proportional to the turning radius of the vehicle. The nonholonomic constraint of the vehicle is given by

$$\dot{y}\cos\theta - \dot{x}\sin\theta = 0. \tag{9.6}$$

It is worthy to take note of some cases. When $v = 0$, $\dot{\theta} = 0$ and the vehicle is unable to turn or change orientation. Further, if the steering angle $\phi = \frac{\pi}{2}$, the front wheel is orthogonal to the back wheel, and the vehicle is unable to move forward (see Figs. 9.2 and 9.3).

For reference purposes, we review some basic motion algorithms for the vehicle as presented in [1]. When it is desired to move to a particular point defined in a plane by (x_r, y_r), the following controller can be designed for the system (see Figs. 9.4, 9.5, and 9.6):

$$v^* = K_v\sqrt{(x_r - x)^2 + (y_r - y)^2}, \tag{9.7}$$

$$\theta^* = \tan^{-1}\left(\frac{y - y^*}{x^* - x}\right), \tag{9.8}$$

$$\phi = K_\theta(\theta^* - \theta). \tag{9.9}$$

The vehicle may also exhibit a line following behavior. Suppose it is desired that the vehicle follows a line defined by

$$ax + by + c = 0. \tag{9.10}$$

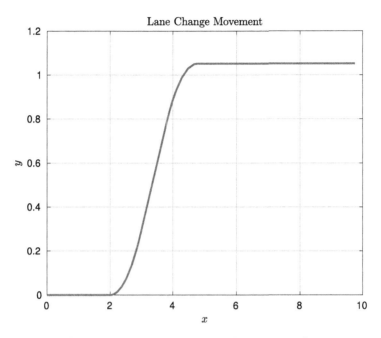

Figure 9.2 Lane change maneuver in x–y plane under constant velocity input.

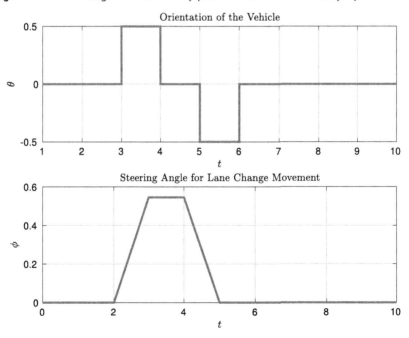

Figure 9.3 Steering angle and orientation of the vehicle during lane change maneuver in the x–y plane under constant velocity input.

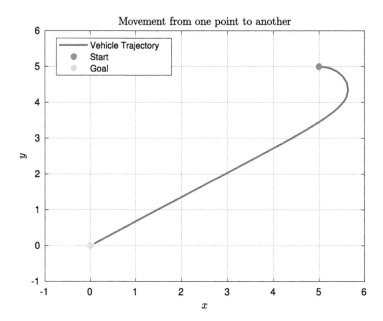

Figure 9.4 Lane change maneuver in x–y plane under constant velocity input.

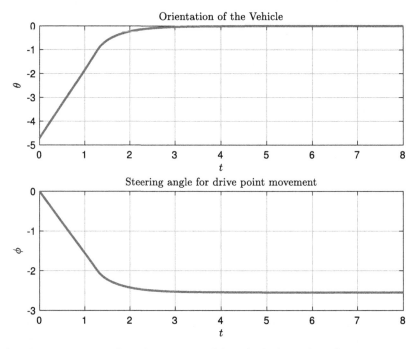

Figure 9.5 Steering angle and orientation of the vehicle during lane change maneuver in the x–y plane under constant velocity input.

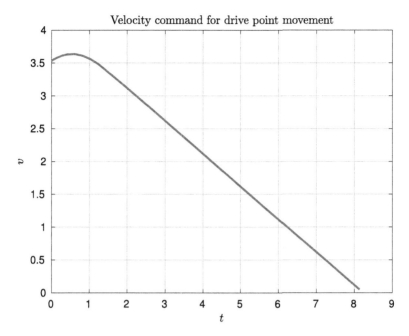

Figure 9.6 Steering angle and orientation of the vehicle during lane change maneuver in the *x–y* plane under constant velocity input.

A line controller can be designed to minimize the normal distance from a line given by

$$d = \frac{(a, b, c) \cdot (x, y, 1)}{\sqrt{a^2 + b^2}}. \tag{9.11}$$

Thus a simple line following controller can be designed as

$$\phi = -k_d d + k_h(\theta^* - \theta), \tag{9.12}$$

where $k_d > 0$ and $k_h > 0$ are proportional gains, and θ^* is defined as

$$\theta^* = \arctan \frac{-a}{b}. \tag{9.13}$$

The car-like vehicle can also may also exhibit a path following behavior. For this purpose, a simple path following PI controller may be designed as follows (see Figs. 9.7 and 9.8):

$$v^* = k_v e + k_i \int e \, dt, \tag{9.14}$$

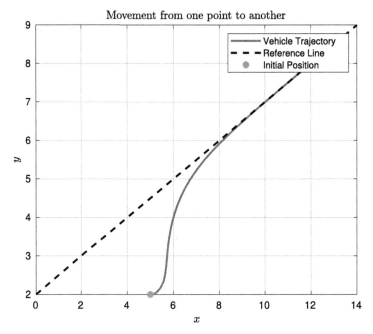

Figure 9.7 Lane change maneuver in the x–y plane under constant velocity input.

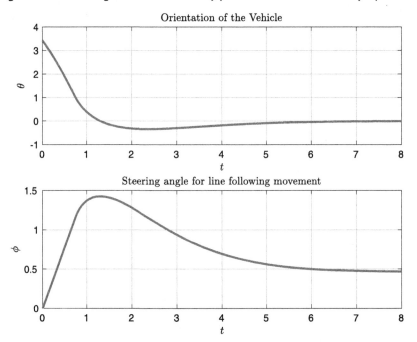

Figure 9.8 Steering angle and orientation of the vehicle during lane change maneuver in the x–y plane under constant velocity input.

$$\theta^* = \arctan \frac{y^* - y}{x^* - x}, \tag{9.15}$$

where e is the error in the 2D Cartesian plane defined as

$$e = \sqrt{(x^* - x)^2 + (y^* - y)^2} - d^*, \tag{9.16}$$

and the steering controller is defined as

$$\alpha = K_h(\theta^* - \theta). \tag{9.17}$$

Finally, a pose controller may be defined for the vehicle as follows. Suppose it is desired to maintain a geometric pose

$$\begin{bmatrix} \dot{x} \\ \dot{y} \\ \dot{\theta} \end{bmatrix} = \begin{bmatrix} \cos(\theta) & 0 \\ \sin(\theta) & 0 \\ 0 & 1 \end{bmatrix} \begin{bmatrix} \nu \\ \gamma \end{bmatrix}. \tag{9.18}$$

The system can now be transformed to polar coordinates using the transformation:

$$\rho = \sqrt{\Delta_x^2 + \Delta_y^2}, \tag{9.19}$$

$$\alpha = \arctan\left(\frac{\Delta_y}{\Delta_x}\right) - \theta, \tag{9.20}$$

$$\beta = -\theta - \alpha. \tag{9.21}$$

Thus the system dynamics in polar coordinates becomes

$$\begin{bmatrix} \dot{\rho} \\ \dot{\alpha} \\ \dot{\beta} \end{bmatrix} = \begin{bmatrix} -\cos(\alpha) & 0 \\ \dfrac{\sin(\alpha)}{\rho} & -1 \\ -\dfrac{\sin(\alpha)}{\rho} & 0 \end{bmatrix} \begin{bmatrix} \nu \\ \gamma \end{bmatrix}, \tag{9.22}$$

for $\alpha \in [\frac{\pi}{2}, \frac{\pi}{2}]$. Finally, a simple linear controller can be designed as:

$$\nu = k_\rho \rho, \tag{9.23}$$

$$\gamma = k_\alpha \alpha + k_\beta \beta. \tag{9.24}$$

9.3 Global path planning

Path planning problems can be formulated as global optimization problems where it is assumed that the vehicle has a global view of its environment. Common algorithms for solving a global path planning problem include the family of graph-based algorithms such as the Dijkstra, A* and D* algorithms, and sampling based algorithms like the RRT and PRM. Global algorithms attempt to find feasible or optimal paths while having a global view of the environment. The path planning problem requires developing algorithms that generate collision-free paths in a sample workspace. In the recent past, different researchers have studied this problem following diverse approaches. Path planning algorithms can be broadly classified based on the search approach into grid-based, sampling-based, bio-inspired, and interpolation-based techniques. This classification is based on how the problems are modeled and on the solution approaches employed in exploring the search space for feasible and/or optimal solutions.

9.3.1 Graph-based planners

Graph-based planners is a general terminology used to refer to path-planning algorithms that rely on the notion of graph theory and discretized grids to compute feasible paths between source and destination nodes

9.3.1.1 Dijkstra algorithm

The Dijkstra algorithm proposed by [37] is a graph-based path–planning algorithm that finds the shortest path between a source and destination using a configuration space approximated by discrete cell–grid spaces and lattices. The Dijkstra algorithm is a breadth-first-search algorithm for finding the shortest path from a source vertex to all other vertices. Consider an initial node v_0 and let d_y represent the distance from node v_0 to node v_y. In the Dijkstra's algorithm, some initial node values will be assigned and the algorithm improves the value step-by-step:

- Mark all the nodes unvisited. Create a set of all the unvisited nodes called the unvisited set.
- A distance value is assigned to every node and the distance value of the value of the initial node will be set to zero while every other node is assigned infinity.
- For the current node, we consider all its unvisited neighbors and evaluate their tentative distances through the current node.

- Upon considering all the unvisited neighbors of the present node, the node is moved into a set of visited nodes and it is removed from the visited set.
- If the destination node has been marked visited or if the smallest tentative distance among the node in the unvisited set is infinity, then stop.
- Otherwise, select the unvisited node that is marked with the smallest tentative distance, set it as the new current node, and repeat the third step.

A generalized formulation for the Dijkstra algorithm is given in [45] as follows. Consider a graph $\mathbb{G} = (\mathbb{V}(\mathbb{G}), \mathbb{E}(\mathbb{G}))$. For vertices $u, v \in \mathbb{V}(\mathbb{G})$, we let the distance metric $\text{dist}_{(\mathbb{G},c)}(u, v)$ denote the minimum total length of a path in $\mathcal{G}$ from u to v with respect to c, or ∞ is v is not reachable from u. For a given nonempty sources set $S \subseteq \mathbb{V}(\mathbb{G})$, we define a function $d : \mathbb{V}(\mathbb{G}) \to \mathbb{Z}_{\geq 0} \cup \{\infty\}$:

$$d(v) := \min\{\text{dist}_{(\mathbb{G},c)}(s, v) | s \in S\}, \tag{9.25}$$

for $v \in \mathbb{V}(\mathbb{G})$. If we are given a target set $T \subset \mathbb{V}(\mathbb{G})$, we want to compute the distance

$$d(T) := \min\{d(t) \mid t \in T\} \tag{9.26}$$

from S to T in $\mathbb{G}$ with respect to c, or ∞ if T is not reachable from S.

Rather than labeling individual vertices with distance related values, we label subgraphs of $\mathbb{G}$ induced by subsets of vertices with distance-related function. Therefore, we assume to be given disjoint subsets U of $\mathbb{V}(\mathbb{G})$ and subsets S and T:

$$\mathbb{V}(\mathbb{G}) = \bigcup_{U \subseteq V} U, \quad S \subseteq V, \quad T \subseteq V. \tag{9.27}$$

It is required that the graph $\mathbb{G}$ with $\mathbb{V}(\mathbb{G}) := V$ and edge set defined by

$$\mathbb{E}(\mathbb{G}) = \{(u, u') | \exists u \in U, u' \in U', (u, u') \in \mathbb{E}(\mathbb{G}) \text{ with } c((u, u')) = 0\} \tag{9.28}$$

is acyclic. Therefore, there exists a topological sequence $U_1, U_2, \ldots, U_{|V|}$ of V with $i < j$ is $(U_i, U_j) \in \mathbb{E}(\mathbb{G})$. For $U \subseteq V$, the index of U is defined as $I(U) = i$ iff $U = U_i$. Throughout the execution of the algorithm and for every $U \subseteq V$, we maintain a function $d_u : U \to \mathbb{Z}_{\geq 0} \cup \{\infty\}$ which is an upper bound on d, that is,

$$d_u(v) \geq d(v) \quad \forall v \in U, \tag{9.29}$$

and a feasible potential on $\mathbb{G}[U]$,

$$d_u(v) \geq d_u(u) + c((u,v)) \quad \forall\ (u,v) \in \mathbb{E}(\mathbb{G}[U]) \tag{9.30}$$

where $\mathbb{G}[U]$ denotes the subgraph of $\mathbb{G}$ induced by U. Initially, we set

$$d_u(v) = \begin{cases} 0, & v \in U \subseteq S, \\ \infty, & v \in U \subseteq \mathbb{V}\setminus S. \end{cases} \tag{9.31}$$

Because we are interested in a specific structure of the graph, we distinguish between two labeling operations as follows. We require a partition of $\mathbb{V}$ into $N \geq 1$ sets, $V_1, \ldots, V_N$, called blocks and a function $B : U \to \{V_1, \ldots V_N\}$ such that

$$\forall\ 1 \leq i \leq N, \quad \emptyset \neq V_i \subseteq \mathbb{V}, \tag{9.32}$$

$$\forall\ U \subseteq \mathbb{V}, \quad U \subseteq B(U), \tag{9.33}$$

$$\forall\ 1 \leq i \leq j \leq N,\ V_i \cap V_j = \emptyset. \tag{9.34}$$

9.3.1.2 A*-based algorithms

The A* algorithm [46] is known as an informed search algorithm or a best-first search algorithm that is formulated as solution to a weighted graph problem. The aim of the A* algorithm is to find a minimum-cost path from a start node to a goal node, and this is done by maintaining a tree of paths originating at the start node and extending those paths one edge at a time until a termination criterion is fulfilled. At each iteration, the algorithm determines the path to extend based on the associated cost of the path, specifically, the A* algorithm selects the path that minimizes

$$f(n) = g(n) + h(n), \tag{9.35}$$

where n is the next node on the path, $g(n)$ is the cost of the path from start node to n, and $h(n)$ is a heuristic function which minimizes the cost of the cheapest path from n to the next goal.

The D* algorithm is an extension of the A* algorithm originally proposed by [47] which considers a dynamic variation in cost parameters during the search process. Other variations of the A* algorithm include the Field D* [48], Theta* [49], Anytime repairing A* (ARA*), and Anytime D* (AD*) [50], and so on.

In [56], a multiobjective D*-lite algorithm was proposed based on the D*-lite for path planning applications where objectives other than the path length objective were considered. In [32], three A*-based algorithms were proposed for path planning, namely the scaled A*, A*++, and A*++ with reconstruction. In particular, an iterative segment planner, which generates a set of spatially diverse paths by iteratively planning segments of all paths in a sequential manner was proposed. To prevent the generation of spatially close paths, the iterative segment planner utilizes a penalty function to increase the cost of paths that lie near the route(s) assigned to other vehicles. By varying the penalty gains, different degrees of path diversity can be achieved. A simultaneous localization and mapping (SLAM) algorithm based on D* was proposed for path planning in [25] considering negative edge weights in unknown environments with moving obstacles.

9.3.2 Sampling-based methods

Sampling-based methods for path planning explore the configuration spaces using sampling methods. That is, connectivity of the sampling space is captured by sampling the entire configuration space. Common sampling-based path-planning algorithm include Randomized potential planner (RPP), Rapidly exploring random trees (RRT), Probabilistic roadmap (PRM) to mention a few. Given any configuration space or C-space which contains disjoint sets representing the obstacle-free space, C-free, and the obstacle-space, C-obst, we are required to obtain a path from some initial configuration q_0 to a final configuration q_f. The sampling-based path-planning algorithm solves the path-planning problem using some common algorithmic steps outlined below:

1. (Sampling) First a configuration is selected randomly or quasirandomly in either the free or obstacle configuration space.
2. (Metric evaluation) Given any two configurations q_a and q_b, a metric is used to evaluate the effort required to reach q_b from q_a.
3. (Nearest neighbor) Next, a search algorithm is used to determine the nearest neighbor to the newly sampled configuration based on the predefined metric function.
4. (Select parent) In this phase, an existing node is selected to be the parent of the newly sampled node. This process is executed a bit differently in various sampling-based path-planning algorithms.
5. (Local planning) Given two configurations q_a and q_b, a connection will be established between them in this phase of the sampling-based path planning algorithm.
6. (Collision checking) Finally, the connection path will be examined for possible intersections with the obstacle space. This step is implemented

as a Boolean operation that returns true or false depending on whether the object is in the obstacle space or not.

Sampling-based path-planning algorithms are said to satisfy the weaker notion of the completeness property known as the probabilistic completeness property. This means that sampling-based methods are not guaranteed to generate feasible paths deterministically but are able to do so in a probabilistic sense. Next, we present a brief overview of the common sampling-based algorithms.

9.3.2.1 Rapidly exploring random trees (RRT)

The RRT algorithm was originally introduced by [38] for path planning in vehicular systems to handle nonholonomic constraints and high degrees of freedom. Contrary to the probabilistic roadmap approach, the RRT algorithm does not require point-to-point convergence, rather, it applies control inputs that drive the agents slightly towards randomly selected points. The general steps for the RRT algorithm are given as follows [40]:

- The search is initialized from a q_0.
- A node q_r is selected at random from the configuration space using the sampling procedure.
- If q_r is contained in the obstacle space C_{obst}, it is discarded.
- q_n is obtained based on a nearest neighbor search as the closest node to q_r based on the predefined metric.
- The local planner is employed to connect q_r to q_n. However, the planner may return q_{new} if the node q_n is not reachable.
- Next, the path between q_n and q_{new} is checked for possible collisions. If a collision-free path exists, q_{new} is added to the search tree.
- The algorithm terminates when $q_{new} = q_f$ or the number of iterations or specified time elapses.

An improved informed sampling based RRT-based path-planning algorithm was proposed in [9] to reduce inefficiency of the well-known RRT algorithm. The RRT algorithm finds paths to every state in the search domain in an incremental fashion. This approach is plagued by the curse of dimensionality as the exact knowledge of states requires solution of the problem which increases with state dimensions. The presented approach has less theoretical dependence on state dimension and allows for linear convergence. In [36], a systematic online path-planning algorithm is presented based on the RRT algorithm to solve cooperative path replanning for multiple UAVs. A time-cooperative reconstruction strategy is proposed to handle the time cooperativity resulting from path replanning. In [29], a cost-aware fast sampling based RRT algorithm was proposed for path planning in a complex and realistic environment using cross-entropy which combines RRT* and a stochastic optimization method. The proposed algorithm constructs two RRT trees, the first is a standard RRT tree used to determine the nearest node in the tress to be extended to a randomly chosen point, while the second contains the first tree with additional long extensions.

9.3.2.2 Probabilistic roadmap (PRM)

The PRM methods generate a probabilistic roadmap following two main procedures. First, through learning phase, the configuration space is sampled for a certain amount of time. Samples or configurations in the free space are maintained while those in the obstacle space are discarded. In the second phase, known as the query phase, the start and target configurations are established and are connected via roadmaps defined by forest graphs. Because the start and target configurations are defined at a later stage of the problem, PRM is able to solve different instances of the path-planning problem in the same environment. The general steps for the PRM algorithm are given as follows [40]:

- A node q_{rand} is selected randomly from the configuration space using the sampling procedure.
- If the randomly selected point q_{rand} is in the obstacle space, it is discarded.
- Else, q_{rand} is added to the roadmap.
- Next, we find all nodes within a specified range to q_{rand}.
- A local planner is employed to connect all neighboring nodes to q_{rand}.
- Then, all collision path are excluded from the roadmap.
- The process will be repeated iteratively until a certain amount of nodes are sampled.

9.3.3 Interpolating curve planners

Interpolation refers to the process of constructing and inserting new sets of data within the range of a previously known points. Mathematical algorithms take as input some set of points, generate new set of data which are smooth in the interest of trajectory continuity while considering vehicle and environmental constraints. Different mathematical models can be used to define interpolation between any two points in the environment. Paths between two points in an environment can be established by straight lines and circles. Polynomial curves are commonly implemented to meet constraints needed in the points they interpolate. Path planning routines can be generated from Bezier curves. Bezier curves are parametric curves defined using Bernstein polynomials. A Bezier curve is defined as

$$P_{[t_0,t_1]}(t) = \sum_{i=0}^{n} B_i^n(t_i)P_i,$$
(9.36)

where P_i are control points such that $P(t_0) = P_0$ and $P(t_1) = P_n$, and $B_i^n(t)$ is a Bernstein polynomial given by

$$B_i^n(t) = \binom{n}{i}\left(\frac{t_1 - t}{t_1 - t_0}\right)^{n-i}\left(\frac{t - t_0}{t_1 - t_0}\right)^{i}, \quad i \in \{0, 1, \ldots, n\}.$$
(9.37)

Some useful properties of Bezier curves include:

- They pass through P_0 and P_n.
- They are always tangent to the lines connecting $P_0 \rightarrow P_1$ and $P_n \rightarrow P_{n-1}$ at P_0 and P_n, respectively.
- They always lie within the convex hull consisting of their control points.

Two Bezier curve-based path-planning algorithms are proposed in [63] for autonomous vehicles with way-points and corridor constraints. Spline curves are piecewise polynomial parametric curves divided into subintervals. The junction between each subsegment is called a knot, and they commonly possess high degree of smoothness constraints at the joints between the pieces of the spline.

9.3.4 Bio-inspired methods

The path-planning problem can be posed as an optimization problem, therefore solutions can be obtained through bio–inspired optimization algorithms such as genetic algorithm (GA), particle-swarm optimization (PSO), support vector machines, grey-wolf optimization to mention a few. The path-planning problem can be posed as a multiobjective problem modeling objectives like path length, path smoothness, obstacle avoidance, and so on. Some of the objectives as described by earlier researchers are highlighted subsequently.

The path length objective is used to get the shortest possible path in the configuration space. Given two points $P_1 = (x_1, y_1)$ and $P_2 = (x_2, y_2)$, the

Euclidean distance between the two points is calculated as

$$d(P_1, P_2) = \sqrt{(x_1 - x_2)^2 + (y_1 - y_2)^2}. \tag{9.38}$$

Therefore, the length of the path is

$$p_l = \sum_{i=1}^{n-1} d(P_i, P_{i+1}), \tag{9.39}$$

where n is the size of the list representing the path and P_i and P_{i+1} are the ith and the $(i+1)$th points in the list of inputs. An adaptive genetic algorithm is proposed based on the distance metric (9.38) to solve the path-planning problem. The issue of a trap in a local minima and premature convergence is handled using a specialized selection operator. The fitness function employed in that study is defined as:

$$f = \begin{cases} \sum_{i=1}^{n-1} d(p_i, p_{i+1}), & \text{for feasible paths,} \\ \sum_{i=1}^{n-1} d(p_i, p_{i+1}) + w(t) \times \text{penalty}, & \text{otherwise,} \end{cases} \tag{9.40}$$

where $w(t)$ is an adaptive penalty weight defined as

$$w(t) = C \times k^t \tag{9.41}$$

and t denotes the number of generations.

The distance metric is given in [52] based on PSO as:

$$\text{dist} = \sum_{i=1}^{n} \left(\sqrt{(x_i^{\text{curr}} - x_i^{\text{goal}})^2 + (y_i^{\text{curr}} - y_i^{\text{goal}})^2} \right) \\ + \left(\sqrt{(x_i^{\text{curr}} - x_i^{\text{goal}})^2 + (y_i^{\text{curr}} - y_i^{\text{goal}})^2} \right),$$

where x_i^{next}, y_i^{next} are given by

$$x_i^{\text{next}} = x_i^{\text{curr}} + v_i^{\text{curr}} \cos \theta_i \Delta t, \tag{9.42}$$

$$y_i^{\text{next}} = y_i^{\text{curr}} + v_i^{\text{curr}} \cos \theta_i \Delta t. \tag{9.43}$$

Path safety is another important objective in modeling path-planning problems for bio-inspired algorithms. In [51], the path safety objective was defined based on the occupancy of the visited cells of the grid as the robot navigates its environment. Under the assumption that the robot occupies the same size as the partition of the map, the path safety objective is defined as [51]

$$p_s = \sum_{i=1}^{n-1} \Pi_{i,i+1},$$

(9.44)

where n is the size of the sorted list representing the navigation path and $\Pi_{i,i+1}$ defines the sum of the occupancy of the visited cells by the robot in the segment formed by the path between the ith and the $(i+1)$th point in the sorted list. The obstacle avoidance objective in [52] was defined as follows:

$$F_{\text{obst}}(x_p) = \begin{cases} \dfrac{k}{\gamma}\left(\dfrac{1}{d_{\min}(x_p)} - \dfrac{1}{n_0}\right), & d_{\min}(x_p) \le n_0, \\ 0, & \text{otherwise}, \end{cases}$$

(9.45)

where n_0 is the influence range of the obstacle, k is a positive constant, and $\gamma \ge 2$ shapes the radial profile of potential; $d_{\min}(x_p)$ is the minimum distance of x_p from the obstacles.

Another important objective in describing the path-planning problem is the path smoothness, which is directly proportional to the amount of energy consumed in traversing the path. The path smoothness is defined as [51]

$$p_s = \sum_{i=1}^{n_a-1} (180° - \alpha_i),$$

(9.46)

where α_i is the ith angle of the path; n_a is the number of angles of the path. The path smoothness objective (9.47) is modeled differently in [52] where PSO was used to solve the path-planning problem:

$$p_s = \arccos \frac{\alpha_1}{\alpha_2},$$

(9.47)

where α_1 and α_2 are defined as:

$$\alpha_1 = (x_i^{\text{curr}} - x_i^{\text{goal}})(x_i^{\text{gbest}} - x_i^{\text{goal}}) + (y_i^{\text{curr}} - y_i^{\text{goal}})(y_i^{\text{gbest}} - y_i^{\text{goal}}),$$

$$\alpha_2 = \sqrt{(x_i^{\text{curr}} - x_i^{\text{goal}})^2 + (y_i^{\text{curr}} - y_i^{\text{goal}})^2}\sqrt{(x_i^{\text{curr}} - x_i^{\text{goal}})^2 + (y_i^{\text{curr}} - y_i^{\text{goal}})^2},$$

where x_i^{curr}, x_i^{gbest}, y_i^{curr}, and y_i^{gbest} represent the current and global best computed by the particle swarm algorithm.

Furthermore, in [52] a metric for predicting dynamic obstacle position is proposed as

$$p_r = \sum_{i=1}^{n} \sqrt{(x_p - x_i^{\text{goal}})^2 + (y_p - y_i^{\text{goal}})^2}. \tag{9.48}$$

In [54], a bacterial foraging based optimization algorithm was proposed to find the shortest feasible path in an unknown environment. The optimization cost function is based on the estimated distance to target and a Gaussian cost function of the particle. In this approach, the robot explores its environment using a simple robot sensor. The study considers the following optimization objective:

$$J = J_{\text{goal}} + J_{\text{obst}}, \tag{9.49}$$

where J_{goal} and J_{obst} are defined as

$$J_{\text{goal}} = -H_{\text{goal}} \exp(-W_{\text{goal}}(\theta_i(t) - P_g(t))^2), \tag{9.50}$$

with H_{goal} and W_{goal} being the height and width of the attractant. Also

$$J_{\text{obst}} = \begin{cases} H_{\text{obst}} \exp(-W_{\text{obst}}(\theta_i(t) - P_0(t))^2), & |P_0 - C(t)| \leq \beta, \\ 0, & \text{otherwise,} \end{cases} \tag{9.51}$$

where H_{obst} and W_{obst} are constant values defining height and width of the repellant; $P_0(t)$ denotes the obstacle position detected by the sensor and β is the sensor range.

A multiobjective shuffled frog-leaping algorithm was employed for solving the path-planning problem in [51] considering objectives such as path length, path smoothness, and obstacle avoidance. An improved particle swarm optimization and gravitational search algorithm was hybridized in [52] to solve the path-planning problem under the following assumptions:

- Current position/initial position and goal position/target position of all the robot are known in a prior coordinate system.
- At any instant of time, the robot can decide any action from a set of predefined actions for its motion.
- Each robot is performing its action until reaching its respective target position in several steps.

The authors used several strategies to handle the listed assumptions:
- For determining the next position from its current position, the robot tries to align its heading direction towards the goal position.
- The alignment may cause a collision with the robots/obstacles (which are static in nature) in the environment. Hence, the robot turns its heading direction with a certain angle either to the left or to the right for determining its next position from its current position.
- If a robot can align itself with a goal without collision, then it will move to that determined position.
- If the heading direction is rotated to the left or to the right, then it is required for the robot to rotate by the same angle about its z-axis; if it is same for more than one angle, and then it can decide randomly.

The authors of [53] proposed a modified genetic algorithm for path-planning problems based on the Bezier curve interpolation method. In the proposed method, the robot path is dynamically generated based on the obstacle location.

Specifically, the optimization problem is posed as

$$\min \tilde{D} \tag{9.52}$$

subject to:

$$D \geq r, \tag{9.53}$$

$$\forall \theta \in \delta \implies \theta \notin \tilde{P}, \tag{9.54}$$

$$\forall p \in \beta \implies p \notin \delta, \tag{9.55}$$

$$s, t \notin \delta, \tag{9.56}$$

$$\tilde{p} = \phi, \tag{9.57}$$

where $\tilde{D}$ is the sum of distances between d and the adjacent points $(p_1, p_2, \ldots, p_i)$ on the Bezier curve; D is the distance between two control points; and δ is a set of obstacle θ. Also $\tilde{D}$ is defined as

$$\tilde{D} = \sum_{i=1}^{\eta} d(p_i, p_{i+1}), \tag{9.58}$$

where η represents the count of the Bezier curve points, and $\tilde{D}$ can be calculated by the integral of Bezier curve as follows:

$$\tilde{D} = \oint_l P(t) = \oint \sum_{i=1}^{n} p_i B_{i,n}(t) = \int_0^1 \sqrt{x_t'^2 + y_t'^2}\, dt. \tag{9.59}$$

The fitness function for the modified genetic algorithm is defined as

$$f = \frac{1}{\overline{D}}.$$ (9.60)

In [57], a path-planning algorithm was proposed based on Q-learning and neural network algorithm to develop collision-free trajectories for robots in an environment with static and dynamic obstacles. In the Cartesian system representing the environment considered, an angle θ is constructed between the line formed by the robot and the target and the line formed by the robot and the closest obstacle, where θ is defined as

$$\theta = \arcsin\left(\frac{D}{d_0}\right),$$ (9.61)

where

$$D = \frac{|(y_{\text{rob}} - y_t)x_0 + (x_t - y_{\text{rob}})y_0 + x_{\text{rob}}y_t -_t y_{rob}|}{\sqrt{(y_{\text{rob}} - y_t)^2 + (x_t - y_{\text{rob}})^2}}$$ (9.62)

and

$$d_0 = \sqrt{(x_{\text{rob}} - x_0)^2 + (y_{\text{rob}} - y_0)^2}.$$ (9.63)

The following assumptions were considered in [57]:
- All dynamic parameters are known at each time instant, including the Cartesian position and velocity of the robot, obstacles, and target.
- The robot has a velocity higher than the velocity of the target, a common sense condition that enforces convergence.
- The robot is activating in an unknown working space. The collision detection is achieved in a simulated environment based on the Euclidean distance from the margin of a virtual sphere that includes the robot structure, to the margin of a virtual sphere that includes the obstacle. In real environments, collision is detected either by laser or sonar sensors, or by bumpers, once the robot physically touches the obstacle.

A coevolutionary improved genetic algorithm was proposed in [58] for global path planning of multiple robots to determine near optimal or optimal collision-free paths. The following assumptions were considered in the development of the algorithm:
- The environment of a robot's motion is a 2D workspace, where obstacles are represented by their absolute coordinates, and the height of obstacles is not considered.

- The information of the static obstacles is completely known.
- There are no dynamic obstacles in the environment.
- Each robot can be regarded as point-sized and occupies only one location at a time, and the size of obstacles can be scaled according to the size of the robot.
- Robots moves with a fixed speed.

The evaluation function for feasible paths is computed as

$$f_f(p) = \frac{1}{w_d d(p) + w_s s(p)} + w_k h(p), \tag{9.64}$$

where $d(p)$ is defined as the path length, $s(p)$ is the safety of the path, and $h(p)$ is the smoothness of the path; w_d, $w(s)$, and w_k are appropriate positive weight constants for length, safety, and smoothness, respectively. The path length $d(p)$ of path p is defined as

$$d(p) = \sum_{i=1}^{N-1} \sqrt{(x_{i+1} - x_i)^2 + (y_{i+1} - y_i)^2}, \tag{9.65}$$

where x_i, y_i are the coordinates of the ith node. Path safety is defined as

$$s(p) = \sum_{i=1}^{N-1} \alpha(l_i). \tag{9.66}$$

The cost function for evaluating infeasible paths is given as

$$f_i(p) = \frac{1}{w_d d(p) + w_s s(p) + w_c c(p)}, \tag{9.67}$$

where c_p represents the proportion of path length crossing the obstacles. The proportion of infeasible lines is described as

$$l(p) = \frac{L_{\text{infeasible}}}{L_{\text{total}}} \tag{9.68}$$

where $L_{\text{infeasible}}$ is the number of infeasible lines and L_{total} is the number of total lines in path p. The proportion of path length crossing the obstacles is computed similarly.

9.4 Multirobot path planning

Multirobot path-planning problems are primarily concerned with finding feasible paths for teams of robots in a configuration space such that some team objectives are fulfilled.

9.4.1 Independent objectives

Consider a team of N robots in a workspace bounded by the subset of $\mathbb{R}^2$ or $\mathbb{R}^3$. The position and orientation of each robot A_i in the workspace are specified by a point in an n-dimensional configuration space C^i. Some obstacles exist in the workspace and forbid certain configurations of the robot. However, in the workspace, the robot has some valid configurations defined by C_v^i.

Let the state space representing the configuration of all robots be represented by X as:

$$X = C_v^1 \times C_v^2 \times \cdots \times C_v^N, \tag{9.69}$$

where $\times$ denotes Cartesian product. If A_i^0 represents the interior of A_i, then the set of states, X_{ij}^{col}, where any two sets of robots collide is

$$X_{ij}^{\text{col}} = \{x \in X | A_i^0(x_i) \cap A_i^0(x_i) \neq \emptyset\}. \tag{9.70}$$

Therefore, the collision subset $X^{\text{col}} \subset X$ is defined by the open set:

$$X^{\text{col}} = \bigcup_{i \neq j} X_{ij}^{\text{col}}. \tag{9.71}$$

A state is therefore in the collision subset if the interiors of two or more robots intersect. Thus, the valid subset for the robots in the workspace is $X^v = X \backslash X^{\text{col}}$. The objective is to bring each robot from an initial configuration $x_i^0 \in X_i$ to a final configuration $x_i^f \in X_i$. For each robot, a performance function is defined as [38]

$$L^i(x_i^0, x_i^f, u^1, u^N) = \int_0^T g_i(t, x_i(t), u_i(t))dt + \sum_{j=1} c_{ij}(x(\cdot)) + q_i(x_i(T)), \tag{9.72}$$

where c_{ij} is a function that penalizes collision given by

$$c_{ij}(x(\cdot)) = \begin{cases} 0, & \text{if } x(t) \in X_v, \\ \infty, & \text{otherwise,} \end{cases} \tag{9.73}$$

and q_i represents the goal in terms of performance

$$q_i(x_i(T)) = \begin{cases} 0, & \text{if } x_i(T) = x_i^f, \\ \infty, & \text{otherwise.} \end{cases} \quad (9.74)$$

The function g_i is a standard continuous optimal control cost function which models energy consumption. At the final (or goal) state, it is expected that

$$g_i(t, x_i(t), u_i(t)) = 0. \quad (9.75)$$

The solution concept proposed in [38] presents an approach to find a state feedback control law based on some strategies available to each robot A_i denoted by $\gamma_i \in \Gamma$, where Γ is the set of possible (or allowable) strategies.

It is assumed that the robots can chose from some strategy space Γ. Therefore, given an initial state x_0 and strategy $\gamma = \{\gamma_1, \gamma_2, \dots, \gamma_N\}$, it is possible to determine the sequence of configurations $x(t)$ which drives the robot to the final trajectory x^f.

9.4.2 Mixed path planning

Mixed path-planning algorithms combine different path-planning approaches to induces different behaviors for robots, especially in dynamic environments. Mixed path planning was discussed by [39] in an approach that combines graph search and potential field algorithms to solve a path-planning problem for robots operating within a hospital environment. Their approach combines both corridor and room path planning. In the corridor path planning algorithm, each robot plans its optimal path and send a request to a central module, which eliminates possible conflicts in the path of the robots. According to the authors, the path planning algorithm is capable of real-time planning and guaranteeing the safety of pedestrians to a large extent. For the potential field path planning module, a potential field function was defined as:

$$U_{\text{att}}(q) = \frac{1}{2} k_{\text{att}} (x_g - x)^2 + (y_g - y)^2 \quad (9.76)$$

where k_{att} is a positive control parameter which maximizes the attraction and U_{att} defines the attraction value at position q. The repulsion potential field is defined by

$$
U_{\text{rep}}(q) = \begin{cases} \dfrac{1}{2} k_{\text{rep}} \left(\dfrac{1}{\delta(q)} - \dfrac{1}{\rho} \right)^{0.5}, & \delta(q) \le \rho, \\ 0, & \delta(q) > \rho, \end{cases} \tag{9.77}
$$

where $\delta(q) = ((x - x_0)^2 + (y - y_0)^2)^{0.5}$, therefore the resultants of the potential fields is

$$
U(q) = U_{\text{att}}(q) + U_{\text{rep}}(q). \tag{9.78}
$$

In the mixed path planning of [39], there are two main modules, namely the robot module and the central module. The robot module directs the activity of each robot while the central module coordinates the activity of the entire robot team. At initialization, the path point set path_i, task set ϕ_i, static potential field $U_{\text{sta}}(q)$.

9.4.3 Distributed task assignment

In some path-planning problems, the objective could be more than moving from one point to another in the configuration space. Agents could be given some tasks to perform at some specified locations in the configuration space. Tasks may include moving objects from a source to a destination in a cyclic or event-triggered manner. Thus in each cycle, the robot may be required to plan its path repeatedly. In such path-planning problems, it is required that the agent demonstrate some form of learning to improve its efficiency in terms of resource allocated and possibly fuel or energy consumed.

Game-theory based distributed path-planning via distributed learning was investigated in [2] for agents working cooperatively to execute some tasks optimally within some specified time window. In the described configuration, each task is assigned a value, and the task is completed if sufficiently many robots are present at the required location within a pre-specified time window. Tasks are issued over cyclic time periods; however, the specification of the tasks, including number of robots, location, time window, and value, is considered unknown a priori. Each agent maximizes the value obtained from completed tasks by planning its trajectories for upcoming cycles based on past experiences. Further, each agent has recharging and maintenance needs, and as such each robot is required to start and end each cycle at the assigned stations. A formal presentation of the distributed task assignment [2] for a team of mobile robots goes as follows.

Consider a team of n homogeneous robots $R = \{r_1, r_2, \ldots, r_n\}$ that are required to optimize their trajectories in a given environment while performing some cyclic tasks that are assigned within specified periods. The environment representing the workspace of the robots is described by a discretized 2D grid, denoted by $P = \{1, 2, \ldots, \bar{x}\} \times \{1, 2, \ldots, \bar{y}\}$, where $\bar{x}, \bar{y} \in \mathcal{N}$ denote the number of cells along corresponding directions. In the discretized environment, some of the cells may be occupied by some obstacles $P_0 \subset P$ and the robots are free to navigate over the feasible cell space $P_f = P \backslash P_0$. In the feasible cell space, there are some m stations $S = \{s_1, \ldots, s_m\}$ where the robots can dock and recharge to prepare for the next operating cycle. Each robot is assumed to be assigned to a specific docking station and must return to its station every cycle.

The trajectory of each robot r_i at time is denoted by $p_i = \{p_i^0, p_i^1, \ldots, p_i^T\}$, where T denotes the time step in each operating cycle. The robots can either maintain their current position or move to any of the feasible neighboring cells within a time step. Specifically, if a robot is at some cell $p = (x, y) \in P_f$, at the next time step it has to be within p's neighborhood on the grid $N(p) \subseteq P_f$, that is,

$$N(p) = \{(x', y') \in P_f \mid |x' - x| \leq 1, |y' - y| \leq 1\}. \tag{9.79}$$

For any robot $r_i \in R$, the set of feasible trajectories P_i is defined as

$$P_i = \{p_i \mid p_i^0 = p_i^T = \sigma_i, \ p_i^{t+1} \in N(p_i^t), \ t = 0, \ldots, T - 1\}, \tag{9.80}$$

where $\sigma_i \in S$ denotes the station of robot r_i. Accordingly, the combined set of feasible trajectories is denoted by $P = P_1 \times \cdots \times P_n$. Given such an environment and a set of mobile robots, we consider a set of k tasks, $\tau = \{\tau_1, \tau_2, \ldots, \tau_k\}$, each of which is defined as a tuple $\tau_i = \{c_i^*, l_i, t_i^a, t_i^d, v_i\}$, where $c_i^* \in N$ is the required number of robots, $l_i \in P_f$ is the location, $t_i^a < t_i^d \in \{0, \ldots, T\}$ are the arrival and departure times, and $v_i \in \mathbb{R}_+$ is the value.

Accordingly, the task is completed if at least c_i^* robots simultaneously spend one time step at l_i within the time window $[t_i^a, t_i^d]$. More specifically, given the trajectories of all robots, P, the set of completed tasks $\tau^*(p) \in \tau$ is defined as

$$\tau^*(p) = \{\tau_i \in \tau \mid \in t \in [t_i^a, t_i^d - 1], c_i(p, t) \geq c_i^*\}, \tag{9.81}$$

where $c_i(p, t) \in \{0, \ldots, n\}$ is the counter denoting the number of robots that stayed at $l_i \in P_f$ from time t to $t + 1$ in that cycle, that is,

$$c_i(p, t) = |\{r_i \in R | p_i^t = p_i^{t+1} = l_i\}|. \tag{9.82}$$

Each task can be completed within one time step if sufficiently many robots simultaneously stay at the corresponding location within the specified time window. Some typical examples of tasks may be placing a box on a shelf where the weight of the box determines the number of robots needed, or aerial monitoring where the required ground resolution determines the number of drones needed. The robots are assumed to know how to achieve the lower level coordination. The core of the problem lies in having sufficient number of robots in the right place and time window. Thus, the function can be formulated to quantify the performance resulting from the trajectories of all robots p, as the total value of completed tasks:

$$f(p) = \sum_{\tau_i \in \tau^*} v_i. \tag{9.83}$$

Consider a scenario where the tasks are unknown a priori. Robots are expected to improve their overall performance by updating their trajectories over the cycles based on their observations. In such a setting, $p(t)$ denotes the trajectories at the tth cycle for $t \in \{0, 1\}$ and it is required to optimize the long-run performance defined by the following performance index:

$$\lim_{t^* \to \infty} \frac{1}{t^* + 1} \sum_{t=0}^{t^*} f(p(t)). \tag{9.84}$$

9.5 Some typical applications

Path planning problems may also appear in complex 3D environments involving manipulation of sophisticated objects. The authors of [3] considered automatic path planning for a dual-crane lifting problem in a complicated environment. Based on their claims, the novel path planner can produce optimized paths in complex 3D environments. Their approach utilizes a comprehensive and computationally efficient mathematical model of the dual-crane system. The path-planning algorithm utilizes a novel multiobjective parallel genetic algorithm to generate optimized paths for lifting the objects while relying on an efficient algorithm for continuous collision detection.

Another important application of path-planning algorithms is in disassembly problems. Both assembly and disassembly planning are important problems that appear in manufacturing engineering. In the disassembly problem, an assembled object is provided and it is required to compute an optimal disassembly path by carrying out some assembly maintainability study. Existing computational approaches to solving the problem rely on relation graph models of the assembly or precedence graphs and utilize graph theory and AI for obtaining disassembly routines. Sampling-based path-planning algorithms are considered very efficient tools for computing optimal disassembly paths due to their efficiency and ease of implementation. A disassembly path-planning algorithm based on a modified RRT algorithm was proposed for complex articulated objects in [5]. An important feature of the proposed method is the ability to handle objects with a high number of mobile parts and automatically identify DOFs for the assembly tasks.

Path-planning problems usually consider a configuration space which may feature some complexity in terms of the obstacles present in the environment. In [7], automatic path planning was discussed for a mobile robot considering an environment featuring obstacles of arbitrary shape. The first phase of the proposed algorithm involves obtaining a graph which defines all collision-free paths in the environment. The next phase of the algorithm determines an optimal path for the algorithm using a fast distance transformation (FDT) method.

In [8], the notion of cooperative route planning is discussed within the framework of Internet-of-Vehicles (IoV). Cooperative route planning is defined by [8], as a concept for optimizing global vehicular routing based on data about planned route from interconnected vehicles present in the network. Cooperative route planning is beneficial in the sense that the user benefits from minimizing traffic; however, this induces some security risks. Privacy constraints are introduced into cooperative route planning without significant sacrifices in cost whilst providing anonymity for users of the network.

Local path-planning algorithms consider the problem of finding optimal paths using local information and ensuring that the robot is not lost. Key challenges for local path-planning algorithms are evaluating localizability of a path and resulting impact on the path planning process. Secondly, it is important to balance localizability of the path and path-planning criteria such as obstacle avoidance, short path length, and exploration time. To overcome this problem, a novel path evaluation method was proposed in [10] to deal with uncertainty resulting from dead-reckoning and map matching. A function is proposed to evaluate the impact of localizability of path planning with consideration for traditional path-planning criteria.

A path-planning problem was investigated for a network of distributed robots deployed for surveillance from a remote station to detect some unknown static targets. The approach presented demonstrated cooptimization of sensing and communication during the motion planning process. A communication-constrained motion-planning algorithm was proposed while considering path loss, shadowing, and multipath fading problems.

A 3D volume based coverage path-planning (VCPP) algorithm was developed for robotic evacuation of intracerebral hemorrhage in [13]. The proposed algorithm differs from existing algorithms in that it removes the need to decompose the volume area into a series of 2D planning problems. The algorithm minimizes the configuration space distance traveled.

Path planning in three dimensional spaces for nonholonomic parallel orienting robots employs algorithms that generate maneuvers comprising a sequence of moves interlinked by points of zero velocity. The existing methods limit the uses of the robots employing these techniques to just a few applications. A novel geometric path-planning algorithm without maneuvers was developed in [14] for nonholonomic parallel robotic systems. The research shows that there are infinitely differentiable paths connecting two points in 3D special orthogonal planes which can be used to develop a practical path planner for nonholonomic parallel orienting robots that generate single-move maneuvers.

A centralized and decoupled algorithm was proposed in [15] for solving multirobot path-planning problems defined by grid graphs considering applications in on-demand and automated warehousing. The study investigates both the traditional problem of moving some set of robots from an initial location to a predefined goal location and a more complicated problem which models frequent replanning to accommodate some adjustments in goal configurations. A heterogeneous ant colony optimization algorithm was proposed in [17] for solving a global path-planning problem which addresses the problem of accumulated pheromone and intensity of heuristic value as the ants approach the goal point by introducing a bilateral cooperative exploration (BCE) method. The BCE method performs the search task by switching the goal point into the starting point, and vice versa.

A path-planning algorithm was proposed for UAVs based on genetic algorithms in [18]. The fitness function for the path planning algorithm was formulated considering the fitness function defined using the total distance traveled by the UAVs, clearance distance, turning angles, areas covered by multiple UAVs, and the number of repetitive routes of multiple UAVs. The planning algorithm was designed following the Bezier curve interpolation method. Cooperative path-planning problem was studied for multiple underactuated autonomous surface vehicles in [19] moving along a parameterized path. The modular cooperative path-planning algorithm was developed combining line-of-sight guidance scheme, tracking differentiators, and path variable containment scheme. The dynamics of the vehicle was subject to uncertain kinematics and unknown kinetics induced by model uncertainties and ocean disturbances.

A novel strategy for online planning of optimal motions paths was presented in [22] for wilderness search and rescue applications. The proposed strategy considered planning time-optimal and piecewise polynomial paths for all robots, implementation and regular evaluation of paths through some series of checks that gauge the feasibility of path completion within the available time. Also, that work discussed online path replanning wherever it was deemed necessary. In [24], path planning was discussed for a team of cooperating vehicles for package delivery applications. In that work, the cooperating team comprised two vehicle types, a truck to navigate the street networks and a microaerial vehicle to perform deliveries. The problem was formulated on a graph with the objective of finding shortest cooperative route enabling the quadrotor to deliver items at requested locations.

Data gathering in large-scale sensor networks is another typical application area for unmanned vehicles. An efficient strategy for data collection in autonomous vehicles should consider cooperation amongst sensors within communication range, advanced coding, and data storage to ease cooperation, while route planning should be content and cooperation aware. In [30], the efficient data gathering problem is formulated as a cooperative route optimization problem with communication constraints. Given the complexity of the problem, the authors of [30] use heuristic optimization techniques such as particle swarm optimization to calculate the AV's route and the times for communication with each sensor and/or cluster of sensors.

9.6 Notes

In this chapter, path-planning problems and algorithms were presented for multivehicle systems. Some typical path planning problems in two-dimensional space were discussed and existing solution approaches were reviewed.

References

[1] P. Corke, Robotics, Vision and Control: Fundamental Algorithms in MATLAB, arXiv, Springer Tracts in Advanced Robotics, vol. 73, Springer, 2011.

[2] R. Bhat, Y. Yazıcıoğlu, D. Aksaray, Distributed path planning for executing cooperative tasks with time windows, arXiv:1908.05630v1, 2019.

[3] P. Cai, I. Chandrasekaran, J. Zheng, Y. Cai, Automatic path planning for dual-crane lifting in complex environments using a prioritized multi-objective PGA, IEEE Trans. Ind. Inform. 14 (3) (2018) 829–845.

[4] R.A. Conn, M. Kam, On the moving-obstacle path-planning algorithm of Shih, Lee and Gruver, IEEE Trans. Syst. Man Cybern. 27 (1) (1997) 136–138.

[5] J. Cortes, L. Jaillet, T. Simeon, Disassembly path planning for complex articulated objects, IEEE Trans. Robot. 24 (2) (2008) 475–481.

[6] D. Devaurs, T. Simeon, J. Cortex, Optimal path planning in complex cost spaces with sampling-based algorithms, IEEE Trans. Autom. Sci. Eng. 13 (2) (2016) 415–423.

[7] J.L. Diaz, J.H. Sossa, Automatic path planning for a mobile robot among obstacles of arbitrary shape, IEEE Trans. Syst. Man Cybern. 28 (2) (1998) 467–472.

[8] M. Florian, S. Finster, I. Baumgart, Privacy-preserving cooperative route planning, IEEE Int. Things J. 1 (6) (2014) 590–599.

[9] J.D. Gammell, T.D. Barfoot, S.S. Srinivasa, Informed sampling for asymptotically optimal path planning, IEEE Trans. Robot. 34 (4) (2018) 966–984.

[10] Y. Gao, J. Liu, M.Q. Hu, H. Xu, K.P. Li, H. Hu, A new path evaluation method for path planning with localizability, IEEE Access 7 (2019) 162583–162597.

[11] A. Ghaffarkhah, Y. Mostofi, Path planning for networked robotic surveillance, IEEE Trans. Signal Process. 60 (7) (2012) 3560–3575.

[12] D. Gonzalez, J. Perez, V. Milanes, F. Nashashibi, A review of motion planning techniques for automated vehicles, IEEE Trans. Intell. Transp. Syst. 17 (4) (2016) 1135–1145.

[13] J. Granna, I.S. Godage, R. Wirz, K.D. Weaver, R.J. Webster, J. Burgner-Kahrs, A 3-D volume coverage path planning algorithm with application to intracerebral hemorrhage evacuation, IEEE Robot. Autom. Lett. 1 (2) (2016) 876–883.

[14] P. Grosch, F. Thomas, Geometric path planning without maneuvers for nonholonomic parallel orienting robots, IEEE Robot. Autom. Lett. 1 (2) (2016) 1066–1072.

[15] S.D. Han, J. Yu, DDM: fast near-optimal multi-robot path planning using diversified-path and optimal sub-problem solution database heuristics, IEEE Robot. Autom. Lett. 5 (2) (2020) 1350–1357.

[16] Y. Jin, Y. Liao, A.A. Minai, M.M. Polycarpou, Balancing search and target response in cooperative unmanned aerial vehicle teams, IEEE Trans. Syst. Man Cybern. 36 (3) (2006) 571–587.

[17] J. Lee, D. Lee, J. Lee, Global path planning using improved ant colony optimization algorithm through bilateral cooperative exploration, in: Proc. IEEE International Conference on Digital Ecosystems and Technologies, 2011, pp. 109–113.

[18] H. Li, Y. Fu, K. Elgazzar, L. Paull, Real-time trajectory generation for the cooperative path planning of multi-vehicle systems, in: Proc. the 41st IEEE Conference on Decision and Control, 2002, pp. 3766–3769.

[19] L. Liu, D. Wang, Z. Peng, T. Li, Modular adaptive control for LOS-based cooperative path maneuvring of multiple underactuated autonomous surface vehicles, IEEE Trans. Syst. Man Cybern. Syst. 47 (7) (2017) 1613–1624.

[20] C. Liu, S. Zhang, A. Akbar, Ground feature oriented path planning for unmanned aerial vehicle mapping, IEEE J. Sel. Top. Appl. Earth Obs. Remote Sens. 12 (4) (2019) 1175–1187.

[21] X. Yu, L. Liu, G. Feng, Trajectory tracking for nonholonomic vehicles with velocity constraints, IFAC-PapersOnLine 48 (11) (2015) 918–923.

[22] A. Macwan, J. Vilela, G. Nejat, B. Benhabib, A multirobot path-planning strategy for autonomous wilderness search and rescue, IEEE Trans. Cybern. 45 (9) (2015) 1784–1797.

[23] P. Maini, K. Sundar, M. Singh, S. Rathinam, P.B. Sujit, Cooperative aerial-ground route planning with fuel constraints for coverage applications, IEEE Trans. Aerosp. Electron. Syst. 55 (6) (2019) 3016–3028.

[24] N. Mathew, S.L. Smith, S.L. Waslander, Planning paths for package delivery in heterogeneous multirobot teams, IEEE Trans. Autom. Sci. Eng. 12 (4) (2015) 1298–1308.

[25] I. Maurovic, M. Seder, K. Lenac, I. Petrovic, Path planning for active SLAM based on the D* algorithm with negative edge weights, IEEE Trans. Syst. Man Cybern. 48 (4) (2018) 1321–1331.

[26] D. Drake, S. Koziol, E. Chabot, Mobile robot planning with a moving goal, IEEE Access 6 (2018) 12800–12814.

[27] H. Nam, Data-gathering protocol-based AUV path-planning for long-duration cooperation in underwater, IEEE Sens. J. 18 (21) (2018) 8902–8912.

[28] T. Oral, F. Polat, MOD Lite: an incremental path planning algorithm taking care of multiple objectives, IEEE Trans. Cybern. 46 (1) (2016) 245–257.

[29] J. Suh, J. Gong, S. Oh, Fast sampling-based cost-aware path planning with nonmyopic extensions using cross entropy, IEEE Trans. Robot. 33 (6) (2017) 1313–1326.

[30] P.B. Sujit, D.E. Lucani, J.B. Sousa, Bridging cooperative sensing and route planning of autonomous vehicles, IEEE J. Sel. Areas Commun. 30 (5) (2012) 912–922.

[31] S. Upadhyay, A. Ratnoo, Continuous-curvature path planning with obstacle avoidance using four parameter logistic curves, IEEE Robot. Autom. Lett. 1 (2) (2016) 609–616.

[32] J. Votion, Y. Cao, Diversity-based cooperative multivehicle path planning for risk management in cost-map environment, IEEE Trans. Ind. Electron. 66 (8) (2019) 6117–6127.

[33] W. Yao, N. Qi, C. Yue, N. Wan, Curvature-bounded lengthening and shortening for restricted vehicle path planning, IEEE Trans. Autom. Sci. Eng. 17 (1) (2020) 15–28.

[34] Y. Gao-Yang, Z. Shao-Lei, Y. Shi, Cooperative path planning of multi-missiles, in: Proc. IEEE Chinese Guidance, Navigation and Control Conference, 2014, pp. 767–772.

[35] Y. Zhing, Y. Hsueh, W. Lee, Y. Jhang, Efficient cache-supported path planning on roads, IEEE Trans. Knowl. Data Eng. 28 (4) (2016) 951–964.

[36] Z. Zhen, C. Gao, Q. Zhao, R. Ding, Cooperative path planning for multiple UAVs formation, in: IEEE Int. Conference on Automation, Control and Intelligent Systems, 2014, pp. 469–473.

[37] E.W. Dijkstra, A note on two problems in connection with graphs, Numer. Math. 1 (1959) 269–271.

[38] S.M. Lavalle, S.A. Hutchinson, Optimal motion planning for multiple robots having independent goals, IEEE Trans. Robot. Autom. 14 (6) (1998) 912–925.

[39] X. Huang, Q. Cao, X. Zhu, Mixed path planning for multi-robots in structured hospital environment, J. Eng. 14 (6) (2019) 512–516.

[40] M. Elbanawi, M. Simic, Sampling-based robot motion planning, IEEE Access 2 (2014) 56–77.

[41] L. Kavraki, J.C. Latombe, Randomized preprocessing of configuration space for path planning: articulated robots, in: Proc. IEEE/RSJ/GI Int. Conf. IROS, vol. 3, 1994, pp. 1764–1771.

[42] N.M. Amato, Y. Wu, A randomized roadmap method for path and manipulation planning, in: Proc. IEEE Int. Conf. Robot. Autom., vol. 1, 1996, pp. 113–120.

[43] L.E. Kavraki, P. Svestka, J.C. Latombe, M.H. Overmars, Probabilistic roadmaps for path planning in high-dimensional configuration spaces, IEEE Trans. Robot. Autom. 12 (4) (1996) 566–580.

[44] P. Svestka, M.H. Overmars, Motion planning for car-like robots using a probabilistic learning approach, Int. J. Robot. Res. 16 (2) (1997).

[45] S. Peyer, D. Rautenbach, J. Vygen, A generalization of Dijkstra's shortest path algorithm with applications to VLSI routing, J. Discret. Algorithms 7 (2009) 377–390.

[46] P.E. Hart, N.J. Nilsson, B. Raphael, A formal basis for the heuristic determination of minimum cost paths, in: IEEE Trans. System Science and Cybernetics, vol. ssc4, no. 2, 1968, pp. 100–107.

[47] A. Stentz, Optimal and efficient path planning for partially-known environments, in: Proc. IEEE Int. Conf. Robot. Autom., 1994, pp. 3310–3317.

[48] D. Ferguson, A. Stentz, Using interpolation to improve path planning: the field D* algorithm, J. Field Robot. 23 (2) (2006) 79–101.

[49] A. Nash, K. Daniel, S. Koenig, A. Felner, Theta*: any-angle path planning on grids, in: Proc. Nat. Conf. Artif. Intell., 2007, pp. 1177–1183.

[50] M. Likhachev, D. Ferguson, G. Gordon, A. Stentz, S. Thrun, Theta*: anytime search in dynamic graphs, Artif. Intell. 172 (14) (2008) 1613–1643.

[51] A. Hidalgo-Paniagua, M. Vega-Rodriguez, J. Ferruz, N. Pavon, MOSFLA-MRPP: multi-objective shuffled frog-leaping algorithm applied to mobile robot path planning, Eng. Appl. Artif. Intell. 44 (2015) 123–136.

[52] P.K. Das, H.S. Behara, B.K. Panigrahi, A hybridization of an improved particle swarm optimization and gravitational search algorithm for multi-robot path planning, Swarm Evol. Comput. 28 (2016) 14–28.

[53] M. Elhoseny, A. Tharwaat, A.E. Hassanien, Bezier curve based path planning in dynamic field using modified genetic algorithm, J. Comput. Sci. 25 (2018) 339–350.

[54] M. Hossain, I. Ferdous, Autonomous robot path planning in dynamic environment using a new optimization technique inspired by bacterial foraging technique, Robot. Auton. Syst. 64 (2015) 137–141.

[55] A.H. Karami, M. Hasanzadeh, An adaptive genetic algorithm for robot motion planning in 2D complex environments, Comput. Electr. Eng. 43 (2015) 317–329.

[56] T. Oral, F. Polat, MOD* Lite: an incremental path planning algorithm taking care of multiple objectives, IEEE Trans. Cybern. 46 (1) (2016) 245–257.

[57] M. Duguleana, G. Mogan, Neural networks based reinforcement learning for mobile robots obstacle avoidance, Expert Syst. Appl. 62 (2016) 104–115.

[58] H. Qu, K. Xing, T. Alexander, An improved genetic algorithm with co-evolutionary strategy for global path planning of multiple mobile robots, Neurocomputing 120 (2013) 509–517.

[59] R. Al-Jarrah, A. Shahzad, H. Roth, Path planning and motion coordination for multi-robots system using probabilistic neuro-fuzzy, IFAC-PapersOnLine 48 (10) (2015) 46–51.

[60] B. Tang, Z. Zhanxia, J. Luo, A convergence-guaranteed particle swarm optimization for mobile robot global path planning, Assem. Autom. (2016) 114–129.

[61] Q. Yang, S. Yoo, Optimal UAV path planning: sensing data acquisition over IOT sensor networks using multi-objective bio-inspired algorithms, Assem. Autom. 6 (2018) 13671–13684.

[62] Y. Zhang, D. Gong, J. Zhang, Robot path planning in uncertain environment using multi-objective particle swarm optimization, Neurocomputing 103 (2018) 13671–13684.

[63] J. Choi, R. Curry, G. Elkaim, Path planning based on Bézier curve for autonomous ground vehicles, in: Advances in Electrical and Electronics Engineering – IAENG Special Edition of the World Congress on Engineering and Computer Science, 2008, pp. 1–9.

CHAPTER 10

Path planning in autonomous aerial vehicles

10.1 Introduction

Unmanned aerial vehicles are employed in numerous real life applications such as payload delivery, traffic monitoring, moving objects in seemingly dangerous environment, and surveillance. The use of UAVs in any of these applications necessitates the planning of feasible and optimal trajectories for the motion of the vehicles. Path-planning algorithms for UAV flights differ from ground vehicles in that the planning problems need to be solved in a three-dimensional configuration space. Compared to two-dimensional spaces, these environments are subject to higher degrees of uncertainty and moving obstacles. UAVs will be required to interact dynamically with other flying or static objects which may appear in their flight paths, therefore making global path planning almost impossible, as it is almost impossible to fully map out the configuration space. In some applications, UAV and UGV path-planning problems are combined for joint task execution in complex environments. UAV path planning involves designing a flight path directed towards a target with minimal comprehensive costs, i.e., minimal probability of being destroyed while meeting the UAV performance constraints. In general, path planning for UAVs has the following attributes:

- (Stealth) This aspect concerns the safety of UAVs. UAVs are usually required to carry out missions in threatening environments. Thus, it is very important to minimize the probability of detection by a hostile radar and other UAVs.
- (Physical feasibility) This refers to the physical limitations in the use of UAVs, which include the maximum path distance and the minimum path leg length.
- (Performance of the mission) This refers to whether a path can satisfy the requirements of a specified mission. To complete the mission, various requirements must be met when we design a path. These requirements usually include the maximal turning angle, the maximum climbing/diving angle, and the minimal flying height.
- (Real-time implementation) This refers to the efficiency of path planning. The flight environments of UAVs are usually constantly changing.

Advanced Distributed Consensus for Multiagent Systems
https://doi.org/10.1016/B978-0-12-821186-1.00018-0

Therefore, our path-planning algorithm must be computationally efficient. Replanning ability is critical for adapting to unforeseen threats.

Algorithms for path planning in a two-dimensional configuration space have been modified to accommodate and solve path-planning problems as they appear in three-dimensional spaces for UAVs. Popular search grid-based path-planning algorithms include the A*, D*, RRT algorithms and have been proposed to solve the path-planning problem. More recently, researchers [1,2,6–9] have employed computational intelligence (CI) based techniques to solve more complex path-planning problems.

10.2 UAV and its environment

Path-planning algorithms are designed with special consideration for the models of the UAV, the environment where the UAV will be deployed and operate, and possible objects in the environment that pose some threats to the operation of the vehicle.

10.2.1 UAV models

Different kinematic models have been studied in the UAV literature for path planning in a three-dimensional space. In [14], the following kinematic model was studied:

$$\dot{x} = V \cos\gamma \cos\phi, \tag{10.1}$$

$$\dot{y} = V \cos\gamma \sin\phi, \tag{10.2}$$

$$\dot{z} = V \sin\gamma, \tag{10.3}$$

$$\dot{\phi} = \omega, \tag{10.4}$$

where $P = (x, y, z)$ is the position of the UAV in the three-dimensional space, $(u_x, u_y, u_z) = (\dot{x}, \dot{y}, \dot{z})$ defines the velocity of the UAV, V is the airspeed, γ is the flight path angle, ϕ is the heading angle, and ω is the turn rate. The motion of the vehicle is subject to some constraints in the workspace:

$$\|\phi_{k+1} - \phi_k\| \leq \omega_{\max} \Delta T, \tag{10.5}$$

$$\gamma_{\min} \leq \gamma_k \leq \gamma_{\max}, \tag{10.6}$$

$$v_{\min} \leq v_k \leq v_{\max}, \tag{10.7}$$

$$h_{\min} \leq z_k \leq h_{\max}, \tag{10.8}$$

where ω_{max} is the maximum UAV turn rate, ΔT is the sampling time, γ_{min} and γ are the maximum and minimum flight angles, v_{min} and v_{max} are the minimum and maximum air speed, h_{min} and h_{max} are the maximum and minimum altitude.

The following kinematic model was considered in the multi-UAV path-planning problem for precision agriculture in [2]:

$$\dot{p}_i = v_i, \tag{10.9}$$

$$\dot{v}_i = -ge_3 + \frac{1}{m_i} T_i R_i e_3, \tag{10.10}$$

$$\dot{R}_i = R_i S(\Omega_i), \tag{10.11}$$

$$I_i \dot{\Omega}_i = -\Omega_i \times I_i \Omega_i + \tau_i, \tag{10.12}$$

where $e_3 = [0, 0, 1]^T$, $T_i \in \mathbb{R}$ is the collective mass normalized thrust input, $S(\Omega_i)$ is the skew-symmetric matrix formed by Ω_i and $\tau_i \in \mathbb{R}^3$ is the vector of aerodynamic torque input applied to the ith UAV. Also R_i is defined as

$$R_i = \begin{bmatrix} c_\theta c_\psi & s_\phi s_\theta c_\phi - c_\psi s_\phi & c_\phi s_\theta c_\phi + s_\phi s_\psi \\ c_\theta s_\pi & s_\phi s_\theta c_\phi + c_\psi c_\phi & c_\phi s_\theta s_\phi - s_\phi c_\psi \\ -s_\theta & s_\phi c_\theta & c_\phi c_\theta \end{bmatrix} \tag{10.13}$$

where ψ, θ, ϕ are roll, pitch, and yaw angles; c_x and s_x are $\cos x$ and $\sin x$, respectively.

10.2.2 Environment modeling

The workspace or environment of the UAV is an important consideration in solving path-planning problems. Obstacles in the workspace are approximated as convex polyhedron obstacles such as spheres, ellipsoids, and cuboids. An expression defining an obstacle in the workspace is given as [13]

$$\Omega(\rho_i) = \left(\frac{x_i - x_0}{a}\right)^2 \left(\frac{y_i - y_0}{b}\right)^2 \left(\frac{y_i - y_0}{c}\right)^2, \tag{10.14}$$

where (x_0, y_0, z_0) represents the three-dimensional position of the obstacle center, $\rho_i = (x_i, y_i, z_i)$ represents the position of the ith UAV in the space. The coefficient a, b, c model the shape and size of the obstacle. The no-fly zone, or the danger zone, is represented by the inside and the surface of the convex polyhedron

$$R^\omega(\rho_i) = \{\rho_i \mid \Omega(\rho_i) \leq 1\}. \tag{10.15}$$

10.2.3 Threat modeling

In [9], the environment is described by Voronoi diagrams, the threat along any given edge (i, j) is given as

$$J_{ij} = a\sum_{n=1}^{N_r} J_{ij,r}(n) + b\sum_{n=1}^{N_t} J_{ij,t}(n) + c\sum_{n=1}^{N_m} J_{ij,m}(n) + dJ_{ij,l}, \tag{10.16}$$

where a, b, c, d are chosen such that $a + b + c + d = 1$; N_r, N_t, and N_m are number of radars, terrain obstacles, and guided missiles, respectively; $J_{ij,r}(n), J_{ij,t}(n)$, and $J_{ij,m}$ denote the nth threat to edge (i, j) from radars, terrain obstacles, and guided missiles, respectively; $J_{ij,l}$ denotes the fuel cost of edge (i, j).

The radar threat is due to the fact that UAV been detected by the enemy radar. The radar transmitted power is considered a threat level, and the radar threat to UAV is defined by

$$J_{ij,r} = \int_0^{\frac{I_{ij}}{V}} \frac{Q_n}{R_n^4(t)} dt, \tag{10.17}$$

where V is the velocity of the UAV, I_{ij} is the length of the edge (i, j), R_n is the distance between UAV and the nth radar at time t, and Q_n us the detecting power of the nth radar expressed as

$$Q = \frac{P_t G A_e \sigma}{(4\pi)^2 R^4}, \tag{10.18}$$

where P_t is the power of the transmitter, G is the gain of the antenna, A_e is the effective acreage of the antenna, σ is the section acreage of radar and R is the distance from radar to object. It is also quite possible to compute the threat cost at different locations along an edge via a computationally efficient cost function defined by

$$J_{ij,r}(n) = l_{ij}\left(\frac{1}{d_{1/8,ij,n}^4} + \frac{1}{d_{3/8,ij,n}^4} + \frac{1}{d_{5/8,ij,n}^4} + \frac{1}{d_{7/8,ij,n}^4}\right). \tag{10.19}$$

The terrain threat expresses the threats posed to the UAV when penetrating low altitude, mountains, and other complex terrain factors. The influence of terrain threats is defined as

$$J_{ij,t}(n) = L_n e^{-k d_{ij}(n)},$$ (10.20)

where L_n is the threat level of the nth terrain factor, k is a coefficient and $d_{ij}(n)$ is the Euclidean distance between the midpoint of edge (i, j) and the nth terrain threat.

The threat cost of a guided missile is defined by

$$J_{ij,m}(n) = B(1 - A)p_{ij}(n),$$ (10.21)

where B is the attack ability of the guided missile, $p_{ij}(n)$ is the detection probability of the nth guided missile. Finally, the fuel cost is expressed as

$$J_{ij,l}(n) = \lambda l_{ij},$$ (10.22)

where λ is the proportionality coefficient and l_{ij} is the length of the edge (i, j). In summary, the overall threat cost can be defined as

$$J_{ij} = a \sum_{n=1}^{N_T} \left[l_{ij} \left(\frac{1}{d_{1/8,ij,n}^4} + \frac{1}{d_{3/8,ij,n}^4} + \frac{1}{d_{5/8,ij,n}^4} + \frac{1}{d_{7/8,ij,n}^4} \right) \right]$$
$$+ b \sum_{n=1}^{N_t} L_n e^{-k d_{ij}(n)} + c \sum_{n=1}^{N} B(1 - \alpha)p_{ij}(n) + d\lambda l_{ij}.$$ (10.23)

10.3 UAV path planning

This section highlights some path planning algorithms and techniques for UAVs.

10.3.1 Potential fields

Artificial potential fields are based on the defining functions that generate some attraction or repulsion to the mobile robots. These functions are formulated such that each UAV considers other UAVs as obstacles, and as such a repulsive field is generated between them. Consider a multi-UAV path-planning problem involving n UAVs, the potential field between the *i*th UAV and other UAVs may be formulated as [13]

$$
U_{ij}^{\rho} = \begin{cases} 0, & \text{if } \rho_{ij} \notin S, \\ \displaystyle\sum_{j=1}^{n} \left\| \frac{1}{\rho_{ij}} - \frac{1}{\rho_0} \right\| (\rho_{ig})^{0.5}, & \text{if } \rho_{ij} \in S, \end{cases} \tag{10.24}
$$

where ρ_{ij} is the relative position between any two robots i and j; η_i is a positive repulsion factor; $S \in [\rho_{ij}^{\min}, \rho_{ij}^{\max}]$ is the distance-based scope, $\rho_{ij}^{\min}$ and $\rho_{ij}^{\max}$ are the minimum and maximum obstacle avoidance area, respectively. A modified potential field force function is proposed in [13] as

$$
\begin{aligned}
F_{ij}^{\text{rep}} = &\sum_{j=1, i \neq j}^{n} \frac{\eta_{ij}}{e^{\| \rho_{ij} \|}} \left(\left\| \frac{1}{\rho_{ij}} - \frac{1}{\rho_0} \right\| \rho^{0.5} \left(\frac{1}{\rho_{ij}} \right)^2 \nabla \rho_{ij} \right. \\
&+ \frac{1}{2} \sum_{j=1, j \neq i}^{n} \left(\left\| \frac{1}{\rho_{ij}} - \frac{1}{\rho_0} \right\| \rho^{-0.5} \left(\frac{1}{\rho_{ij}} \right)^2 \nabla \rho_{ij} \right),
\end{aligned} \tag{10.25}
$$

where η_{ij} is the weight factor between UAVs i and j. A distance factor $e^{\| \rho_{ij} \|}$ is introduced such that when ρ_{ij} is far, the influence of the ith UAV on the jth UAV is small, and vice versa. The target tracking strategy considers the following function:

$$
\begin{aligned}
U_i(\rho_i) &= U_{ig}^{\text{att}}(\rho_i) + \sum_{j=1, j \neq i}^{n} U_{ij}^{\text{rep}}(\rho_i) \\
&= (\eta_i \| \rho_{ig} \| + \kappa_i \| v_{ig} \|) + \sum_{j=1, j \neq i}^{n} U_{ij}^{\text{rep}}(\rho_i).
\end{aligned} \tag{10.26}
$$

An hybrid potential field approach was proposed in [14] combining Lyapunov guidance vector field and improved interfered fluid dynamical system (IIFDS) to solve the problems of target tracking and obstacle avoidance in three-dimensional cooperative path planning for multiple unmanned aerial vehicles. First, LGVF method is improved for UAV cooperative target tracking in 3D environment by introducing vertical component, with two guidance layers containing steering control and speed control. Secondly, IIFDS method is presented for UAVs to avoid obstacles or threats in a complicated environment, where the local minimum problem is well resolved. Moreover, some cooperative strategies are added into the IIFDS framework to satisfy the constraints of obstacle avoidance and cluster maintenance. Finally, the missions of tracking a target and avoiding obstacles can be performed simultaneously, by replacing the original sink fluid of IIFDS with the vector field of LGVF. Besides, the reactive parameters of IIFDS can be adjusted by the rolling optimization strategy to enhance the path quality.

10.3.2 Grid-based approaches

A multistep A*-based search algorithm is proposed in [17] for planning multiobjective 4D motion of vehicles in uncertain environments. For UAVs operating in large dynamic uncertain 4D environments, the motion plan consists of a sequence of connected linear tracks (or trajectory segments). The track angle and velocity are important parameters that are often restricted by assumptions and a grid geometry in conventional motion planners.

A decentralized motion planning method was proposed in [18] for multiple vehicles moving through 3D polygonal obstacles in an urban-like environment. The proposed algorithm combines a prioritized A* algorithm for high-level planning, along with a coordination method based on barrier functions for low-level trajectory generation and vehicle control. Furthermore, the authors augment the low-level trajectory generation and control with a prioritized A* path planning algorithm, in order to compute way points and paths that force the agents of lower priority to avoid the paths of the agents of higher priority, thus reducing congestion.

10.3.3 Intelligent approaches

As with any computational intelligence based solution approach, CI methods for path planning begin by defining an appropriate cost function that describes the problem. In [2], an improved fruit-fly optimization based algorithm is proposed to solve the UAV path-planning problem. The proposed algorithm combines the phase angle-encoded strategy and mutation adaptation mechanism into the basic FOA and is referred to as θ-MAFOA. Mutation adaptation mechanism is adopted to enhance the balance of FOA in terms of the exploitation and exploration ability, while the phase angle-based encoded strategy for fruit-fly locations helps achieve the high performance in the convergence process.

The approach in [2] represents the path as a B-spline curve. Given $n+2$ control points with coordinates defined by

$$(x_c(0), y_c(0), z_c(0)), \dots, (x_c(k), y_c(k), z_c(k)), \dots,$$
$$(x_c(n+1), y_c(n+1), z_c(n+1)),$$

the B-spline curve with coordinates $(x_p(t), y_p(t), z_p(t))$ can be constructed as:

$$x_p(t) = \sum_{i=0}^{n+1} x_{c,i} B_{i,K}(t), \tag{10.27}$$

$$y_p(t) = \sum_{i=0}^{n+1} y_{c,i} B_{i,K}(t), \tag{10.28}$$

$$z_p(t) = \sum_{i=0}^{n+1} z_{c,i} B_{i,K}(t), \tag{10.29}$$

where $B_{i,K}(t)$ is the blending function defined recursively as

$$B_{i,1}(t) = \begin{cases} 1, & \text{if } \mathrm{knot}(i) \le t \le \mathrm{knot}(i+1), \\ 1, & \text{if } \mathrm{knot}(i) \le t \le \mathrm{knot}(i+1) \text{ and } t = n - K + 3, \\ 0, & \text{otherwise}, \end{cases} \tag{10.30}$$

$$B_{i,k}(t) = \frac{(t - \mathrm{knot}(t)) B_{i,K}(t)}{\mathrm{knot}(i+k-1) - \mathrm{knot}(i)} + \frac{(\mathrm{knot}(i+k) - t) B_{i+1,k-1}(t)}{\mathrm{knot}(i+k) - \mathrm{knot}(i+1)} \tag{10.31}$$

and

$$\mathrm{knot}(i) = \begin{cases} 0, & \text{if } i < K, \\ i - K + 1, & \text{if } K \le i \le n + 1, \\ n - K + 3, & \text{if } n < i. \end{cases} \tag{10.32}$$

Typical objective functions for the UAV path-planning optimization problems based on the computational intelligence techniques include path length, collision avoidance, minimum threat, flight safety altitude, and so on. The path length objective can be computed as

$$f_l = \sum_{k=0}^{N} \sqrt{(\tilde{x})^2 + (\tilde{y})^2) + (\tilde{z})^2)}, \tag{10.33}$$

where $\tilde{x}(k)$, $\tilde{y}(k)$, and $\tilde{z}(k)$ are defined as:

$$\tilde{x} = x_p(k+1) - x_p(k), \tag{10.34}$$

$$\tilde{y} = y_p(k+1) - y_p(k), \tag{10.35}$$

$$\tilde{x} = z_p(k+1) - z_p(k). \tag{10.36}$$

The probability of radar detection, P_r, is a function of the radar cross-section of the UAV and the distance between radar and UAV expressed by

$$P_r = \begin{cases} 0, & \text{if } d > R_{r,max}, \\ \dfrac{1}{1 + \eta_2 \left(\dfrac{d^4}{RCS}\right)^{\eta_1}}, & \text{otherwise}, \end{cases} \qquad (10.37)$$

where η_1 and η_2 are inherent parameters of the radar; $R_{r,max}$ denotes the maximum detection distance and RCS depends on the orientation of the UAV with respect to the radar and can be calculated as

$$RCS = \frac{\pi a^2 b^2 c^2}{\sqrt{(a\alpha_z\beta_\phi)^2 + (b\alpha_z\alpha_\phi)^2 + (c\beta_z)^2}}, \qquad (10.38)$$

with $\alpha_z = \sin\psi^e$, $\beta_z = \cos\psi^e$, $\alpha_\phi = \sin\phi^e$, and $\beta_\phi = \cos\phi^e$, where ψ^e is the angle between the velocity vector of the UAV and the segment joining the UAV and radar positions, and $\phi^e = \phi - \arctan(\tan\theta/\sin\psi)$, where ϕ, θ and ψ denote the roll, elevation, and azimuth between the positions of the UAV and the radar.

The probability of a kill, P_M, by an antiaircraft missile is expressed as

$$P_M = \begin{cases} 0, & d > R_{M,max}, \\ \dfrac{R_{M,max}^4}{d^4 + R_{M,max}^4}, & d \le R_{M,max}. \end{cases} \qquad (10.39)$$

The probability of kill, P_G, by an antiaircraft gun is expressed as

$$P_G = \begin{cases} 0, & d > R_G, \\ \dfrac{1}{d}, & R_G \le d \le R_{G,max}, \\ 0, & \text{otherwise}, \end{cases} \qquad (10.40)$$

where R_G and $R_{G,max}$ denote the effective and maximum hit radius, respectively, and d is the distance between UAV and the antiaircraft gun. Flight safety altitude is also an important objective for the UAV. The flight safety objective is expressed as

$$f_{safety} = \sum_{k=1}^{N} f_{safety}^*, \qquad (10.41)$$

where

$$f^*_{\text{safety}} = \begin{cases} 0, & \text{if } z_k < H_{\text{map}}(x_p(k), y_p(k)), \\ z_k - H_{\text{map}}(x_p(k), y_p(k)), & \text{otherwise,} \end{cases} \tag{10.42}$$

with $H_{\text{map}}(x_p(k), y_p(k))$ denoting the altitude of the terrain at the waypoint $(x_p(k), y_p(k))$.

The maneuvering constraint of the UAV is expressed as

$$g_1 = \sum_{k=1}^{N} g_k^1, \tag{10.43}$$

where

$$g_k^1 = \begin{cases} 0, & \text{if } \phi_k \leq \phi_k^{\text{max}}, \\ C, & \text{otherwise,} \end{cases} \tag{10.44}$$

with ϕ_k denoting the turning angle of the path at the waypoint p_k and C being a penalty coefficient. The UAV maneuverability is limited to a maximum turning angle ϕ_k^{max} at each waypoint p_k given by

$$\phi_k^{\text{max}} = \frac{n_{\text{max}} g}{V^2} \sqrt{(x_p(k+1) - x_p(k))^2 + y_p(k+1) - y_p(k))^2}, \tag{10.45}$$

where g is the gravitational acceleration, V is the flight velocity, and n_{max} is the maximum lateral overload. Climbing and gliding constraint expresses the minimum gliding slopes and maximum climbing. The UAV slope S_k at a waypoint p_k should not exceed the maximum climbing slope α_k and the minimum gliding slope β_k which are functions of the altitude z_k. A penalty function g_2 is thus constructed as

$$g_2 = \sum_{k=1}^{N} g_k^2, \tag{10.46}$$

where

$$g_k^2 = \begin{cases} 0, & \text{if } \beta_k \leq S_k \leq \alpha_k, \\ C, & \text{otherwise,} \end{cases} \tag{10.47}$$

with

$$\alpha_k = -1.5377 \times 10^{10} z_p(k)^2 - 2.6997 \times 10^{-5} z_p(k) + 0.4211, \tag{10.48}$$

$$\beta_k = 2.5063 \times 10^{-9} z_p(k)^2 - 6.3014 \times 10^{-6} z_p(k) - 0.325, \tag{10.49}$$

$$S_k = \frac{z_p(k+1) - z_p(k)}{\sqrt{(x_p(k+1) - x_p(k))^2 + (y_p(k+1) - y_p(k))^2}}. \tag{10.50}$$

UAVs cannot fly through some terrains, so the terrain constraint is formulated as

$$g_3 = \sum_{k=1}^{N} g_k^3, \tag{10.51}$$

with

$$g_k^3 = \begin{cases} 0, & \text{if } z_k \geq H_{\text{map}}(x_p(k), y_p(k)) + H_{\text{safe}}, \\ C, & \text{otherwise}, \end{cases} \tag{10.52}$$

and H_{safe} denoting the minimal safe flight height.

In [25], the evaluation function for the UAV flight-planning problem is given by

$$f = f_l + f_t + f_r + f_c, \tag{10.53}$$

where f_l is the minimal flying path length cost, f_t is the terrain cost, f_r is the radar detection cost, and f_c is the collision cost among the flying UAVs. Mathematically, f_l is defined as

$$f_l = \frac{\sum_{i=2}^{N_w} \sqrt{(x_i - x_{i-1})^2 + (y_i - y_{i-1})^2 + (z_i - z_{i-1})^2}}{\sqrt{(x_{N_w} - x_1)^2 + (y_{N_w} - y_1)^2 + (z_{N_w} - z_1)^2}}, \tag{10.54}$$

where N_w is the total number of waypoints including the starting and destination waypoints. The cost function for mountain obstacles is

$$f_t = \sum_{i=2}^{N_w} \sum_{j=1}^{n} A_{i,j}, \tag{10.55}$$

where

$$A_{i,j} = \begin{cases} 1, & \text{if } z_{i,j} \leq z_{k,i,j}^m, \\ 0, & \text{otherwise}, \end{cases} \tag{10.56}$$

with A_{ij} being a binary value and $z^m_{k,i,j}$ the return height for horizontal position $(x_{i,j}, y_{i,j})$ of the kth mountain. The radar threat cost is defined as

$$f_r = \sum_{i=2}^{N_w} \sum_{j=1}^{n} B_{i,j}, \tag{10.57}$$

where

$$B_{i,j} = \begin{cases} \left(\dfrac{\delta}{D_{i,j}}\right)^4, & \text{if } D_{i,j} \leq R_k, \\ 0, & \text{otherwise}, \end{cases} \tag{10.58}$$

$$D_{i,j} = \sqrt{(x_{i,j} - x^r_k)^2 + (y_{i,j} - y^r_k)^2 + (z_{i,j} - z^r_k)^2}. \tag{10.59}$$

The collision cost function is designed as

$$f_c = \sum_{i=2}^{N_w} \sum_{j=1}^{n} C_{i,j}, \tag{10.60}$$

with

$$C_{i,j} = \begin{cases} 1, & \text{if } d_{i,j} \leq \bar{d}, \\ 0, & \text{otherwise}, \end{cases} \tag{10.61}$$

where $d_{i,j}$ is the shortest distance between $(x_{i,j}, y_{i,j})$, and $z_{i,j}$.

Optimal path planning for UAV considering energy efficiency for surveillance purposes was studied in [27] while anticipating disturbances and energy consumptions. A trained and tested energy consumption regression model is used as the cost function of an optimal path-planning scheme, which is designed from a clustered 3D real pilot flight pattern with the proposed K-agglomerative clustering method, and is processed via A* and set-based particle-swarm-optimization (S-PSO) algorithm with adaptive weights. An online adaptive neural network (ANN) controller with varied learning rates is designed to ensure the control stability while having a reliably fast disturbance rejection response.

10.4 Multi-UAV path-planning problems

Different path-planning problem formulations have been proposed in the literature depending on the applications. In this section, we review some notable common formulations presented by some earlier researchers.

10.4.1 Cooperative search–attack mission

UAV path planning is an important task in cooperative search and attack mission subject to some environment, collision, and threat constraints. Some constraints as outlined in [5] are defined as follows:

- (Maneuverability) Each UAV's turning angle and maximum turning angle θ_{max} satisfy the following inequality:

$$\theta_i - \theta_{max} \leq 0. \tag{10.62}$$

- (Collision avoidance) Each UAV should maintain a safe distance from a possible neighbors as defined by:

$$d_{min} - d_{ij}(k) \leq 0 \quad \forall\, i, j \in \{1, \ldots, N\}, \tag{10.63}$$

where $d_{ij}(k)$ is the distance between any ith and jth UAV and d_{min} is the safe distance between the UAVs.

- (Threat avoidance) The safe distance from potential threats or possible collisions is defined by the following inequality:

$$R_T^l - d_T^{il}(k) \leq 0 \quad \forall\, i = 1, 2, \ldots, N,\ l = 1, 2, \ldots, N_T, \tag{10.64}$$

where d_T^{il} is the distance between the ith UAV and the lth threat and R_T^l is the radius of the lth threat.

- (Range) The range constraint describes the flight distance any given UAV can cover before running out of fuel. Therefore, the distance traversed by each UAV should be less than its range expressed by

$$L_{past}^i(k) - L_{max}^i \leq 0, \tag{10.65}$$

where L_{max}^i is the maximum range of the ith UAV and L_{past}^i is the distance traversed by the UAV.

The cooperative mission evaluation function is described in [5] as:

$$J = \sum_{i=1}^{N} \mu_i J_i, \tag{10.66}$$

$$J_i = \omega_i J_{si} + (1 - \omega_i) J_{ai}, \tag{10.67}$$

where μ_i is the weight of the ith UAV, and J_{si} and J_{ai} represent evaluation functions related to search and attack missions. Thus, the distributed cooperative search and attack mission optimization model is formulated as

$$U^* = \arg\max_U J \qquad (10.68)$$

subject to:

$$\theta_i - \theta_{\max} \leq 0, \qquad (10.69)$$

$$d_{\min} - d_{ij}(k) \leq 0 \quad \forall\, i, j \in \{1, \ldots, N\}, \qquad (10.70)$$

$$R_T^l - d_T^{il}(k) \leq 0 \quad \forall\, i = 1, 2, \ldots, N,\ l = 1, 2, \ldots, N_T, \qquad (10.71)$$

$$L_{\text{past}}^i(k) - L_{\max}^i \leq 0. \qquad (10.72)$$

Metrics such as surveillance coverage rate $P(k)$ and attack benefit $Q(k)$ are used to define the evaluation functions J_{si} and J_{ai}, respectively:

$$P(k) = \frac{\sum_{x=1}^{N_L} \sum_{y=1}^{N_w} }{N_L \times N_w} \text{grid}_{(x,y)}(k), \qquad (10.73)$$

where $\text{grid}_{(x,y)}(k) \in \{0, 1\}$ is an indicator function that determines whether the concerned grid has been searched. The attack benefit $Q(k)$ is defined as

$$Q(k) = \sum_{m=1}^{N_{\text{target}}(k)} \text{value}_m, \qquad (10.74)$$

where value_m defines the value of the targets that have been attacked and N_{target} is the total number of targets attacked.

An intelligent self-organized algorithm is proposed in [5] to solve the cooperative search–attack mission problem for multiple UAVs using a distributed control architecture which divides the global optimization problem into several local optimization problems. Considering the range constraint of UAV, a new state transition rule is designed to guide UAV back to its initial point within the maximum flight range. In the second phase, Dubins curve is employed to smoothly connect the waypoints generated by the ACO. As for the unexpected threats during the flight, an online threat avoidance method is proposed to replan the paths. A probabilistic approach to the cooperative search and attack problem was considered in [12]. The search was carried out in a probabilistic map environment. The search environment was divided into $L_x \times L_y$ grid with the following assumptions:

- The UAVs take-off in the same flight attitude to enter the search environment.

- Each UAV flies at a constant speed.
- UAVs can only fly to the adjacent cell in one time step.

Further, the flight is constrained to a maximum steering angle of 45°. The UAVs can choose from eight possible steering angles corresponding to the North, Northeast, Southeast, South, Southwest, West, and Northwest directions. Two probability functions, $p_k(i,j) \in [0,1]$ and $\chi(i,j) \in [0,1]$, model target existence and uncertainty of each search cell. Thus, the target search evolves according to

$$p_{k+1}(i,j) = \frac{p_d p_k(i,j)}{p_d p_k(i,j) + p_f(1 - p_k(i,j))}, \tag{10.75}$$

if a target is found in a cell (i,j), and

$$p_{k+1}(i,j) = \frac{(1 - p_d)p_k(i,j)}{(1 - p_d)p_k(i,j) + (1 - p_f)(1 - p_k(i,j))}, \tag{10.76}$$

if no target is found. The objective function based on the probabilistic model of [12] for a flight path with L_e steps is

$$R_w = \sum_{k=1}^{L_e} (\omega_1 R_e + \omega_2 R_t + \omega_3 R_d), \tag{10.77}$$

where the weight constants ω_1, ω_2, ω_3 are chosen such that

$$\omega_1 + \omega_2 + \omega_3 = 1. \tag{10.78}$$

Also R_e is the environmental search reward defined by

$$R_e = \chi_k(i,j) - \chi_{k+1}(i,j), \tag{10.79}$$

R_t is the target search reward function given by

$$R_t = (p_d - p_f)p_{k+1}(i,j) + p_f, \tag{10.80}$$

and the search distance reward is the Euclidean distance of the UAV from step k to step $k+1$, namely

$$R_d = \sqrt{(i_{k+1} - i_k)^2 + (j_{k+1} - j_k)^2}. \tag{10.81}$$

Collision avoidance is established by considering the Euclidean distance between any two UAVs at each time step:

$$D_d = \sqrt{(i_{k+1}(n) - i_{k+1}(m))^2 + (j_{k+1}(n) - j_{k+1}(m))^2}, \tag{10.82}$$

where $(i_{k+1}(n), j_{k+1}(n))$ is the position of the nth UAV at time step $k + 1$. The flight path fitness function is thus defined as

$$F_t = \begin{cases} R_w, & D_d > \sqrt{2}, \\ R_w - w_f D_d, & 0 < D_d < \sqrt{2}, \\ 0, & D_d = 0, \end{cases} \qquad (10.83)$$

where w_f is the penalty factor.

The closed trajectory problem for multiple UAVs was solved in two phases in [12]. In the first phase, an immune genetic algorithm was proposed to solve the target search efficiency of UAV in an uncertain environment. The second phase employed a divide-and-conquer approach and a deterministic path optimization algorithm to generate optimal paths for each UAV from the position of the event trigger to the time instant to the nearest return base, with initial and terminal velocity vector constraints.

An intelligent cooperative mission-planning scheme based on hybrid artificial potential field and ant colony optimization was proposed in [16]. In the search–attack mission environment of UAV swarm under the dynamic topology interaction, a time-sensitive target probability map was established. Based on the HAPF, the target attraction field, threat repulsive field, and repulsive field were constructed for the environmental cognition. A distributed ACO algorithm was designed to improve the UAVs' global searching capability. For this mission planning problem, four time-sensitive moving target types and four constraint types of UAV swarm were considered, which contributed to the practical applications of the HAPF-ACO.

A cooperative search–attack mission-planning problem with dynamic targets and threats was solved in [15] based on a distributed intelligent self-organized mission-planning algorithm for multiple UAVs. The DISOMP algorithm can be divided into four modules: a search module designed based on the distributed ant colony optimization (ACO) algorithm, an attack module designed based on the parallel approach (PA) scheme, a threat avoidance module designed based on the Dubins curve (DC), and a communication module designed for information exchange between the multi-UAV system and the dynamic environment.

10.4.2 Coverage path planning

Coverage path planning is a subproblem in path planning where a group of vehicles are required to cover a volume of interest. Coverage path planning has practical applications in precision agriculture, robot cleaning, and underwater applications. In [29], a multirobot path-planning problem was studied for 2D and 3D coverage planning. Typical objectives in coverage path planning include minimization of energy usage, maximization of coverage area, and collision avoidance.

Coverage planning problems can be classified as continuous or discrete, depending on the way the robot covers the structure. In complete coverage planning problems, it is desired to generate paths that allow vehicles cover the total volume given. In [28], the coverage planning problem was formulated as a multiobjective problem considering objectives like energy usage minimization and coverage area maximization. Practical applications of the 3D coverage planning can be found in volumetric inspections of piping and building structures for defects, robotic surgery, and in farm operations to locate defective crops. Coverage path planning can be decomposed into two subproblems:

- Coverage sampling, which deals with decomposition of the coverage area into the smallest set of viewpoints that provides full coverage, and
- Multigoal path planning, which is concerned with finding an optimal sequence and collision free paths that connects all viewpoints.

Coverage sampling strategies are needed to obtain necessary viewpoints that form the admissible sets. Numerous coverage strategies have been developed by some earlier researchers. A random art gallery algorithm was proposed in [20] for coverage sampling where the admissible set is generated by randomly sampling the workspace. The proposed sampling strategy does not, however, guarantee full coverage of the inspection space. The sampling strategy proposed in [21] incorporates an RRT planning algorithm for coverage sampling. The study in [19] investigates the coverage path planning for robotic single-sided dimensional inspection of free-form surfaces. That study reveals that a nonrandom targeted viewpoint sampling strategy significantly contributes to solution quality of the resulting planned coverage path. By deploying optimization during the viewpoint sampling, an optimal set of admissible viewpoints can be obtained, which consequently significantly shortens the cycle-time for the inspection task.

A 3D volume coverage path planning (VCPP) algorithm for robotic intracerebral hemorrhage evacuation was considered in [23]. In contrast to existing 3D planning techniques, the proposed algorithm generates 3D paths without first decomposing the volume into series of 2D planning problems. It considers the morphology of the volume to be covered and minimizes the configuration or task space distance traveled. The algorithm merges elements from existing grid-based and wavefront approaches and accommodates kinematic and environmental constraints, as well as obstacle avoidance.

The next-best view (NBV) problem is another important subproblem in coverage planning [22]. The problem addressed in [22] is to plan the next sensor's position, called the "next best view" (NBV). The NBV is the best view for the reconstruction process from a set of candidate views. Determining the NBV is a complex problem because the determined view must fulfil the following constraints:

- The NBV must see unknown surfaces in order to completely observe the object.
- The view must be reachable by the robot and there must be enough space around the view location for the robot's placement.
- The surfaces to be seen must be within the camera's field of view and depth of view; in addition, the angle formed between the sensor's orientation and the surface normal must be smaller than a given angle defined by the sensor in question.
- To ensure that the new scan will be merged with the previous ones, there must be an overlap between them.

The utility function for the NBV problem is

$$f_{\text{utility}} = f_{\text{area}}(f_{\text{quality}} + f_{\text{navigation}} + f_{\text{occlusion}}), \tag{10.84}$$

where f_{area} represents the area factor and serves to perceive unseen areas in the workspace, f_{area} is mathematically defined as

$$f_{\text{area}} = \sum_{i=1}^{2} f_i, \tag{10.85}$$

where each f_i evaluates the percentage of a certain voxel-type. Each f_i is bounded as $0 \leq f_i \leq 1$, reaching a maximum when the percentage x_i of voxels i reaches the desired percentage. Each function is constrained as follows:

$$f(0) = 0, \quad f'(0) = 0, \tag{10.86}$$
$$f(\alpha) = 1, \quad f'(\alpha) = 0, \tag{10.87}$$
$$f(1) = 0, \quad f'(1) = 0, \tag{10.88}$$
$$f(x) > 0, \quad \forall x \in (0, 1), \tag{10.89}$$
$$f'(x) > 0, \quad \forall x \in (0, \alpha), \tag{10.90}$$
$$f'(x) < 0, \quad \forall x \in (\alpha, 1). \tag{10.91}$$

Further, $f_{occlusion}$ is a function representing the occlusion factor which aims to determine the occluded areas and is defined as

$$f_{occlusion} = \frac{n_{op}}{\text{rows} \times \text{cols}}, \tag{10.92}$$

where n_{op} is the number of occluded plane voxels, and rows and cols are the numbers of rows and columns of the range image generated by ray tracing. The quality factor is captured by $f_{quality}$ and it defines the quality of perception of the sensor:

$$f_{quality} = \frac{\sum_{i=1}^{n_{oc}} \cos(\alpha_i)}{n_{oc}}. \tag{10.93}$$

The navigation factor is defined as:

$$f_{navigation} = (\rho - 1)x^2 + 1, \tag{10.94}$$

where x is the normalized orthodromic distance and ρ is the smallest value of the function. A model for path planning problem was proposed in [26] for aerial vehicles under QoS constraints. That work leverages on integrated full stack simulation framework to study end-to-end performance of UAV data transfer over a 5G mmWave network. In [30], a system for complete coverage path planning for outdoor surveillance is presented. The proposed system manages the presence of no-fly zones by decomposing the area into smaller subregions via a Morse-based decomposition approach. Furthermore, negotiation for UAV-subregion assignment is managed, taking into account environment geometry and both starting and desired final positions for the UAVs, once the coverage mission is completed. The back-and-forth motion is chosen as coverage pattern inside each subregion.

10.4.3 Planning for cooperative sensing

Cooperative sensing is another important function UAVs are capable of performing. Path planning is an essential step for cooperative sensing. The following utility functions were proposed in [5] for cooperative sensing:

$$U_p = \omega_s \frac{U^s}{U_{max}^s} + \omega_e \frac{U^e}{U_{max}^e} + \omega_t \frac{U^t}{U_{max}^t} + \omega_r \frac{U^r}{U_{max}^r}, \tag{10.95}$$

where

$$\omega_s + \omega_e + \omega_t + \omega_r = 1. \tag{10.96}$$

The optimal path p^* for cooperative sensing is thus defined as

$$p^* = \arg\max_p U_p, \tag{10.97}$$

subject to:

$$U_{p^*}^e \leq E^{\text{lim}}, \tag{10.98}$$

$$U_{p^*}^t \leq T^{\text{lim}}. \tag{10.99}$$

The sensing utility U^s is the sum of the sensing information in all the SIG cells defined as:

$$U^s = \sum_{n=1}^{N_{\text{sig}}} U_n^s, \tag{10.100}$$

$$U_n^s = \sum_{q=1}^{Q} s_n^q v_n^q(t), \tag{10.101}$$

where N_{sig} is the number of SIG cells, $v_n^q(t)$ is an update rule. The energy utility function models the total energy consumption as the sum of the communication energy utility U^{ce}, staying energy utility U^{se}, and the flying energy utility U^{fe}. The communication energy utility U^{ce} is defined as:

$$U^{ce} = -\sum_{n=1}^{N_{\text{sig}}} U_n^{ce}, \tag{10.102}$$

$$U_n^s = P^c N_n^P \sum_{q=1}^{Q} t_e^q s_n^q, \tag{10.103}$$

where P^c is the unit of communication energy for a unit time, N_n^P is the required number of transmissions to successfully transfer a sensing packet in cell n, t_e^q is the single sensing packet exchange time, s_n^q is the number of sensor nodes of type q in cell n; while N_n^P is defined as

$$N_n^P = \frac{1}{1 - p_n^e}. \tag{10.104}$$

The staying energy utility is defined as:

$$U^{se} = -\sum_{n=1}^{N_{\text{sig}}} U_n^{se}, \tag{10.105}$$

$$U_n^{se} = P^s N_n^P \sum_{q=1}^{Q} t_e^q s_n^q, \tag{10.106}$$

where P^s is the unit staying energy provided by UAV power system. The flying energy utility is defined as:

$$U^{fe} = - \sum_{n=1}^{N_{sig}+1} U_n^{fe}, \tag{10.107}$$

$$U_n^{fe} = \sum_{C_{fly(n)}^k \in S_{fly}^n} e_{fly(n)}^{k,k+1}, \tag{10.108}$$

$$e_{fly(n)}^{k,k+1} = t_{fly(n)}^{k,k+1} P^f, \tag{10.109}$$

$$t_{fly(n)}^{k,k+1} = \frac{d_{fly(n)}^{k,k+1}}{v_{fly(n)}^{k,k+1}}. \tag{10.110}$$

The total energy utility is thus

$$U^e = U^{ce} + U^{se} + U^{fe}. \tag{10.111}$$

The time utility for the planning problem is defined as

$$U^t = - \left\{ \sum_{n=1}^{N_{sig}} \left(N_n^P \sum_{q=1}^{Q} t_e^q s_n^q \right) + \sum_{n=1}^{N_{sig}+1} \sum_{C_{fly(n)}^k \in S_{fly}^n} t_{fly(n)}^{k,k+1} \right\}. \tag{10.112}$$

The risk utility for the planning problem is defined as

$$U^r = - \left\{ \sum_{n=1}^{N_{sig}} \left(r_n N_n^P \sum_{q=1}^{Q} t_e^q s_n^q \right) + \sum_{n=1}^{N_{sig}+1} \sum_{C_{fly(n)}^k \in S_{fly}^n} r_{fly(n)}^{k,k+1} t_{fly(n)}^{k,k+1} \right\}. \tag{10.113}$$

10.5 Typical applications

In [10], the authors considered cooperative path planning for multi-missiles. The solution was divided into three phases. The first phase was that of producing flyable paths with minimum length, the second was to choose the reference path and calculate the cooperative time for each missile, and the third was to produce paths of equal length or different length with same interval. In the first phase, Dubins paths were used to produce flyable paths with minimum length for each missile, the second phase was to calculate the cooperative time for each missile by choosing the reference path. Finally, in the third phase, paths of equal length or different length with same interval were achieved by manipulating these paths.

UAVs path-planning problem for remote sensing and photogrammetry was discussed in [11]. A ground feature oriented path-planning method was proposed for UAV mapping. The proposed method first estimates the distribution of the ground feature points from a lower-resolution image. Then, image footprints are selected by applying a three-step optimization. The flight path for the UAV is then generated by solving the "grouped traveling salesman" problem. This approach ensures the geo-registration of images during orthoimage stitching while maximizing the orthoimage coverage. Two cases, including a simulation and a real-world case, together with standard path-planning modes with different overlaps, are selected to evaluate the proposed method.

Some applications usually require low-cost aerial vehicle systems owing to their light weight, stealth, and maneuverability. However, these vehicle systems need more frequent refueling depending on the applications. In [3], a two-stage strategy for coupled route planning for UAV and GV was proposed for a coverage mission. In the first stage, refueling sites that ensure reachability of points of interest by the UAV are computed, as well as the feasible routes for both UAV and GV. The second stage oversees solving a mixed-integer linear programming problem to plan exact routes for the vehicles in the network.

An incremental moving path-planning algorithm was proposed in [24] for unmanned aerial vehicle systems with moving goals. The proposed algorithm leverages on previously planned paths to update in cases where the target moves. A cooperative path-planning algorithm to track a moving target was proposed in [31] for applications involving UAV and UGV. The novelty of the algorithm is that it takes into account vision occlusions due to obstacles in the environment. The algorithm uses a dynamic occupancy grid to model the target state, which is updated by sensor measurements using a Bayesian filter. Based on the current and predicted target behavior, the path planning algorithm for a single vehicle (UAV/UGV) is first designed to maximize the sum of the probability of detection over a finite look-ahead horizon. The algorithm is then extended to multiple vehicle collaboration scenarios, where a decentralized planning algorithm relying on an auction scheme is designed to plan finite look-ahead paths that maximize the sum of the joint probability of detection over all vehicles.

10.5.1 Simulation example

Consider a quadrotor system defined by the following nonlinear dynamics:

$$\dot{x} = f(x, u) = \begin{bmatrix} \dot{\phi} \\ \dot{\theta}\dot{\psi}a_1 + \dot{\theta}a_2\Omega_r + b_1 U_2 \\ \dot{\theta} \\ \dot{\phi}\dot{\psi}a_3 - \dot{\phi}a_4\Omega_r + b_2 U_3 \\ \dot{\psi} \\ \dot{\theta}\dot{\phi}a_5 + b_3 u_4 \\ \dot{z} \\ g - (\cos\phi\cos\theta)\dfrac{1}{m}U_1 \\ \dot{x} \\ u_x\dfrac{1}{m}U_1 \\ \dot{y} \\ u_y\dfrac{1}{m}U_1 \end{bmatrix}, \tag{10.114}$$

where x and U are state and control inputs to the system. The state vector is $x = [\phi \; \dot{\phi} \; \theta \; \dot{\theta} \; \psi \; \dot{\psi} \; z \; \dot{z} \; x \; \dot{x} \; y \; \dot{y}]$. The control input is $U = [U_1 \; U_2 \; U_3 \; U_4]$. The control inputs are defined as follows:

$$U_1 = b(\Omega_1^2 + \Omega_2^2 + \Omega_3^2 + \Omega_4^2), \tag{10.115}$$
$$U_2 = b(-\Omega_2^2 + \Omega_4^2), \tag{10.116}$$
$$U_3 = b(\Omega_1^2 - \Omega_3^2), \tag{10.117}$$
$$U_4 = d(-\Omega_1^2 + \Omega_2^2 - \Omega_3^2 + \Omega_4^2). \tag{10.118}$$

The parameters a_1, a_2, a_3, a_4, a_5 and b_1, b_2, b_3 are:

$$a_1 = \frac{(I_{yy} - I_{zz})}{I_{xx}}, \tag{10.119}$$

$$a_2 = \frac{J_r}{I_{xx}}, \tag{10.120}$$

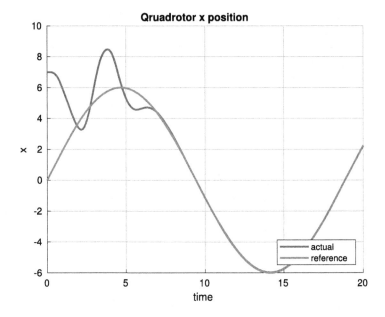

Figure 10.1 Quadrotor x-position.

$$a_3 = \frac{(I_{zz} - I_{xx})}{I_{yy}}, \tag{10.121}$$

$$a_4 = \frac{Jr}{I_{yy}}, \tag{10.122}$$

$$b_1 = \frac{l}{I_{xx}}, \tag{10.123}$$

$$b_2 = \frac{l}{I_{yy}}, \tag{10.124}$$

$$b_3 = \frac{1}{I_{zz}}, \tag{10.125}$$

while u_x and u_y are defined as:

$$u_x = (\cos\phi \sin\theta \cos\psi + \sin\phi \sin\psi), \tag{10.126}$$

$$u_y = (\cos\phi \sin\theta \sin\psi - \sin\phi \cos\psi). \tag{10.127}$$

Figs. 10.1–10.6 give the state responses of the system with respect to a path-tracking objective. The control input of the system are shown in Figs. 10.7–10.10, and Figs. 10.11 and 10.12.

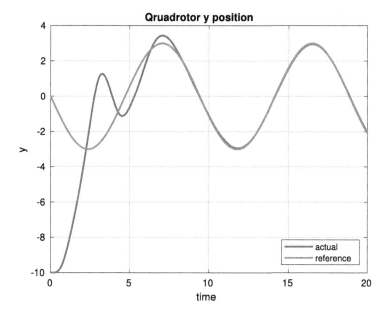

Figure 10.2 Quadrotor *y*-position.

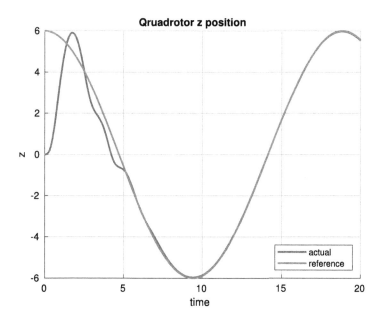

Figure 10.3 Quadrotor *z*-position.

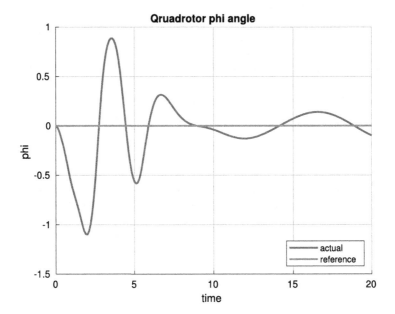

Figure 10.4 Quadrotor ϕ angle.

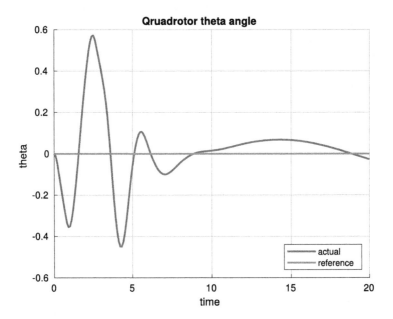

Figure 10.5 Quadrotor θ angle.

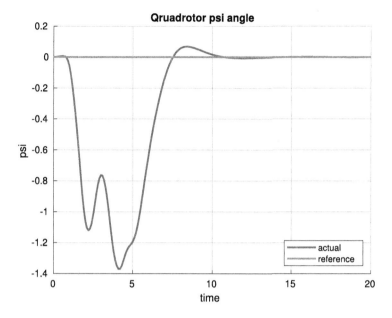

Figure 10.6 Quadrotor ψ angle.

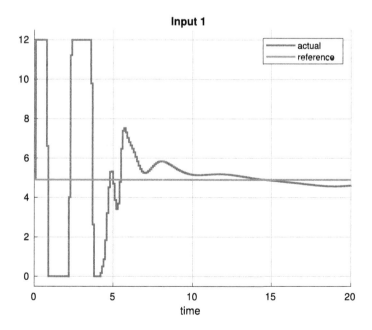

Figure 10.7 Quadrotor input U_1.

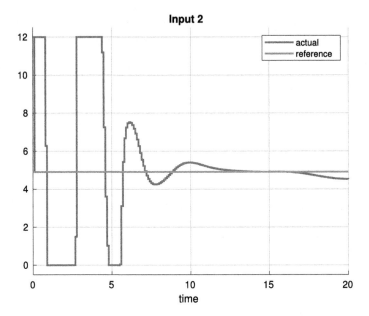

Figure 10.8 Quadrotor input U_2.

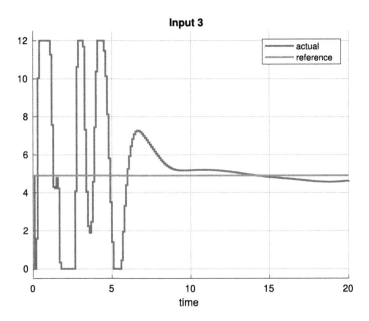

Figure 10.9 Quadrotor input U_3.

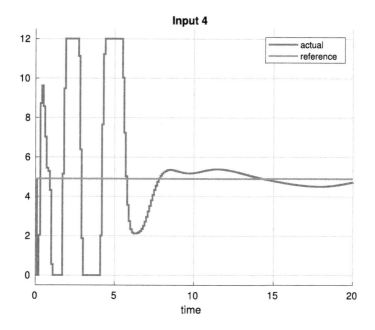

Figure 10.10 Quadrotor input U_4.

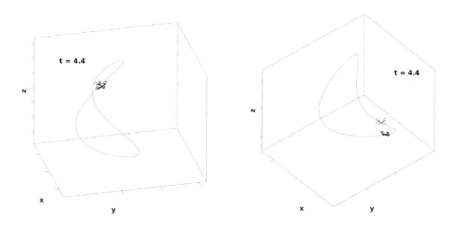

Figure 10.11 Quadrotor path-tracking simulation at $t = 4.4$ s.

10.6 Notes

In this chapter, path-planning problems and algorithms were presented for unmanned aerial vehicle systems. Some typical path-planning problems in

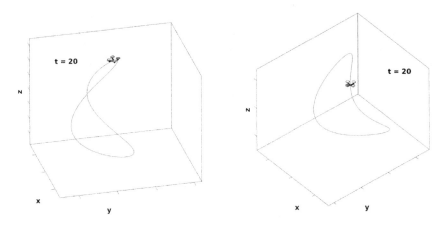

Figure 10.12 Quadrotor path-tracking simulation at $t = 20$ s.

a three-dimensional space were discussed, and existing solution approaches were reviewed.

References

[1] J. Zhang, J. Yan, P. Zhang, X. Kong, Design and information architectures for an unmanned aerial vehicle cooperative tracking controller, IEEE Access 6 (3) (2018) 45821–45833.

[2] X. Zhang, X. Lu, S. Jia, X. Li, A novel phase angle-encoded fruit fly optimization algorithm with mutation adaptation mechanism applied to UAV path planning, Appl. Soft Comput. 70 (2018) 371–388.

[3] P. Maini, K. Sundar, M. Singh, S. Rathinam, P.B. Sujit, Cooperative aerial-ground route planning with fuel constraints for coverage applications, IEEE Trans. Aerosp. Electron. Syst. 55 (6) (2019) 3016–3028.

[4] D. Drake, S. Koziol, E. Chabot, Mobile robot planning with a moving goal, IEEE Access 6 (2018) 12800–12814.[uncited]

[5] Z. Ziyang, X. DongJing, G. Chen, Cooperative search-attack mission planning for multi-UAV based on intelligent self-organized algorithm, Aerosp. Sci. Technol. 76 (2018) 402–411.

[6] C. Qu, W. Gai, J. Zhang, M. Zhong, A novel hybrid grey wolf optimizer algorithm for unmanned aerial vehicle (UAV) path planning, Knowl.-Based Syst. 194 (2020) 105530.

[7] M. Mah, H. Lim, A. Tan, Secrecy improvement via joint optimization of UAV relay flight path and transmit power, Veh. Commun. 23 (2020) 1–13.

[8] G. Jain, G. Yadav, D. Prakash, A. Shukla, R. Tiwari, MVO-based path planning scheme with coordination of UAVs in 3D environment, J. Comput. Sci. 37 (2019) 1–15.

[9] L. Huang, H. Qu, P. Ji, X. Liu, Z. Fan, A novel coordinated path planning method using k-degree smoothing for multi-UAVs, Appl. Soft Comput. 48 (2016) 182–192.

[10] Y. Gao-yang, Z. Shao-lei, Y. Shi, Cooperative path planning of multi missiles, in: Proc. 2014 IEEE Chinese Guidance, Navigation and Control Conference, 2014, pp. 767–772.

[11] Y. Liu, X. Zhang, Y. Zhang, X. Guan, Collision free 4D path planning for multiple UAVs based on spatial refined voting mechanism and PSO approach, Chin. J. Aeronaut. 32 (6) (2019) 1504–1519.

[12] Z. Zhou, D. Luo, J. Shao, Y. Xu, Y. You, Immune genetic algorithm based multi-UAV cooperative target search with event-triggered mechanism, Phys. Commun. 41 (August 2020) 101103, https://doi.org/10.1016/j.phycom.2020.101103.

[13] X. Zheng, F. Wang, Z. Li, B. Han, Obstacle avoidance model for UAVs with joint target based on multi strategies and follow-up vector field, Proc. Comput. Sci. 170 (2020) 257–264.

[14] P. Yao, H. Wang, Z. Su, Cooperative path planning with applications to target tracking and obstacle avoidance for multi-UAVs, Aerosp. Sci. Technol. 54 (2016) 10–22.

[15] Z. Zhen, P. Zhu, Y. Xue, Y. Ji, Distributed intelligent self-organized mission planning of multi-UAV for dynamic targets cooperative search-attack, Chin. J. Aeronaut. 32 (12) (2020) 2706–2716.

[16] Z. Zhen, Y. Chen, L. Wen, B. Han, An intelligent cooperative mission planning scheme of UAV swarm, Aerosp. Sci. Technol. 100 (2020) 1–16.

[17] P. Wu, D. Campbell, T. Merz, Multi-objective four dimensional vehicle motion planning in large dynamic environments, IEEE Trans. Syst. Man Cybern., Part B, Cybern. 41 (3) (2011) 621–634.

[18] X. Ma, Z. Jiao, Z. Wang, D. Panagou, 3D decentralized prioritized motion planning and coordination for high-density operations of micro-aerial vehicles, IEEE Trans. Control Syst. Technol. 26 (3) (2018) 939–953.

[19] E. Gloriuex, P. Franciosa, D. Ceglarek, Coverage path planning with targeted viewpoint sampling for robotic free-form surface inspection, Robot. Comput.-Integr. Manuf. 61 (2020) 1–11.

[20] E. Gloriuex, P. Franciosa, D. Ceglarek, A randomized art gallery for sensor placement, in: Proc. the Seventeenth Annual Symposium on Computational Geometry, 2001, pp. 232–240.

[21] A. Bircher, M. Kamel, K. Alexis, H. Oleynikova, R. Siegwart, Receding horizon path planning for 2D exploration and surface inspection, Auton. Robot. 42 (2) (2018) 291–306.

[22] J.I. Vasquez-Gomez, L.E. Sucar, R. Murrieta-Cid, E. Lopez-Damian, Volumetric next-best-view planning for 3D object reconstruction with positioning error, Int. J. Adv. Robot. Syst. 11 (159) (2014) 1–13.

[23] J. Granna, I.S. Godage, R. Wirz, K.D. Weaver, R.J. Webster, J. Burgner-Kahrs, A 3-D volume coverage path planning algorithm with application to intracerebral hemorrhage evacuation, IEEE Robot. Autom. Lett. 1 (2) (2016) 876–883.

[24] H.Y. Zhang, W.M. Lin, A.X. Chen, Path planning for the mobile robot: a review, Symmetry 10 (2018) 450.

[25] S. Shao, Y. Peng, C. He, Y. Du, Efficient path planning for UAV formation via comprehensively improved particle swarm optimization, ISA Trans. 97 (2020) 415–430.

[26] L. Shi, S. Xu, H. Liu, Z. Zhan, QoS-aware UAV coverage path planning in 5G mmWave network, Comput. Netw. 175 (2020) 1–10.

[27] R. Wai, A.S. Prasetia, J. Zhang, Adaptive neural network control and optimal path planning of UAV surveillance system with energy consumption prediction, IEEE Access 7 (2019) 126137–126153.

[28] K.O. Ellefsen, H.A. Lepikson, J.C. Albiez, Multiobjective coverage path planning: enabling automated inspection of complex, real-world structures, Appl. Soft Comput. 61 (2017) 264–282.

[29] N. Baras, M. Dasygenis, N. Ploskas, Multi-robot coverage path planning in 3-dimensional environment, in: Proc. the 8th International Conference on Modern Circuits and Systems Technologies (MOCAST), 2019, pp. 1–4.

[30] D.C. Guastella, L. Cantelli, G. Giammello, C.D. Melita, G. Spatino, G. Muscato, Complete coverage path planning for aerial vehicle flocks deployed in outdoor environments, Comput. Electr. Eng. 75 (2019) 189–201.

[31] H. Yu, K. Meier, M. Argyle, R.W. Beard, Cooperative path planning for target tracking in urban environments using unmanned air and ground vehicles, IEEE/ASME Trans. Mechatron. 20 (2) (2015) 541–552.

Index

A

Ackermann vehicle, 299
Adaptive
  control, 60, 193
  control gains, 64
  control laws, 33, 56, 69
  distributed controller, 32
  group consensus problem, 46
Adaptive neural network (ANN), 342
Adjacency matrix, 3–5, 53, 86, 140, 141,
    165, 183, 213, 252, 256, 261
Adversarial agents, 173
Aerial vehicle, 272, 273, 275, 286, 287, 349
Agents
  control inputs, 9, 44, 168, 200
  cooperative, 173, 174, 177, 178
  dynamics, 9, 12, 15, 98, 99, 138, 151
  fitness function, 250
  followers, 26, 57, 82–84, 102, 107
  heterogeneous, 46, 60
  mobile, 1, 2
  model, 12, 268
  model dynamics, 76
  motions, 10
  network, 1, 20, 23, 24, 54, 56, 167, 168
  nonfaulty, 172
  pinning, 65
  position communication topologies, 83
  state, 10, 59, 174, 181, 200, 212, 213
  systems, 2
  trusted, 169, 170, 253
  trusting, 253
Agreement
  set, 219
  space, 17
  states, 53
Algebraic
  connectivity, 6, 141, 146, 150
  graph theory, 147, 156, 199
Allocation rule, 245, 246
Ant colony optimization (ACO), 325, 344,
    346

Antagonistic
  interaction, 53, 68, 178
  interaction networks, 53
  networks, 68
Asynchronous
  consensus, 97
  control, 120
  networks, 100
  switching, 118, 120–125, 129, 131, 132,
    153, 154, 156
  switching control, 118, 119, 121
Attitude
  containment control, 138
  dynamics, 269
  dynamics quadrotor, 290
Autonomous
  agents networks, 1
  ground vehicle, 278
  multiagent systems, 12
  vehicles, 282, 287, 297, 312, 326

B

Battery energy storage system (BESS), 29,
    31, 33, 35
  network, 31
  system, 34
Battery packages, 31–34
Bilateral cooperative exploration (BCE),
    325
Biological
  agents, 272
  multiagent systems, 200
Bipartite consensus, 53–57, 94, 95, 100,
    104, 108, 109, 178, 179
  algorithm, 55
  pinning, 94
  problem, 68
  protocol, 56, 58, 68, 94, 108
  resilient, 178
  tracking, 69, 99
Bipartite output consensus problem, 68
Bluetooth communication networks, 272

C

Cell decomposition methods, 297
Communication
  delays, 42, 46, 56, 68, 99, 156
  graph, 31, 34, 54, 68, 100, 101, 106,
    109, 139, 140, 151, 180, 245, 246
  network, 2, 117, 120, 161, 197, 200, 251
  path, 251, 252
  topology, 24, 27, 62, 76, 97, 100, 142,
    146, 148, 150, 156, 199, 252
Competition network, 54
Complete fuzzy graph, 211
Computational intelligence (CI), 332, 337,
    338
Connectivity assumptions, 28, 97, 100
Consensus
  algorithms, 1, 46, 107, 170, 180, 181,
    190, 193
  analysis, 200
  asynchronous, 97
  control, 12, 139, 147
  dynamics, 183
  errors, 134, 137
  exponential, 117, 118, 121, 124, 153
  filtering, 99, 110
  function, 176, 178
  game, 200, 229, 252
  in multiagent systems, 12, 15, 109
  in networks, 68
  interventional bipartite, 54–58, 68
  leaderless, 17–19, 21, 27, 76, 77, 79,
    101, 106, 212
  multiagent, 251
  multiagent systems, 118, 124
  network, 252
  objective, 15, 26, 82, 213, 256, 263
  performance problem, 139
  pinning, 61–65
  problem, 1, 17, 26, 29, 36, 46, 58, 68,
    73, 101, 106, 109, 118, 138–140,
    142, 146, 147, 150, 153, 177, 198,
    213, 215, 251
    for multiagent systems, 76, 139
    in multiagent systems, 17
  protocol, 8, 12, 20, 40, 42, 45, 64, 77,
    85, 87, 96, 101–104, 106, 173,
    176–179

  region, 139
  resilient, 68, 163, 165, 167, 168, 172,
    263
  state, 261
  tracking, 97, 98, 138
  value, 36, 181, 183, 193
Containment control problems, 25
Continuous-time resilient asymptotic
    consensus (CTRAC), 173, 175
  problem, 175
Control
  actions, 287
  adaptive, 60, 193
  architectures, 267
  asynchronous, 120
  consensus, 12, 139, 147
  cooperative, 267
  coordination, 161, 198
  gains, 32, 34, 45, 56, 64
  input, 10, 12, 15, 24, 26–29, 31, 57, 58,
    64, 73, 79, 141, 165–167, 176, 179,
    212, 282, 289, 310, 353, 354
  law, 32, 166, 169, 170, 193, 205–207
  multivehicle, 209
  objective, 31
  policies, 287
  problems, 1, 68, 272, 294
  protocol, 16, 17, 34, 42, 63, 64, 66, 109,
    175, 257, 272
  sequence, 208
  systems, 156
  term, 11
  theory, 139, 199
  vehicle, 337
Controllability, 207
  graph, 207
  property, 207
Controllable graph process, 207, 208
Controller
  pose, 305
  steering, 305
Cooperative
  agents, 173, 174, 177, 178
  control, 267
  frameworks, 278
  games, 234, 235
  mission, 343

nodes, 173
path planning, 351
path replanning for multiple UAVs, 310
route, 326
route planning, 324
search, 343, 344
sensing, 349
states, 174
strategies, 336
surveillance, 286
UAVs, 336
Coordination
control, 161, 198
control problems, 198
game, 233
in multiagent systems, 197
multiagent, 10, 197, 268
multirobot, 268
multivehicle, 268
objective, 252
problem, 198
process, 161
states, 17
variable, 197, 198
Cost
function, 287, 298, 315, 318, 320, 334,
337, 341, 342
functional, 287
parameters, 308
path, 309
savings, 235
Coverage
path planning, 346, 347
sampling, 347
sampling strategies, 347
Crisp graphs, 199, 209, 212, 213

D

Delayed multiagent system, 119
stochastic, 118–122, 124, 125, 129,
132–134, 137, 153, 156
Denial-of-service (DoS) attacks, 88, 179,
192, 193
Differential
drive vehicle, 279
flatness property, 289

Digraph, 8, 9, 76, 79, 96, 161, 170, 171,
246, 252
Digraph game, 246
Directed spanning tree, 3, 19, 65, 80, 100,
140, 147, 162, 212
Discounted payoff game (DPG), 244
Distributed denial-of-service (DDoS)
attacks, 179
Double integrator dynamics, 46
Double-integrator position-based mean
squared subsequence reduced
(DP-MSSR) algorithm, 165
Dubins curve (DC), 344, 346
Dubins paths, 351
Dynamics
agents, 9, 12, 98, 99, 138, 151
consensus, 183
heterogeneous, 46, 267
leader, 32, 84, 97, 98
linear, 19, 55, 62, 79, 98, 106
nonlinear, 23, 27, 44, 57, 63, 79, 94, 95,
118, 120, 138, 139, 353
quadrotor, 290
state, 110
vehicle, 9–11, 269

E

Edge flow dynamics, 215
Eigenvalue
Laplacian, 6
maximum, 41, 91, 92
optimization, 201
problem, 22
Energy storage system (ESS), 29
Equilibrium state, 62, 66, 237
Evolutionarily stable state, 237, 261
Evolutionarily stable strategies (EVSS), 229,
235, 237
Evolutionary game theory (EGT), 229,
230, 235
Evolutionary graph (EG), 229, 238, 241,
256
consensus, 249
state-dependent, 256, 263
theory, 238, 241
Exponential consensus, 117, 118, 121, 124,
153

F

Fast distance transformation (FDT), 324
Finite element system (FES), 269
Fitness, 235–239, 247
  relative, 240
  value, 235, 236
Fixation probability, 238–241
Flight
  path, 331, 332, 345, 346, 352
  velocity, 340
Followers
  agents, 26, 57, 82–84, 102, 107
  position, 82
Formation
  consensus problem, 165
  control, 1, 2
Fuel cost, 334, 335
Fuzzy
  bridge, 210, 211
  graph, 208–211, 213, 224
    framework, 213
    theory, 208, 209
  subset, 209

G

Generic linear multiagent systems, 46
Genetic algorithm (GA), 312, 313, 316,
    317, 323, 325
Graph
  controllability, 207
  Laplacian eigenvalue, 6
  Laplacian spectrum, 4
  random, 109, 199
  undirected, 5, 6, 12
  union, 75, 77, 100, 169
Ground vehicle, 286, 331
  agent, 277
  autonomous, 278
  directs, 286
  system, 277, 278
  unmanned, 286
Group consensus, 36, 37, 39, 41–44, 46,
    47, 50, 85, 88, 89, 91, 92, 99, 103,
    104, 170, 171
  for networked mechanical systems, 99
  pinning, 65

  problem, 39, 41, 86, 103
  resilient, 171
  theory, 38

H

Hawk and dove game, 235, 236
Helicopter system, 272, 276, 277
Heterogeneous
  agent, 46, 60
  agent dynamics, 168
  dynamics, 46, 267
  linear MAS, 68
  MAS, 15, 46, 65, 169, 170
  network, 54
Homogeneous
  network, 54
  robots, 322
  stationary state, 63, 64

I

Improved interfered fluid dynamical system
    (IIFDS), 336
Impulsive control strategy, 68
Inelastic massless cable, 289
Infeasible paths, 318
Information exchange, 1, 138, 192, 272,
    346
  protocols, 180
  velocity, 83
Inherent dynamics, 207
Integrator dynamics, 168, 251
Interaction topology, 46, 252
Interventional bipartite consensus, 54–58,
    68
Intruder attacks, 161, 162, 192, 197

L

Lagrangian dynamics, 46, 65
Laplacian, 3, 4, 6, 19, 54, 59, 86, 91, 166,
    198, 213, 252
  eigenvalue, 6
  matrix, 4–7, 18, 25, 46, 47, 80, 81, 87,
    89, 92, 140, 142, 143, 146,
    199–201, 203, 212, 250, 256
  potential, 8
  spectrum, 4, 215

Leader
    agent, 17, 26, 27, 29, 57, 82, 84, 98, 102
    dynamics, 32, 84, 97, 98
    node, 25, 26, 55, 56
    states, 26, 44, 107
Leaderless, 17
    asymptotic consensus, 212
    consensus, 17–19, 21, 27, 76, 77, 79,
        101, 106, 212
    consensus problem, 23, 24, 77, 80, 81,
        102
    output consensus problem, 78
Linear
    dynamics, 19, 55, 62, 79, 98, 106
    dynamics model, 98
    multiagent systems, 138, 139, 156, 193
    multiagent systems consensus problems,
        139
    velocity, 281
Linear matrix inequality (LMI), 98, 132,
    140
Local
    neighborhood, 56
    state feedback control, 105

M

Markovian switching signal, 74, 102
Massless
    links, 293
    rigid links, 291
Matrix, Laplacian, 4, 5, 7, 18, 25, 46, 47,
    80, 87, 92, 140, 142, 143, 146,
    199–201, 203, 212, 250
Max consensus, 15, 16
Maximum
    eigenvalue, 41, 91, 92
    path distance, 331
    position value, 169
Mean payoff game, 243, 244
Mean subsequence reduced (MSSR)
    algorithm, 164
Mean-select-reduced (MSR) algorithm,
    171
Median value consensus, 176
Median-based consensus algorithm (MCA),
    164
Microaerial vehicle (MAV), 286, 326

Mobile
    agents, 1, 2
    multiagent systems, 2
    multiagent systems coordination control,
        2
    robots, 12, 322, 324, 335
Motion
    plan, 337
    planners, 337
    planning, 289, 337
    planning process, 325
    space, 12
    vehicle, 337
Multiagent
    consensus, 251
    coordination, 10, 197, 268
    dynamic systems, 9
    framework, 12, 277
    game, 257
    network, 252, 256
    setting, 200
    synchronization problem, 100
Multiagent system (MAS), 1, 9, 15, 16, 19,
    31, 46, 47, 58, 60, 61, 63, 73, 88,
    102, 103, 117, 125, 128, 137–139,
    144, 149, 153, 172, 173, 197, 200,
    251, 256
    consensus, 118, 124
        criterion, 131
        problem, 117
        seeking, 138
    cooperative control, 1
    for robotic systems, 65
    network, 25, 39, 54, 83, 84, 110, 163,
        168, 170, 173, 267, 268, 272
Multibody dynamics, 269
Multigoal path planning, 347
Multiple
    robots path planning, 317
    UAVs, 326, 344, 346
    vehicle, 337
    vehicle systems, 287
Multiquadrotor systems, 293
Multirobot
    coordination, 268
    path planning, 319
    system, 15, 24

Multivehicle, 267, 268
  applications, 268
  architecture, 268, 272
  control, 209
  cooperative control, 294
  cooperative control framework, 294
  coordination, 268
  coordination problems, 272
  framework, 269
  robot, 15
  systems, 326

N

Navigation path, 314
Neighborhood
  error, 58
  sets, 198, 205, 220, 249, 253, 257
Neighboring agents, 139, 156, 169, 256,
      267, 269
Network
  agents, 1, 20, 23, 24, 54, 56, 167, 168
  algorithm, 197
  analysis, 197
  attacks, 179
  communication, 2, 117, 120, 161, 197,
      200, 251
  communication protocol, 181
  conditions, 199
  connectivity probability, 109
  consensus, 252
  delays, 197
  failures, 161
  heterogeneous, 54
  homogeneous, 54
  MAS, 25, 83, 84, 110, 163, 168, 170,
      173, 267, 268, 272
  multiagent, 252, 256
  multiagent systems, 39, 54
  perspective, 12
  states, 119
  stochastic switched, 131
  structure, 2, 185, 186, 188, 189
  synthesis, 197
  topology, 3, 17, 65, 73, 86, 99–102, 109,
      110, 117, 175, 179, 191, 193, 197,
      268
  vulnerable, 161
  weights, 252

Networked
  control systems, 153, 156
  MAS, 176
  mechanical systems, 46
  multiagent consensus control, 150
  multiagent systems, 139, 193
  predictive control schemes, 99
  system, 150
Neural network, 68, 69, 99, 110, 317
Nonfaulty agents, 172
Nonlinear
  adaptive distributed control protocol, 28
  dynamics, 23, 27, 44, 57, 63, 79, 94, 95,
      118, 120, 138, 139, 353
  MAS, 27, 28, 68
  multiagent systems, 65, 99, 139
  state function, 15, 212

O

Obstacles
  avoidance, 267, 312, 314, 315, 324, 336,
      348
  position, 315
  present, 324
  space, 310, 311
Omnidirectional mobile robots, 10
Optimal
  disassembly path, 324
  motions paths, 326
  path, 306, 320, 324, 346
    planning, 298, 342
    planning problem, 298
Orientation
  angle, 10
  relative, 270

P

Parallel approach (PA), 346
Parameterized path, 326
Particle-swarm optimization (PSO), 312,
      315, 326
Path
  communication, 251, 252
  cost, 309
  diversity, 309
  length, 309, 312, 315, 318, 324, 338
  length cost, 341
  loss, 325

planner for nonholonomic parallel orienting robots, 325
planning, 10, 297, 299, 306, 308–312, 320, 321, 323–326, 331, 332, 336, 337, 346, 347, 349, 352
   algorithms, 306, 320, 335
   problem, 297, 298, 306, 323, 326, 349
   process, 324
   replanning, 310
   robot, 316
   safety, 313, 318
   smoothness, 299, 314
   variable, 326
PI controller, 303
Pin agent states, 61
Pinning
   agents, 65
   bipartite consensus, 94
   cluster consensus control problem, 65
   consensus, 61–65
   consensus problem, 62, 65
   group consensus, 65
   state, 63
Polygonal obstacles, 337
Population dynamics, 229, 238
Pose controller, 305
Position
   communication topologies, 83
   communication topologies agents, 83
   followers, 82
   obstacles, 315
   relative, 167, 336
   values, 166, 169
Positional
   optimal strategies, 243
   strategy, 243, 244
Probabilistic roadmap (PRM), 309–311
Probability
   density, 241, 287
   distribution, 237
   measure, 101, 249
   space, 249
Problem
   bipartite consensus, 68
   coordination, 198
   group consensus, 39, 86, 103
   layout, 41, 62, 63

multiagent systems consensus, 117
path planning, 297, 298, 306, 323, 326, 349
statement, 60, 176, 212

Q

Quadrotor, 272, 289, 291, 293, 326
   aircraft, 273
   array, 289
   attitude dynamics, 290
   attitude kinematics, 289
   dynamics, 290
   system, 273, 275, 276, 288, 293, 353
   UAV, 293

R

Random
   consensus algorithms, 192
   consensus value, 181, 182
   graphs, 109, 199
   networks, 12, 25, 181, 182
Randomized potential planner (RPP), 309
Rapidly exploring random trees (RRT), 309, 310
   algorithm, 310
Reachable neighborhood set, 224
Relative
   fitness, 240
   orientation, 270
   position, 167, 336
   state errors connectivity assumptions, 29
   state value, 174
Replicator dynamics, 236, 237, 247, 253
Resilient
   bipartite consensus, 178
   consensus, 68, 163, 165, 167, 168, 172, 263
   group consensus, 171
   group consensus algorithm, 170
Robot
   cleaning, 346
   configuration, 297
   module, 321
   multivehicle, 15
   path, 316
   sensor, 315

structure, 317
team, 321
Robotic
    evacuation, 325, 347
    surgery, 347
    systems, 325
    systems network, 69
Rotational dynamics, 273
Route planning, 326, 352
    cooperative, 324

S

Secure consensus, 192
Secure consensus problem, 180
Set-based particle-swarm-optimization
    (S-PSO), 342
Shapley function, 253, 257, 259, 261, 263
Signed
    graphs, 94, 100
    networks, 68
Simultaneous localization and mapping
    (SLAM) algorithm, 309
Single
    integrator agents, 178
    integrator dynamics, 24, 41, 88, 110,
        168, 249
    integrator model, 9, 10
    quadrotor, 288–290
Slung load, 288, 291, 293
Social
    interaction networks, 198
    networks, 53, 180, 200, 229
Spanning tree, 4, 9, 41, 42, 76, 77, 80, 86,
        96, 97, 99, 100, 102, 103, 147, 171,
        198, 210
State
    agents, 10, 174, 181
    consensus, 261
    definitions, 15
    dependent weights, 263
    dynamics, 110
    feedback control, 132
    feedback matrix, 102
    information, 17, 63, 68, 251
    relative value, 174
    space, 102, 200, 214, 237
    trajectories, 134, 137, 151, 153

transition rule, 344
value, 172, 181
variable, 101, 104, 105, 176
vector, 9, 108, 118, 181, 237, 289, 353
State of charge (SoC), 29
Static
    Laplacian, 259
    obstacles, 318
Steering
    angle, 12, 280, 281, 300, 345
    controller, 305
    velocity, 280
Stochastic
    delayed multiagent systems, 118–122,
        124, 125, 129, 132–134, 137, 153,
        156
    delayed multiagent systems consensus,
        119, 120, 123
    delayed multiagent systems consensus
        criterion, 154
    linear multiagent systems, 193
    matrix, 184
    switched
        delayed systems, 120
        multiagent systems, 118
        network, 131
        systems, 120, 123, 124, 132, 156
Subdominant eigenvalues, 189, 191, 194
Subgraphs, 41, 42, 75, 103, 170, 171, 215,
        307
Switched
    multiagent systems, 97, 100
    multiagent systems consensus, 97, 100
    networks, 110, 153
    networks consensus, 118
    stochastic systems, 132
Symmetric adjacency matrix, 183
Synchronous networks, 192

T

Target
    position, 315
    state, 2, 352
Terminal velocity vector, 346
Terrain
    cost, 341
    obstacles, 334

Threat cost, 334, 335
Topological states, 199
Topology, 4, 40, 65, 74, 77, 80, 86, 94, 99, 106, 117, 132, 139, 147, 165, 256, 257
  communication, 27, 62, 76, 97, 100, 142, 146, 148, 150, 156, 199
  network, 3, 17, 65, 73, 86, 99–102, 109, 110, 117, 175, 179, 191, 193, 197, 268
  structure, 101
  undirected, 147, 150
Total disagreement, 8
Trusted agent, 169, 170, 253
Trusting agent, 253

U
Uncooperative
  agents, 177
  nodes, 176
Underwater vehicle systems, 285
Undirected
  communication graph, 146
  communication topologies, 29
  evolutionary graphs, 242
  graph, 5, 6, 12, 18, 200, 204, 212, 238
  graph Laplacians, 5
  networks, 15, 18, 96
  topology, 147, 150
Union graphs, 75, 77, 100, 169
Unmanned aerial vehicle (UAV), 291, 325, 326, 331, 332, 335–337, 341, 343, 345, 349, 352, 359
  cooperative, 336
  multiple, 326, 344, 346
  path planning, 331, 335, 343
Unmanned ground vehicle (UGV), 286, 287
  path planning, 331

Unmatched control, 122, 125, 128, 129, 133, 134, 137, 155
Unmatched controller, 137
Unstructured Byzantine nodes, 171, 172

V
Vehicle
  body, 277
  configuration, 300
  control, 337
  dynamics, 9–11, 269
  forward movement, 299
  models, 278, 299
  motion, 337
  systems, 269, 279, 352
  types, 326
Velocity
  agents, 28
  communication topology, 83
  information exchange, 83
  linear, 281
  states, 82
  steering, 280
  vector, 339
Virtual leader, 28, 63
Virtual leader dynamics, 28
Volume coverage path planning (VCPP), 325, 347
Vulnerable networks, 162, 177

W
Weighted mean subsequence reduced (W-MSSR) algorithm, 164, 165
Wireless
  information networks, 272
  networks, 229
Workspace, 306, 317, 319, 322, 332, 333, 347, 348

Printed in the United States
By Bookmasters